Feature Papers in *NeuroSci*: From Consciousness to Clinical Neurology

Feature Papers in *NeuroSci*: From Consciousness to Clinical Neurology

Editors

Lucilla Parnetti
Federico Paolini Paoletti
Xavier Gallart-Palau

MDPI • Basel • Beijing • Wuhan • Barcelona • Belgrade • Manchester • Tokyo • Cluj • Tianjin

Editors

Lucilla Parnetti
University of Perugia
Italy

Federico Paolini Paoletti
Perugia General Hospital and
University
Italy

Xavier Gallart-Palau
Biomedical Research Institute
of LLeida (IRBLLEIDA) -
University Hospital Arnau de
Vilanova (HUAV)
Spain

Editorial Office
MDPI
St. Alban-Anlage 66
4052 Basel, Switzerland

This is a reprint of Topical Collection published online in the open access journal *NeuroSci* (ISSN 2673-4087) (available at: https://www.mdpi.com/journal/neurosci/topical_collections/Feature_NeuroSci).

For citation purposes, cite each article independently as indicated on the article page online and as indicated below:

LastName, A.A.; LastName, B.B.; LastName, C.C. Article Title. *Journal Name* **Year**, *Volume Number*, Page Range.

ISBN 978-3-0365-7846-0 (Hbk)
ISBN 978-3-0365-7847-7 (PDF)

Contents

Preface to "Feature Papers in *NeuroSci*: From Consciousness to Clinical Neurology"

Neuroscience is an exciting and highly evolving field that analyzes the apparition and evolution of the human consciousness to the molecular basis of neurological diseases. In this book compilation, the reader will encounter the latest findings in the field of neuroscience, with exciting scientific contributions that have been carefully hand-picked by the editors to disseminate basic and clinical knowledge in the field.

Lucilla Parnetti, Federico Paolini Paoletti , and Xavier Gallart-Palau
Editors

Editorial

Featured Papers in *NeuroSci*

Xavier Gallart-Palau [1,2,3]

1   Biomedical Research Institute of Lleida Dr. Pifarré Foundation (IRB Lleida), Neuroscience Area, +Pec Proteomics Research Group (+PPRG), University Hospital Arnau de Vilanova (HUAV), 25198 Lleida, Spain; xgallart@irblleida.cat; Tel.: +34-973702201
2   Faculty of Medicine, University of Lleida, 25198 Lleida, Spain
3   Department of Psychology, University of Lleida, 25198 Lleida, Spain

Citation: Gallart-Palau, X. Featured Papers in *NeuroSci*. *NeuroSci* **2023**, 4, 103–104. https://doi.org/10.3390/neurosci4020010

Received: 21 April 2023
Accepted: 25 April 2023
Published: 30 April 2023

In this topical collection, Arsiwalla et al. discuss the viability of the morphospace construct to unite artificial intelligence concepts with biological agents to analyze and explore consciousness [1]. In the same vein, Pereira et al. discuss what makes neuronal organoids sentient from the perspective of sentiomics and the potential involvement of brain stimulation [2]. Liwicki et al. assess the replicability of subject-dependent and -independent methods of inner speech decoding [3].

Kieran Greer proposes an evolutionary model to explain the neural correlates that may sustain intelligence and adaptation, from invertebrates to the human brain [4]. Using comparative neuroscience, Peguero et al. investigate how neuronal tissue from zebrafish can be maintained ex vivo and how it responds molecularly to neuronal insult, with applicability in human neuronal regeneration [5]. Ison et al. use crustacean neurons to investigate the effects of Doxapram on neuronal channels, a platform that can potentially be extrapolated to investigate the effects of several other drugs on neuronal channels [6]. Carvalho et al. explore the alcohol addictive effects in rats that have experienced early life stressful events [7]. Kiffmeier et al. analyze the sex-dependent implications of the cerebellum on the learning process in autism model BTBR mice [8].

Relevant studies involving human subjects and specific clinical populations have also been included in this book. Valdez et al. investigate single neuron response to human face image processing [9], dePace et al. investigate autonomic neuropathies [10], Lentoor et al. explore neurocognitive domains in obesity [11], Gallo et al. look at bilateral facial palsy at the onset of neurosarcoidosis [12], Colombo et al. analyze the neuromodulation of dysautonomia in long COVID patients [13], Tan el al. assess the safety of the use of fluorescein sodium in pediatric neurosurgery [14], and Tariciotti et al. design a deep learning model with clinical applicability in brain tumor classification [15]. Finally, Handle et al. perform an interesting literature review to identify the most relevant findings linking handwriting product patterns to the specific cognitive and behavioral idiosyncrasies of subjects affected by autism spectrum disorder [16].

Conflicts of Interest: The author declares no conflict of interest.

References

1. Arsiwalla, X.D.; Solé, R.; Moulin-Frier, C.; Herreros, I.; Sánchez-Fibla, M.; Verschure, P. The Morphospace of Consciousness: Three Kinds of Complexity for Minds and Machines. *NeuroSci* **2023**, *4*, 79–102. [CrossRef]
2. Pereira, A.; Garcia, J.W.; Muotri, A. Neural Stimulation of Brain Organoids with Dynamic Patterns: A Sentiomics Approach Directed to Regenerative Neuromedicine. *NeuroSci* **2023**, *4*, 31–42. [CrossRef]
3. Simistira Liwicki, F.; Gupta, V.; Saini, R.; De, K.; Liwicki, M. Rethinking the Methods and Algorithms for Inner Speech Decoding and Making Them Reproducible. *NeuroSci* **2022**, *3*, 226–244. [CrossRef]
4. Greer, K. Neural Assemblies as Precursors for Brain Function. *NeuroSci* **2022**, *3*, 645–655. [CrossRef]
5. Peguero, R.L.; Bell, N.A.; Bimbo-Szuhai, A.; Roach, K.D.; Fulop, Z.L.; Corbo, C.P. Ultrastructural Analysis of a Forming Embryonic Embodiment in the Adult Zebrafish Optic Tectum Surviving in Organotypic Culture. *NeuroSci* **2022**, *3*, 186–199. [CrossRef]
6. Ison, B.J.; Abul-Khoudoud, M.O.; Ahmed, S.; Alhamdani, A.W.; Ashley, C.; Bidros, P.C.; Bledsoe, C.O.; Bolton, K.E.; Capili, J.G.; Henning, J.N.; et al. The Effect of Doxapram on Proprioceptive Neurons: Invertebrate Model. *NeuroSci* **2022**, *3*, 566–588. [CrossRef]
7. Carvalho, M.; Morais-Silva, G.; Caixeta, G.A.B.; Marin, M.T.; Amaral, V.C.S. Alcohol Deprivation Differentially Changes Alcohol Intake in Female and Male Rats Depending on Early Life Stressful Experience. *NeuroSci* **2022**, *3*, 214–225. [CrossRef]
8. Kiffmeyer, E.A.; Cosgrove, J.A.; Siganos, J.K.; Bien, H.E.; Vipond, J.E.; Vogt, K.R.; Kloth, A.D. Deficits in Cerebellum-Dependent Learning and Cerebellar Morphology in Male and Female BTBR Autism Model Mice. *NeuroSci* **2022**, *3*, 624–644. [CrossRef]
9. Valdez, A.B.; Papesh, M.H.; Treiman, D.M.; Goldinger, S.D.; Steinmetz, P.N. Encoding of Race Categories by Single Neurons in the Human Brain. *NeuroSci* **2022**, *3*, 419–439. [CrossRef]
10. DePace, N.L.; Santos, L.; Munoz, R.; Ahmad, G.; Verma, A.; Acosta, C.; Kaczmarski, K.; DePace, N.; Goldis, M.E.; Colombo, J. Parasympathetic and Sympathetic Monitoring Identifies Earliest Signs of Autonomic Neuropathy. *NeuroSci* **2022**, *3*, 408–418. [CrossRef]
11. Lentoor, A.G. Obesity and Neurocognitive Performance of Memory, Attention, and Executive Function. *NeuroSci* **2022**, *3*, 376–386. [CrossRef]
12. Gallo, C.; Mazzini, L.; Varrasi, C.; Vecchio, D.; Virgilio, E.; Cantello, R. Bilateral Facial Palsy as the Onset of Neurosarcoidosis: A Case Report and a Revision of Literature. *NeuroSci* **2022**, *3*, 321–331. [CrossRef]
13. Colombo, J.; Weintraub, M.I.; Munoz, R.; Verma, A.; Ahmad, G.; Kaczmarski, K.; Santos, L.; DePace, N.L. Long COVID and the Autonomic Nervous System: The Journey from Dysautonomia to Therapeutic Neuro-Modulation through the Retrospective Analysis of 152 Patients. *NeuroSci* **2022**, *3*, 300–310. [CrossRef]
14. Tan, A.J.L.; Tey, M.L.; Seow, W.T.; Low, D.C.Y.; Chang, K.T.E.; Ng, L.P.; Looi, W.S.; Wong, R.X.; Tan, E.E.K.; Low, S.Y.Y. Intraoperative Fluorescein Sodium in Pediatric Neurosurgery: A Preliminary Case Series from a Singapore Children’s Hospital. *NeuroSci* **2023**, *4*, 54–64.
15. Taricciotti, L.; Ferlito, D.; Caccavella, V.M.; Di Cristofori, A.; Fiore, G.; Remore, L.G.; Giordano, M.; Remoli, G.; Bertani, G.; Borsa, S.; et al. A Deep Learning Model for Preoperative Differentiation of Glioblastoma, Brain Metastasis, and Primary Central Nervous System Lymphoma: An External Validation Study. *NeuroSci* **2023**, *4*, 18–30. [CrossRef]
16. Handle, H.C.; Feldin, M.; Pilacinski, A. Handwriting in Autism Spectrum Disorder: A Literature Review. *NeuroSci* **2022**, *3*, 558–565. [CrossRef]

NeuroSci

Article

Ultrastructural Analysis of a Forming Embryonic Embodiment in the Adult Zebrafish Optic Tectum Surviving in Organotypic Culture

Ricardo L. Peguero, Nicole A. Bell, Andras Bimbo-Szuhai, Kevin D. Roach, Zoltan L. Fulop and Christopher P. Corbo *

Laboratory of Developmental Brain Research & Neuroplasticity, Department of Biological Sciences, Wagner College, Staten Island, NY 10301, USA; ricardo.peguero@wagner.edu (R.L.P.); nicole.bell@wagner.edu (N.A.B.); andras.bimbo-szuhai@wagner.edu (A.B.-S.); kevin.roach@wagner.edu (K.D.R.); zoltan.fulop@wagner.edu (Z.L.F.)
* Correspondence: ccorbo@wagner.edu

Abstract: It has been shown that adult zebrafish are capable of regenerating regions of the central nervous system (CNS) after insult. Unlike in higher-order vertebrates where damage to the CNS leads to glial scar formation and permanent functional deficits, damage to the adult zebrafish CNS is transient and followed by nearly complete reconstitution of both function and anatomy. Our lab's previous work has shown that explants of zebrafish optic tectum can survive in organotypic culture for up to 7 days, and that at 96 h in culture, regenerating cells of the tectum begin to form structures that resemble the embryonic neural tube seen in vertebrate development. The current project aims to elucidate the cellular and ultrastructural components of the formation of this neural tube-like structure using scanning and transmission electron microscopy. Our results show that after injury and cultivation for 96 h, the explants contained differentiating cells that were undergoing several cellular events, such as neovascularization, and rosette/cisternae formation, leading to the formation of a structure resembling the embryonic neural tube. Additionally, we demonstrate healthy cellular ultrastructures in both degenerated and regenerated areas of the explant.

Keywords: zebrafish; plasticity; brain development; optic tectum; brain regeneration

Citation: Peguero, R.L.; Bell, N.A.; Bimbo-Szuhai, A.; Roach, K.D.; Fulop, Z.L.; Corbo, C.P. Ultrastructural Analysis of a Forming Embryonic Embodiment in the Adult Zebrafish Optic Tectum Surviving in Organotypic Culture. *NeuroSci* **2022**, *3*, 186–199. https://doi.org/10.3390/neurosci3020014

Academic Editor: Masato Nakafuku

Received: 7 March 2022
Accepted: 29 March 2022
Published: 2 April 2022

Publisher's Note: MDPI stays neutral with regard to jurisdictional claims in published maps and institutional affiliations.

1. Introduction

In the last several decades, zebrafish (Danio rerio) have become a popular laboratory animal for developmental biology, specifically brain development. Zebrafish are a unique vertebrate model organism; they are small, easy to care for, inexpensive, and space efficient. For these reasons, they have been adopted for many areas of research and the use of adult zebrafish has become more prominent [1–8].

Since zebrafish are low on the vertebrate phylogenetic tree, they possess higher regenerative capacity than the more advanced orders [2,3,5,6,9,10]. Previous studies have shown that different regions of the zebrafish central nervous system demonstrate incredibly robust regenerative processes that result in the full resolution of any tissue damage sustained from injury [10–13]. Such regenerative processes are not seen in mammals and the neurogenic capacity of the mammalian brain has also been shown to be lesser than that of zebrafish [14]; while the zebrafish brain contains several areas of neurogenic activity that are present during adulthood, the adult mammalian brain's neurogenic activity is restricted to the hippocampus and olfactory bulb [14,15]. Additionally, the zebrafish CNS provides an environment that is supportive of newly born neurons in adulthood whereas the mammalian CNS does not [14,15]. Presumably, the lack of reactive gliosis and glial scar formation make the zebrafish CNS more permissive to regeneration after injury [10,14].

Neurotrophic factors, such as brain-derived neurotrophic factor (BDNF) and nerve growth factor, have been shown to be upregulated in several areas of the zebrafish CNS after injury and during development [16]. In addition to this, several molecular mechanisms and cellular components implicated in the zebrafish neuroregenerative process have been interrogated; primarily, the roles of signaling pathways, including the IL-6/Stat 3, and Notch pathways [17,18], as well as the role of glial and neural progenitor cells have been described [12]. This work suggests that signaling pathways and molecules related to embryonic development mediate the neuroregenerative response on a molecular level while radial glia and neural progenitor cells are responsible for repopulating and reconstructing any insulted areas.

A study by Tomizawa and colleagues demonstrated that a whole zebrafish brain was able to survive in organotypic culture for seven days [9]. In another study, Kustermann and colleagues showed that neurons from explants of zebrafish retinas were able to survive and regenerate in organotypic culture [19]. Our lab has long been focused on the zebrafish brain cellular anatomy, in particular the structure of the optic tectum [20]. The work by Tomizawa and colleagues as well as Kustermann and colleagues spurred an interest in our group to analyze how pieces of the optic tectum could survive in organotypic culture. The act of removing the pieces would be a model of brain injury in itself.

Our lab showed that zebrafish optic tectum can survive in organotypic culture for up to seven days, even when cut into four separate pieces [3]. While many cells died early in the culture, there were many that survived. After four days in organotypic culture, surviving cells were able to be recruited and form structures resembling the early development of the embryonic neural tube/fold. These forming structures will herein be referred to as embryoid embodiments.

Our initial analysis focused on the cellular events that occurred over the seven days in culture as well as the cellular structure of the embryoid embodiments. This work set out to analyze this time course utilizing transmission and scanning electron microscopy. We investigated regions that appeared to be degenerating as well as the forming embryonic embodiments. We looked into cellular ultrastructure in dying, surviving, and regenerating regions of the cultured tissue pieces.

2. Material and Methods

2.1. Animal Care and Utilization

This project adhered to the guidelines set forth in the Guide for Care and Use of Laboratory Animals, 8th edition as well as euthanasia protocols in The Zebrafish Book, 5th edition. All zebrafish were obtained from a local pet store in Staten Island, NY and were maintained in a 50-gallon aquarium with a regular day/night cycle (14L:10D) and proper aeration and filtration. Zebrafish were fed once per day with dry tetra min flake food for tropical freshwater fish.

This study utilized 12 adult zebrafish of mixed sex. Each fish generated four explant tissue samples, totaling 48. The explants of three animals (totaling 12 cultured pieces) were processed for scanning electron microscopy (SEM) and those of the other nine animals (totaling 36 cultured pieces) were processed for transmission electron microscopy (TEM). SEM samples were fixed at 24, 48, and 96 h of cultivation. All TEM samples were fixed at 96 h of cultivation.

2.2. Culture Media and Surgical Procedure

All surgical techniques were carried out using aseptic conditions. All fish were anesthetized before surgery using a 0.04% tricaine solution. Complete unconsciousness was determined using forceps to pinch the tailfin.

According to our previously published protocol, the brains of the fish were removed, and the tectum was cut into four pieces [3]. Briefly, after anesthesia, the fish were secured, and the skull was removed followed by complete brain extraction. The brain was transferred to organotypic culture media where the optic tectum was cut into four pieces, transferred to Millipore organotypic culture insert (Millipore Sigma, Burlington, MA, USA, cat# PICM03050), and cultivated in the same organotypic culture medium with 5% CO2 at 37 °C.

The organotypic media recipe, adopted from Tomizawa et al., was made up of minimal essential media (MEM) supplemented with 15% horse serum, 15% Hank's Balanced Salt Solution, 0.2 mM L-glutamine, 50 mg/mL of glucose, and 100 units of penicillin/streptomycin for bacterial inhibition (all ingredients for culture media are from Fisher Scientific, Waltham, MA, USA) [9].

2.3. Scanning Electron Microscopy Histotechniques

All tissue explants were fixed in Karnovky's fixative [21] for at least 24 h, post-fixed in 1% osmium tetroxide (Fisher Scientific, Waltham, MA, USA) for two hours, dehydrated through an increasing ethanol series, and further dried using two treatments with propylene oxide followed by complete desiccation with hexamethyldisilazane (HMDS) (Fisher Scientific, Waltham, MA, USA). The processing vial caps were removed and replaced with aluminum foil and one hole was punctured into foil to allow for slow evaporation of the HMDS in the fume hood for at least 24 h.

Once completely dry, the samples were mounted on aluminum SEM stubs using sticky carbon adhesives. The mounted samples were coated using a Hummer IV (LADD research, Williston, VT, USA) with the sputter coater set at 100 mT of vacuum and 10–15 mA for 10 min. Samples were imaged in a Topcon ABT- 32 SEM (Topcon, Livermore, CA, USA) equipped with an Orion digital imaging system (Topcon, Livermore, CA, USA).

2.4. Transmission Electron Microscopy Histotechniques

All tissue explants were fixed in Karnovsky's fixative (4% paraformaldehyde, 2% glutaraldehyde, pH 7.2) for at least 24 h, post-fixed in 1% osmium tetroxide for two hours, dehydrated through an increasing ethanol series, embedded in Durcupan resin (Millipore Sigma, Burlington, MA, USA,), and polymerized overnight at 60 °C.

Blocks were trimmed by hand, and 1 µm light sections were collected using a Reichert OMU-2 ultramicrotome (Reichert, Depew, NY, USA) and glass knives. Sections were stained with 1% toluidine blue and observed using an Olympus BX40 light microscope (Olympus scientific solutions, Waltham, MA, USA). Ultrathin sections were collected using a Sorvall MT-6000 ultramicrotome (Sorvall, Waltham, MA, USA) and a diamond knife. Silver/gold sections were spread with xylene and collected on 2 mm copper slot grids coated with 0.5% formvar. Grids were contrasted with 2% uranyl acetate and 0.1% lead citrate and carefully washed so as not to disturb the formvar coating. Grids were dried prior to imaging on a Philips CM100 TEM (Mount Holyoke, South Hadley, MA, USA) equipped with a Gatan Orius digital imaging system (Gatan, Pleasanton, CA, USA). Montages were assembled in Adobe Photoshop.

3. Result

In this paper, we present an ultrastructural analysis of zebrafish optic tectum maintained in organotypic culture. Scanning electron microscopy was used to analyze surface structures at all time points, while the transmission electron microscopy ultrastructure was focused on samples in culture for 96 h. This time point was selected based on our previous work where we demonstrated that the earliest time when the forming embryoid embodiments could be detected was 96 h. Scanning electron microscopy was used to characterize the surface structure, analyze cells that had migrated to the periphery, and to determine if we could detect any newly forming structure on the surface of the tissue piece.

Figure 1 depicts scanning electron micrographs of surviving samples at 24, 48, and 96 h, respectively. At 24 h, the ependymal layer, the subventricular zone, and the rest of the cortical region [20] up to the pial surface could be recognized and no general anatomical distortion could be detected. At 48 h in culture, only the ependymal layer was recognizable as a distinct morphological entity. The surface tectal piece was covered with densely packed round cells, which were hard to differentiate. However, after 96 h in culture, different, symmetrical protrusions appeared to have formed on the surface. These structures usually have a midline, a raphe-like groove that divides the formations into two mirrored halves. In the presented sample, three of these formations were formed (boxes). We should note that in our many samples, the number of these formations varied greatly from none to several per piece. The formations in these samples represent the surface of the forming embryoid embodiments.

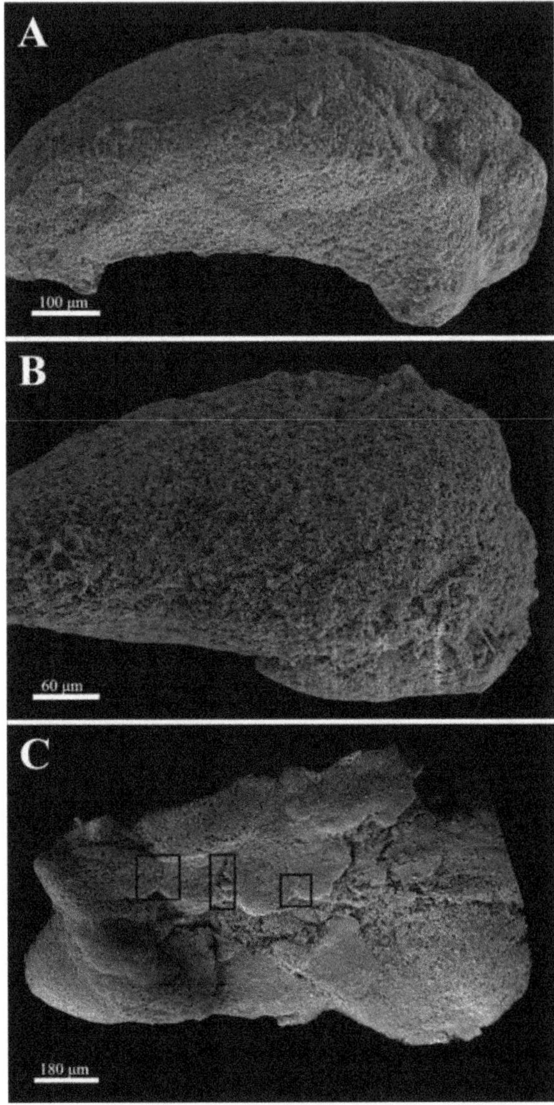

Figure 1. Scanning electron micrographs of three surviving samples representing three time points in culture. At 24 h (**A**), all structural characteristics of the zebrafish optic tectum could be recognized. At 48 h (**B**), only the ependymal layer was recognizable as a distinct morphological entity but the whole surface tectal piece was covered with densely packed round cells. At 96 h (**C**), symmetrical protrusions of regenerating tissue formed on the surface of the explant. These structures usually have a midline, a raphe-like groove (black boxes) that divides the formations into two mirrored halves.

Figure 2 is a panoramic overview of a folding embryoid embodiment in adult zebrafish optic tectum surviving in organotypic culture for 96 h. The image is a montage of 50 electron micrographs captured at approximately 3000× magnification. While the 96-h samples contained larger numbers of advanced formations, they also contained several earlier stages of cell grouping, differentiation, and migration, making it a good time point for detailed morphological analysis. Figure 2 depicts an advanced embryoid embodiment formation in which a forming "ventricular space" can be found. This formation is surrounded by many undifferentiated cellular arrangements and samples of neovascularization.

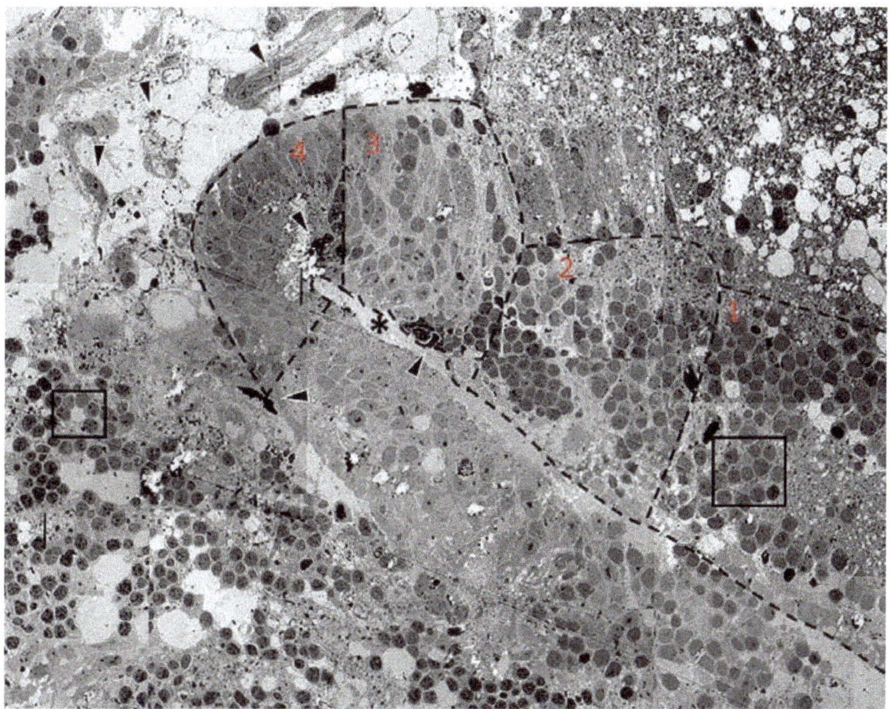

Figure 2. Electron microscopic panoramic overview of a folding embryoid embodiment in adult zebrafish optic tectum surviving in organotypic culture for 96 h. The image is a montage of 50 electron micrographs. Areas 1–4 represent cells differentiated to varying degrees, with area 4 representing the most highly differentiated cells. The central, organized region (asterisk) of the image, which is surrounded with spongiform degenerative regions (upper right corner) intermingled with signs of neovascularization and hematopoiesis (arrowheads) as well as rosettes (black boxes), represents early developmental formations. Inferior to the ventricular space, some mitotic cells can be recognized among differentiating nuclei.

The embryoid embodiment in the middle of Figure 2 is sectioned in the longitudinal plane and the forming ventricular space is visible at the center of the structure (asterisk). A forming ependymal layer could be seen lining the "ventricular space" and microvilli could be detected; these are shown in greater detail in a subsequent figure. Cells in the forming embryoid embodiment showed an increasing level of differentiation toward the tip of the structure. We labeled each level of differentiation with a number (1–4), with 4 being the most differentiated.

The most advanced tissue, found in region 4, contained both advanced ependymal cells and signs of neovascularization (arrowheads) in which newly formed blood cells could also be detected.

Cells in region 3 clearly resembled pseudostratified columnar neuroepithelium typical for the normal development of the healthy vertebrate embryonic brain. The nuclei of these cells could frequently be detected as one elongated cytoplasm that was interconnecting the ependyma and the forming pia mater. This was indicative of the structure and role of the radial glial cells in normal vertebrate brain development. While the pia mater of the forming cortical structure [20] above the ventricular space was clearly recognizable, the pia mater of the forming region below the ventricular space was hardly recognizable. Immediately to the right of this region, region 2 could be seen. This group of cells represents an earlier stage of neuroepithelial formation but, at this position, we did not see a formed pial surface; rather, the cells interacted with a region of spongiform degeneration. The nuclei of these cells were more rounded and several of them were in a stage of cell division. Mitotic figures could still be recognized in several cells.

In region 1, cells forming rosette groupings of rounded cells with dark nuclei could be detected within close proximity to spongiform degeneration. These rosettes are the earliest significant formation leading to the start of brain development in vertebrates [22,23]. These cells, under some chemical influence, group into rosette formations and seem to subsequently form more advanced developing cortical structures [20]. The region of the embryoid embodiment that was inferior to the developing ventricle was not developed, as was the case for its previously described superior counterpart. However, this region did display homogeneous lightly colored nuclei as well as several mitotic cells.

At the upper right corner of the image, an area of spongiform degenerating tissue could be seen with numerous white spots which were locations of former neurons. However, in that region, several small cells with small, dense, dark nuclei could be seen. Among these surviving elements, groups of cells could be seen forming classical rosette arrangements around clear cisternal spaces [3,20]. The nuclei of the cells forming rosettes had different arrangements of chromatin, which is a way to distinguish different cells. Most of the nuclei had a homogenous chromatin with one or two nucleoli. Some of the other nuclei had a distorted shape and dark appearance with recognizable hetero and euchromatin arrangements, similar to the nuclei in other rosettes recognized below the neural fold formation. The left sides of the images displayed several distinct populations of cells grouped together.

Cells in small rosettes could be seen in the lower region of the image (box). These cells had varying densities and were not homogenous, displaying clearly visible hetero and euchromatin. These cells formed the rosette around a cisternal space within the tissue. In the upper left corner, it was possible to see a grouping of cells that were larger than any of the previously mentioned small rosettes. These cells also had varying nuclei colorations, and some cells had nucleoli. Cisternal spaces could be seen at either end of the structure. We hypothesize that this formation was a more advanced form of the simple rosettes below them.

Lastly, there were several events of neovascularization detected both within the embryoid embodiment as well as in the surrounding areas (arrowheads). Some of these vessels were in a more advanced stage of development and, in some cases, showed the presence of newly formed blood cells.

Figures 3–5 present ultrastructural samples from different regions of several tissue blocks representing dying, surviving, and newly formed elements seen after 96 h in organotypic culture. Figure 3A presents a typical sample of spongiform degenerating tissue with a large number of varying size white spaces, likely the location of former cells now dead or migrated away. These spaces were closely associated with small dark dots that were the remnants of former cells. The larger areas were cisternal spaces (c), which were likely assisting in nutrient movement through the tissue piece.

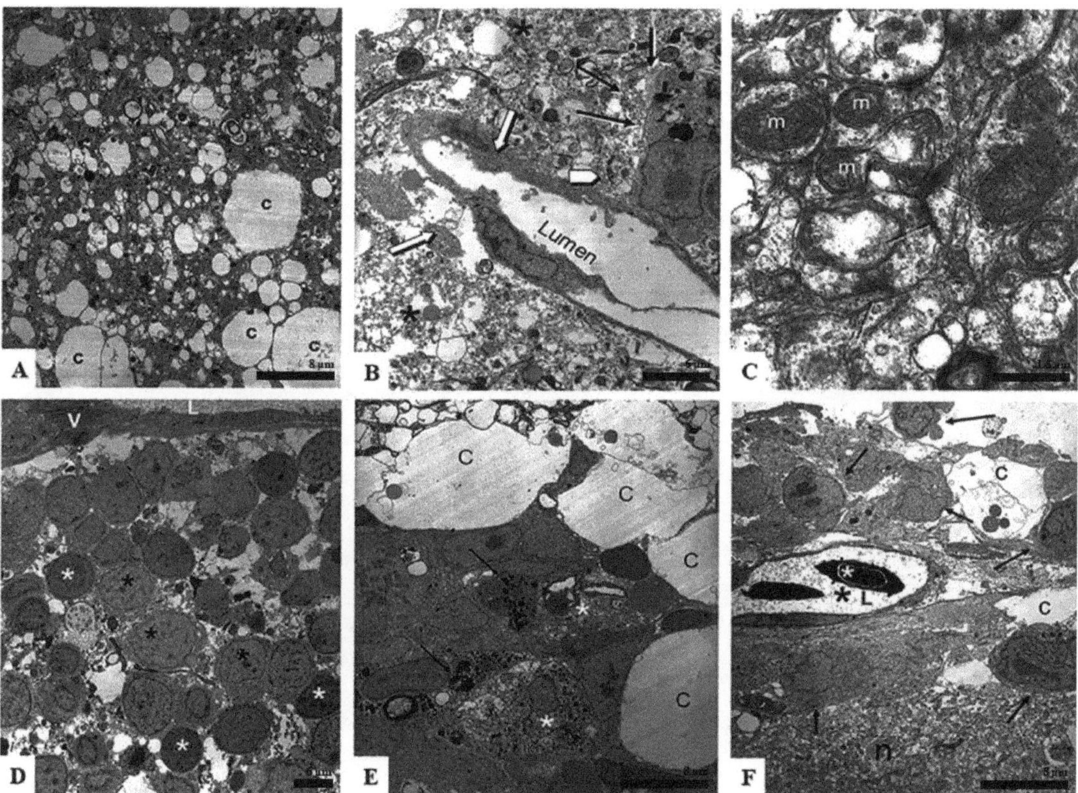

Figure 3. Ultrastructural characteristics of regenerating optic tectum after 96 h in culture. Specifically, (**A**) presents a typical sample of spongiform degenerating tissue with a large number of varying size white spaces, as well as larger cisternal spaces (c). (**B**) shows a surviving blood vessel absent of any blood cells (white arrows). The surrounding area was greatly degenerated (asterisks). The black arrows point to a surviving cell in the area. The white arrowhead points to dense granules. (**C**) shows an area of degrading neuropil. Healthy mitochondria (m) with intact cristae and healthy synapses (black arrows) can be seen. (**D**) is an area of surviving cells that grouped together in a region close to a large vessel seen at the top of the image (V and L). The cells display nuclei of different densities (black asterisk—lighter nuclei; white asterisk—darker nuclei). (**E**) shows an area of early regeneration. Mast cells could be detected (black arrows—granules; asterisk—mast cell body) in this area. The mast cells were surrounded by healthy- appearing cells and cisternae (c). (**F**) shows an area of neuropil (n) that appears more intact than those seen in (**B,C**). One can also see a healthy blood vessel in which new blood cells have formed (white asterisk—red blood cell nuclei; black asterisk—fluid portion of blood vessel; L—lumen of blood vessel). The black arrows are pointing at cells that were active around the vessel.

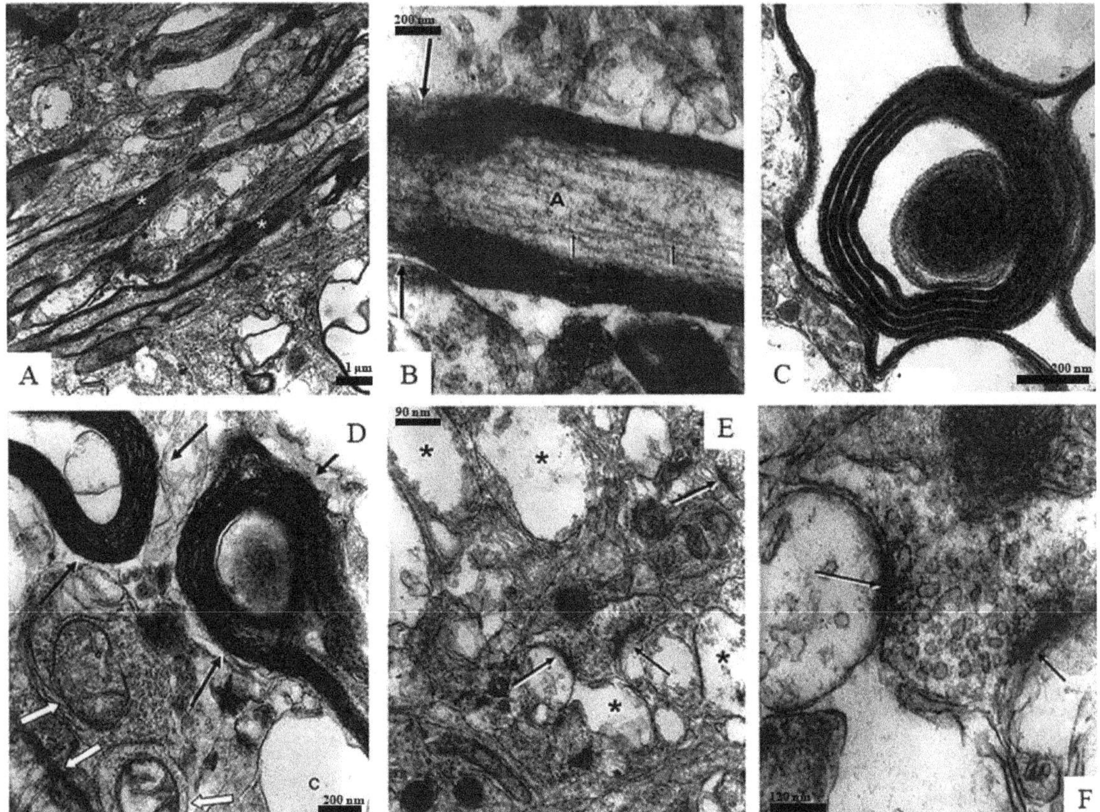

Figure 4. Panel of images depicting different samples of neuronal components found in the spongiform neuropil, such as myelinated axons, synapses, and mitochondria. (**A**) shows healthy, myelinated axons (asterisk). (**B**) is a high-magnification image of (**A**), in which axoplasm (A) and axonal neurofilaments (arrows) can be easily recognized. (**C**) shows an image where both the axon and tissue surrounding the myelin sheath are deteriorating or are completely missing, while the myelin seems to be intact. The myelin in (**D**) also displays a healthy appearance while the axoplasm of the axons (black arrows) is nearly completely degraded. Cisternae (c) as well as mitochondria (white arrows) can also be seen. (**E,F**) focus on surviving synaptic elements (arrows). (**E**) depicts these synaptic elements among degenerative tissue (asterisks). (**F**) is an enlargement of the two synapses seen in the center of (**E**).

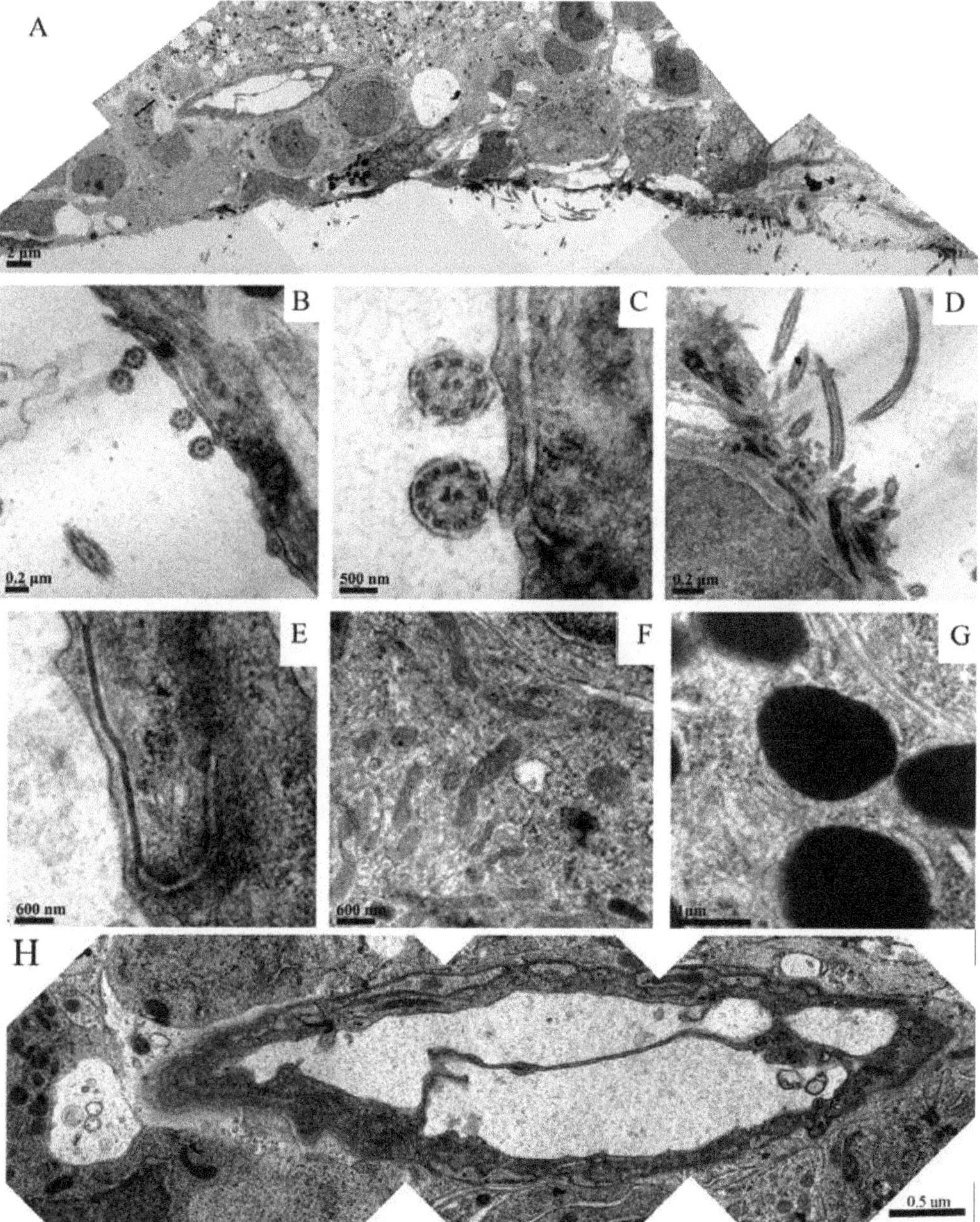

Figure 5. The different regions of the forming ependymal layer in the embryoid embodiment. (**A**) is a montage showing newly formed ependymal cells with their microvilli protruding into the ventricular space. Neovascularization can also be recognized. (**B–D**) depict longitudinal and cross sections of the ependymal microvilli within the neural groove. (**E**) depicts a well-formed tight junction between two neighboring ependymal cells. (**F**) shows that the cytoplasm of these ependymal cells were loaded with healthy mitochondria and other organelles. Some of the forming ependymal cells that were imaged were loaded with dark inclusions, which are shown in (**G**). (**H**) is a montage of the venous blood vessel seen in (**A**) at a higher magnification.

Figure 3B shows a surviving blood vessel absent of any blood cells (white arrows). One can see that the surrounding area was greatly degenerated (asterisks) and the internal lining of the vessel was blebbing off into the luminal space, likely because it was degrading. The black arrows point to a surviving cell in the area. This was likely a macrophage because of its proximity to the blood vessel and the large amount of internal cellular inclusions formed as the cell phagocytized the dying regions. The white arrowhead points to dense granules, likely cellular remnants of apoptosis.

Figure 3C shows an area of degrading neuropil. Interestingly, even in areas of spongiform degeneration, one can find healthy mitochondria (m) with intact cristae and the presence of healthy synapses (black arrows).

Figure 3D is an area of surviving cells that grouped together in a region close to a large vessel seen at the top of the image (V and L). The cells displayed nuclei of different densities (black asterisk—lighter nuclei; white asterisk—darker nuclei).

Figure 3E shows an area of early regeneration. One can detect mast cells (black arrows—granules; asterisk—mast cell body) in this area. The mast cells were surrounded by healthy-appearing cells, one of which was in an active phase of mitosis. As seen in Figure 2, these cells were grouping around large cisternal spaces (c).

Figure 3F shows an area of neuropil (n) that appears to be more intact than those seen in 3B and C. One can also see a healthy blood vessel in which new blood cells formed (white asterisk—red blood cell nuclei; black asterisk—fluid portion of blood vessel; L—lumen of blood vessel). The black arrows are pointing at cells that were metabolically active around the vessel.

Figure 4 is a panel of images depicting different samples of neuronal components found in the spongiform neuropil, such as myelinated axons, synapses, and mitochondria. Figure 4A,B show myelinated axons (asterisks) in longitudinal sections: A is a low-magnification overview and B is a high-magnification image in which axoplasm (A) and axonal neurofilaments (arrows) can be easily recognized. Figure 4C,D depict cross sections of myelinated axons. In Figure 4C, both the axon and tissue surrounding the myelin sheath are deteriorating or are completely missing, while the myelin seems to be intact. The myelin in Figure 4D also displays a healthy appearance while the axoplasm of the axons (black arrows) is nearly completely degraded. It is interesting to note that next to these axons, surviving mitochondria can be seen (white arrows). Figure 4E,F focus on synaptic elements (arrows): E is a low-magnification overview of the region in a spongiform degenerating neuropil (asterisks) while F is an enlargement of the two synapses seen in the center image E. The axon terminals seen in Figure 4F have well-recognizable cell membranes and are loaded with healthy-appearing synaptic vesicles that clearly contain neurotransmitters.

Figure 5 is a panel that focuses on different regions of the forming ependymal layer in the embryoid embodiment. Figure 5A is a montage overviewing a region that clearly shows the specific arrangement of the newly formed ependymal cells with their microvilli protruding into the ventricular space. Signs of neovascularization can also be recognized. Figure 5B–D depict longitudinal and cross sections of the ependymal microvilli within the neural groove. Figure 5E depicts a well-formed tight junction between two neighboring ependymal cells. Figure 5F shows that the cytoplasm of these cells were loaded with healthy mitochondria and other organelles determined through their intact membrane structures. Many of the forming ependymal cells were loaded with dark inclusions, which are shown in a higher magnification in Figure 5G. These inclusions were likely glycogen granules. Figure 5H is a montage of the venous blood vessel seen in 5A at a higher resolution. This image demonstrates healthy endothelial cells and the formation of venous valves. The vessel was surrounded by healthy tissue components that might be either smooth muscle or pericytes.

4. Discussion

The use of organotypic culture of brain tissue is both a powerful and capricious approach to study the plasticity of traumatized mature zebrafish brains. Taking out the brain from the skull, cutting it up into pieces, and placing it into a foreign environment for organotypic culture is a drastic traumatic brain injury (TBI). Each sample provides a unique view of the regenerating embryoid embodiments depending on the orientation in which they were sectioned. Additionally, it is exciting to see such robust regeneration in a brain tissue explant that had been completely removed from the organism, whereby it was removed from continual blood flow and hormone-like signaling molecules. The entirety of the regenerative process occurred with molecules that were present within the tissue sample itself. There were no additives in the growth media to promote regeneration.

Most work conducted in zebrafish brain regeneration is performed in brain injury models using the telencephalon, retina, or forebrain [24,25]. Our model is distinct in two ways: one, our model focuses on the optic tectum, which is the only cortical structure in the zebrafish brain [20], and second, our model uses the surgical procedure of explant generation as a method to introduce brain injury. Additionally, by allowing our explants to survive ex vivo in culture, any influence from factors outside of the tectum on regeneration are removed.

Upon imaging of the external morphology of the tissue explants using SEM, healthy, surviving tissue could be seen up to 24 h in culture, whereas after 48 h in culture, widespread cell death and spongiform degeneration were most prominent. At 96 h in culture, cellular migration and organization could be detected and regenerating, highly organized embryoid embodiments could be seen forming along a raphe. Cells exhibiting migratory characteristics presumably moved from areas within the explant to the surface to access the nutrient-rich media. Additional factors such as BDNF signaling may also have been implicated in the organized migration observed in our explants. These migrating cells also formed structures that resemble the neural tube/fold as it is seen in normal early vertebrate CNS development [26,27].

When analyzing the ultrastructure of tissue maintained in culture for 96 h, several facets of vertebrate neurulation were clearly recognizable, with the most apparent being the formation of neural tube-like structures. However, as we previously showed in our light microscopic analysis, there were other, smaller events that were evident. Examples of these are rosette formation and signs of neovascularization.

The embryoid embodiment was seen in cross section and a ventricular space could be seen surrounded by an ependymal surface made up of cells, some of which appeared mitotic and differentiated to varying extents; presumably, cells in the explant divided, populated, and organized using the spongiform degenerating areas as a scaffold and then differentiated to form embryonic structures that resembled the vertebrate neural tube. Several levels of differentiation could be detected in these explants. Cells forming rosettes around cisternal spaces were among the least differentiated [28], while more differentiated ependymal cells could be seen directly lining the ventricular space and forming an ependymal layer. Within the new parenchyma, more differentiated cells could be seen forming highly organized groupings, and many of these cells exhibited radial glia-like projections. Neovascularization and pial layer formation were also prominent in the tissue explants cultured for 96 h. Even in areas of spongiform degeneration, both healthy and aberrated synapses as well as axons could be seen among other surviving subcellular elements such as mitochondria. Collectively, our results show that the cellular reorganization of the zebrafish optic tectum after TBI is remarkably similar to the organization seen during vertebrate neural development.

The current literature related to the cellular components of the regenerative process in the zebrafish optic tectum is scarce and even in the more heavily interrogated brain regions such as the telencephalon, little work is focused on analyzing the regenerative process from an ultrastructural perspective; our project provides insight into the regenerative process of the zebrafish optic tectum from a perspective that has received little to no attention.

Although our methods were sufficient to describe the ultrastructural characteristics of cell death, neovascularization, cellular organization, and regenerative neurogenesis in the optic tectum after injury, several questions arise from our results, with the most pressing of these questions pertaining to the role of mast cells in the regenerative response. We demonstrated in our previous study that the cells that characteristically appear as mast cells are in fact mast cells due to their metachromasia with toluidine blue staining [3]. Additionally, we characterized the presence of the mast cells at different time points in a separate work [29].

Although we showed that mast cells are present in regenerating tissue explants and are seen associating with other rosette-forming cells and degenerative tissue, little can be said about their function. Since mast cells have been shown to degranulate and, in some cases, mediate regenerative tissue responses including angiogenesis and epithelial tissue repair [30–32], our group hypothesizes that mast cells act in a fashion that can both induce and/or mediate the neuroregenerative process through their degranulation. This hypothesis is in part supported by previous work that showed that mast cell activation can induce BDNF expression in microglial cells [33]. Further support for this hypothesis is seen through previous experiments, which demonstrated that several distinct neuropeptides, including nerve growth factor, can both activate and be released by mast cells [34]. Additionally, it was shown that sterile inflammation is sufficient for the initiation of the regenerative response in zebrafish [35]. Because mast cells, through their degranulation, play a role in inflammatory responses, we reason that there may be a connection between mast cell degranulation and inflammatory processes that are necessary in the initiation of the regenerative response in zebrafish.

In future studies, our group will seek to describe the molecular components of mast cell involvement in neuroregeneration after time in culture. Future work may also concern itself with the characterization of the molecular components of the neovascularization and rosette formation seen in our results to determine the role of newly forming circulatory elements and rosettes. Our group's work suggests that particular attention should be paid to the processes of early vertebrate neurulation when interrogating the regenerative response in zebrafish after TBI.

Author Contributions: Conceptualization, C.P.C. and Z.L.F.; methodology, C.P.C., Z.L.F., R.L.P., A.B.-S., N.A.B. and K.D.R.; formal analysis, C.P.C., Z.L.F., R.L.P., A.B.-S. and N.A.B.; writing—original draft preparation, C.P.C. and R.L.P.; writing—review and editing, C.P.C., R.L.P. and A.B.-S.; supervision, C.P.C. and Z.L.F.; project administration, C.P.C. and Z.L.F.; funding acquisition, C.P.C. and Z.L.F. All authors have read and agreed to the published version of the manuscript.

Funding: This work was funded through an internal donation at Wagner College funding undergraduate research in the sciences.

Institutional Review Board Statement: This project adhered to the guidelines set forth in the Guide for Care and Use of Laboratory Animals, 8th edition, as well as euthanasia protocols in The Zebrafish Book, 5th edition.

Data Availability Statement: The authors confirm that the data supporting the findings of this study are available within the article.

Conflicts of Interest: The authors declare no conflict of interest.

References

1. Baier, H.; Wullimann, M.F. Anatomy and function of retinorecipient arborization fields in zebrafish. *J. Comp. Neurol.* **2021**, *529*, 3454–3476. [CrossRef] [PubMed]
2. Beckers, A.; Vanhunsel, S.; Van Dyck, A.; Bergmans, S.; Masin, L.; Moons, L. Injury-induced Autophagy Delays Axonal Regeneration after Optic Nerve Damage in Adult Zebrafish. *Neuroscience* **2021**, *470*, 52–69. [CrossRef] [PubMed]
3. Corbo, C.P.; Fulop, Z.L. Formation of structures resembling early embryonic neural plate in traumatized adult zebrafish optic tectum maintained in organotypic culture. *In Vivo* **2017**, *38*, 28–43.
4. Eachus, H.; Choi, M.K.; Ryu, S. The Effects of Early Life Stress on the Brain and Behaviour: Insights From Zebrafish Models. *Front. Cell Dev. Biol.* **2021**, *9*, 657591. [CrossRef]

5.	Johnson, K.; Barragan, J.; Bashiruddin, S.; Smith, C.J.; Tyrrell, C.; Parsons, M.J.; Doris, R.; Kucenas, S.; Downes, G.B.; Velez, C.M.; et al. Gfap-positive radial glial cells are an essential progenitor population for later-born neurons and glia in the zebrafish spinal cord. *Glia* **2016**, *64*, 1170–1189. [CrossRef]
6.	Jopling, C.; Sleep, E.; Raya, M.; Martí, M.; Raya, A.; Belmonte, J.C.I. Zebrafish heart regeneration occurs by cardiomyocyte dedifferentiation and proliferation. *Nature* **2010**, *464*, 606–609. [CrossRef]
7.	Marymonchyk, A.; Malvaut, S.; Saghatelyan, A. In vivo live imaging of postnatal neural stem cells. *Development* **2021**, *148*. [CrossRef]
8.	Shimizu, Y.; Kawasaki, T. Differential Regenerative Capacity of the Optic Tectum of Adult Medaka and Zebrafish. *Front. Cell Dev. Biol.* **2021**, *9*, 686755. [CrossRef]
9.	Tomizawa, K.; Kunieda, J.; Nakayasu, H. Ex vivo culture of isolated zebrafish whole brain. *J. Neurosci. Methods* **2001**, *107*, 31–38. [CrossRef]
10.	Cacialli, P. Neurotrophins Time Point Intervention after Traumatic Brain Injury: From Zebrafish to Human. *Int. J. Mol. Sci.* **2021**, *22*, 1585. [CrossRef]
11.	Cosacak, M.I.; Papadimitriou, C.; Kizil, C. Regeneration, Plasticity, and Induced Molecular Programs in Adult Zebrafish Brain. *BioMed Res. Int.* **2015**, *2015*, 769763. [CrossRef] [PubMed]
12.	Kroehne, V.; Freudenreich, D.; Hans, S.; Kaslin, J.; Brand, M. Regeneration of the adult zebrafish brain from neurogenic radial glia-type progenitors. *Development* **2011**, *138*, 4831–4841. [CrossRef] [PubMed]
13.	Kishimoto, N.; Shimizu, K.; Sawamoto, K. Neuronal regeneration in a zebrafish model of adult brain injury. *Dis. Model Mech.* **2012**, *5*, 200–209. [CrossRef] [PubMed]
14.	Ming, G.L.; Song, H. Adult neurogenesis in the mammalian brain: Significant answers and significant questions. *Neuron* **2011**, *70*, 687–702. [CrossRef] [PubMed]
15.	Kizil, C.; Kaslin, J.; Kroehne, V.; Brand, M. Adult neurogenesis and brain regeneration in zebrafish. *Dev. Neurobiol.* **2012**, *72*, 429–461. [CrossRef]
16.	Cacialli, P.; Palladino, A.; Lucini, C. Role of brain-derived neurotrophic factor during the regenerative response after traumatic brain injury in adult zebrafish. *Neural Regen. Res.* **2018**, *13*, 941–944. [CrossRef]
17.	Kim, H.K.; Lee, D.-W.; Kim, E.; Jeong, I.; Kim, S.; Kim, B.-J.; Park, H.-C. Notch Signaling Controls Oligodendrocyte Regeneration in the Injured Telencephalon of Adult Zebrafish. *Exp. Neurobiol.* **2020**, *29*, 417–424. [CrossRef]
18.	Shimizu, Y.; Kiyooka, M.; Ohshima, T. Transcriptome Analyses Reveal IL6/Stat3 Signaling Involvement in Radial Glia Proliferation After Stab Wound Injury in the Adult Zebrafish Optic Tectum. *Front. Cell Dev. Biol.* **2021**, *9*, 668408. [CrossRef]
19.	Kustermann, S.; Schmid, S.; Biehlmaier, O.; Kohler, K. Survival, excitability, and transfection of retinal neurons in an organotypic culture of mature zebrafish retina. *Cell Tissue Res.* **2008**, *332*, 195–209. [CrossRef]
20.	Corbo, C.P.; Othman, N.A.; Gutkin, M.C.; Alonso, A.D.C.; Fulop, Z.L. Use of different morphological techniques to analyze the cellular composition of the adult zebrafish optic tectum. *Microsc. Res. Tech.* **2012**, *75*, 325–333. [CrossRef]
21.	Reese, T.S.; Karnovsky, M.J. Fine structural localization of a blood-brain barrier to exogenous peroxidase. *J. Cell Biol.* **1967**, *34*, 207–217. [CrossRef] [PubMed]
22.	Broccoli, V.; Giannelli, S.G.; Mazzara, P.G. Modeling physiological and pathological human neurogenesis in the dish. *Front. Neurosci.* **2014**, *8*, 183. [CrossRef] [PubMed]
23.	Germain, N.; Banda, E.; Grabel, L. Embryonic stem cell neurogenesis and neural specification. *J. Cell Biochem.* **2010**, *111*, 535–542. [CrossRef] [PubMed]
24.	Adolf, B.; Chapouton, P.; Lam, C.S.; Topp, S.; Tannhäuser, B.; Strähle, U.; Götz, M.; Bally-Cuif, L. Conserved and acquired features of adult neurogenesis in the zebrafish telencephalon. *Dev. Biol.* **2006**, *295*, 278–293. [CrossRef]
25.	Cameron, D.A. Cellular proliferation and neurogenesis in the injured retina of adult zebrafish. *Vis. Neurosci.* **2000**, *17*, 789–797. [CrossRef]
26.	Nikolopoulou, E.; Galea, G.L.; Rolo, A.; Greene, N.D.E.; Copp, A.J. Neural tube closure: Cellular, molecular and biomechanical mechanisms. *Development* **2017**, *144*, 552–566. [CrossRef]
27.	Suzuki, M.; Morita, H.; Ueno, N. Molecular mechanisms of cell shape changes that contribute to vertebrate neural tube closure. *Dev. Growth Differ.* **2012**, *54*, 266–276. [CrossRef]
28.	Harding, M.J.; McGraw, H.F.; Nechiporuk, A. The roles and regulation of multicellular rosette structures during morphogenesis. *Development* **2014**, *141*, 2549–2558. [CrossRef]
29.	Maniscalco, J.; Corbo, C.P.; Fulop, Z. Two Types of Cells are Positively Labeled for Heparin in Surviving Organotypic Culture of the Optic Tectu, of Adult Zebrafish Brain. *In Vivo* **2018**, *39*, 103.
30.	Krystel-Whittemore, M.; Dileepan, K.N.; Wood, J.G. Mast Cell: A Multi-Functional Master Cell. *Front. Immunol.* **2015**, *6*, 620. [CrossRef]
31.	Khramtsova, Y.S.; Artashyan, O.S.; Yushkov, B.G.; Volkova, Y.L.; Nezgovorova, N.Y. Influence of Mast Cells on Reparative Regeneration of Tissues with Different Degree of Immunological Privileges. *Tsitologiia* **2016**, *58*, 356–363.
32.	Hiromatsu, Y.; Toda, S. Mast cells and angiogenesis. *Microsc. Res. Tech.* **2003**, *60*, 64–69. [CrossRef] [PubMed]
33.	Yuan, H.; Zhu, X.; Zhou, S.; Chen, Q.; Zhu, X.; Ma, X.; He, X.; Tian, M.; Shi, X. Role of mast cell activation in inducing microglial cells to release neurotrophin. *J. Neurosci. Res.* **2010**, *88*, 1348–1354. [CrossRef] [PubMed]

34. Kritas, S.K.; Caraffa, A.; Antinolfi, P.; Saggini, A.; Pantalone, A.; Rosati, M.; Tei, M.; Speziali, A.; Pandolfi, F.; Cerulli, G.; et al. Nerve growth factor interactions with mast cells. *Int. J. Immunopathol. Pharmacol.* **2014**, *27*, 15–19. [CrossRef] [PubMed]
35. Kyritsis, N.; Kizil, C.; Zocher, S.; Kroehne, V.; Kaslin, J.; Freudenreich, D.; Iltzsche, A.; Brand, M. Acute inflammation initiates the regenerative response in the adult zebrafish brain. *Science* **2012**, *338*, 1353–1356. [CrossRef]

 *NeuroSci*

MDPI

Article

Alcohol Deprivation Differentially Changes Alcohol Intake in Female and Male Rats Depending on Early-Life Stressful Experience

Marielly Carvalho [1,2,†], Gessynger Morais-Silva [3,4,†], Graziele Alícia Batista Caixeta [1,2], Marcelo T. Marin [3,4] and Vanessa C. S. Amaral [1,2,*]

[1] Laboratory of Pharmacology and Toxicology of Natural and Synthetic Products, State University of Goias, Exact and Technological Sciences Campus, Anapolis 75132-903, CO, Brazil; marielly.carvalho.mc@gmail.com (M.C.); grazielealicia2015@gmail.com (G.A.B.C.)

[2] Graduate Program in Sciences Applied to Health Products (PPGCAPS) UEG, Anápolis 75132-903, GO, Brazil

[3] Laboratory of Pharmacology, School of Pharmaceutical Sciences, Sao Paulo State University (UNESP), Araraquara 14800-903, SP, Brazil; gessynger.morais@unesp.br (G.M.-S.); marcelo.marin@unesp.br (M.T.M.)

[4] Joint Graduate Program in Physiological Sciences (PIPGCF) UFSCar/UNESP, São Carlos, Araraquara 14801-903, SP, Brazil

* Correspondence: vanessa.cristiane@ueg.br

† These authors contributed equally to this work.

Citation: Carvalho, M.; Morais-Silva, G.; Caixeta, G.A.B.; Marin, M.T.; Amaral, V.C.S. Alcohol Deprivation Differentially Changes Alcohol Intake in Female and Male Rats Depending on Early-Life Stressful Experience. *NeuroSci* 2022, 3, 214–225. https://doi.org/10.3390/neurosci3020016

Academic Editors: Charles L. Pickens and François Ichas

Received: 22 February 2022
Accepted: 7 April 2022
Published: 9 April 2022

Publisher's Note: MDPI stays neutral with regard to jurisdictional claims in published maps and institutional affiliations.

Abstract: Experiencing early-life adverse events has enduring effects on individual vulnerability to alcohol abuse and the development of addiction-related behaviors. In rodents, it can be studied using maternal separation (MS) stress. Studies have shown that, depending on the protocol used, MS can affect the mother and pups' behavior and are associated with behavioral alterations later in adulthood, associated with both positive or negative outcomes. However, it is not fully elucidated how MS affects relapse-like behaviors when experienced by female or male individuals. Therefore, the aim of our study was to evaluate the effects of brief and prolonged MS on the alcohol deprivation effect (ADE) in female and male rats. Female and male Wistar rats were exposed to brief (15 min/day) or prolonged (180 min/day) MS from postnatal day (PND) 2 to 10. Later, during adulthood (PND 70), animals were submitted to an ADE protocol. Brief MS exposure prevented the ADE in both females and males, while prolonged MS exposure also prevented the ADE in female rats. Moreover, the ADE was more robust in females when compared to males. In conclusion, we showed that male and female rats are differentially affected by alcohol deprivation periods depending on their early-life experiences.

Keywords: addiction; maternal separation; risk factor; sex differences

1. Introduction

The main feature involved in alcohol addiction is the neuroplasticity that undergoes the neurocircuitry related to motivation, habit, learning, and cognitive control. These neuroplastic events occur due to chronic drug intake and involve complex interactions with the environment and genetic factors [1]. Such long-lasting neuronal alterations are a crucial part of the most challenging step related to the treatment of addiction, namely relapse [2].

Given the pivotal role of neuroplasticity in the development of addiction, it is not surprising that early-life adverse events have enduring effects on individual vulnerability to alcohol abuse and the development of addiction-related behaviors [3]. A history of childhood maltreatment increases the incidence of alcohol use disorder (AUD) in adolescents [4] and adults [5,6] and is related to an earlier onset of drinking and alcohol abuse [7,8] and persistence of alcohol-related disorders through life [9]. Moreover, the greater the severity of the childhood abuse experienced, the greater is the impact on the prevalence and severity of the psychiatric disorders later in life [6,10]. Some studies highlight that such effects could be sex-dependent since they are more prevalent in women [11,12]. Female and male rats exposed to early life stress show an early maturation of the connections between the basolateral amygdala and the medial prefrontal cortex, while this alteration occurs even earlier in females compared to males [13]. Female rats exposed to early life adversity also show an increase in serotonin metabolites in the ventral tegmental area compared to males [14]. Frontal cortex maturation occurs earlier in females than males [15].

Sex differences in alcohol drinking behavior and the incidence of AUD are extensively reported in the literature. In rodents, most studies show that ethanol intake in females is greater than males' [16–20], while men drink more and have a greater prevalence of AUD [21]. Although such data seems to be contradictory to rodent studies, human drinking behavior is largely influenced by social factors [22,23]. Moreover, recent reports are showing that women are drinking more, facing more problems with alcohol drinking, and women who drink excessively are more susceptible to the development of AUD than men [24].

The alcohol deprivation effect (ADE) is a broadly used rodent paradigm to study relapse-like drinking and has been exploited to the screening of possible candidates for the treatment of AUD relapse [25–27]. It consists of the chronic exposition to voluntary ethanol intake followed by periods of ethanol deprivation and re-exposure, resulting in a temporary increase in intake and preference [25]. The FDA (USA Food and Drug Administration) approved medications naltrexone and acamprosate, block the ADE in rodents, and are used for the treatment of AUD in humans [28–30].

Maternal separation (MS) is an animal model widely used to study the effects of early-life stress on alcohol intake and abuse. It relies on the importance of the pups-mother relationship, especially during the first two weeks (at least in rodent models) and, can be roughly separated into two main categories: the brief and prolonged MS [31]. These conditions affect the mother and pups' behavior and are associated with behavioral alterations later in adulthood, associated with both positive and negative outcomes, depending on the protocol used [31–33]. In the brief MS, the pups are kept apart from their dams for short time intervals (usually 3 to 15 min), which mimics the natural environment of these animals (in situations such as dams leaving the nest to go after food, for example). In prolonged MS, the pups are separated from their dams for prolonged time intervals (from 60 min to 24 h) [34]. While the brief MS has been associated with positive behavioral outcomes, reducing the effects of stress exposure later in life [33], the prolonged MS has been associated with negative outcomes, increasing the stress reactivity later in life [35].

In animal models, a history of early-life stress also seems to impact the individual vulnerability to addiction-related behaviors and alcohol abuse. As well as for stress responses, such effects are highly dependent on the protocol used. Most of the studies have shown that the brief MS seems to decrease while the prolonged MS increases ethanol intake in rats when the animals are tested in adult life [16,31,36–38]. Prolonged MS also increases the impulsivity for alcohol consumption in adult rats [39] and the ethanol conditioned place preference (CPP) acquisition in adolescent rats, while it does not impact CPP in adult rats [16]. However, there are no studies examining MS effects on animal models of ethanol intake relapse. Thus, considering the importance of the relapse behavior to understanding ethanol addiction and to the development of treatment strategies, we studied the effects of brief and prolonged MS on the alcohol deprivation effect, a rodent model for the study of ethanol intake relapse [25] in female and male rats.

2. Materials and Methods

2.1. Animals

Female and male Wistar rats from the State University of Goias were used for mating and as experimental animals. Experiments were carried out according to the principles and standards of the National Council for the Control of Animal Experimentation (CONCEA), based on NIH Guidelines for the Care and Use of Laboratory Animals as approved by the Commission on Ethics in Animal Use (CEUA) of the State University of Goias (protocol number 008/2016).

Forty-five adult Wistar rats (females: n = 30; males: n = 15) were used for mating. Primiparous females were mated with adult males in a 1:1 ratio, and gestation day 0 (GD0) was defined as the day when copulatory plugs and/or sperm were found. Pregnant rats were kept alone during gestation and checked daily for newborn pups.

Seventy-six Wistar rats (males: n = 39; females: n = 37) were used as experimental animals. After weaning (postnatal day 21 - PND 21), they were kept in groups of 3–5 animals/cage, until the beginning of the alcohol deprivation effect experiment (at PND 70), when they were individually housed. The females' estrous cycle was not pharmacologically synchronized.

Food and water were available *ad libitum* through the experiments. Animals were maintained in polypropylene cages (41 × 34 × 16 cm), under a 12 h light/dark cycle in a temperature-controlled environment (22 ± 2 °C).

2.2. Maternal Separation (MS)

The brief and prolonged MS was performed as previously described [16]. From PND 2 to 10, pups were daily separated from their dams for 15 minutes (brief MS – MS15) or 180 min (prolonged MS – MS180). During the separation period, the dams were transferred to clean cages and moved to a separate room, away from the pups. The pups were kept in their nests, under a red incandescent light, maintaining the temperature at 32 ± 2 °C. Weaning was normally performed at PND 21. The control group was kept undisturbed with the dams (except for cage cleaning) until weaning. To avoid any bias related to the litter, a maximum of two pups of each sex, from each dam, were used in each experimental group.

2.3. Alcohol Deprivation Effect (ADE)

The ADE protocol was adapted in our laboratory from the previously described [40].

Starting on PND 70, animals were exposed, for two days, to two bottles containing ethanol (6%) as the only source of fluid. Next, animals were given free access to two bottles, one bottle containing crescent ethanol concentrations (6, 8, and 10%) and one containing filtered water, for 4 days each concentration. This phase was called habituation and was intended to initiate alcohol drinking and habituate animals to the taste of the ethanol and experimental procedure. The ethanol consumption of each concentration was analyzed separately and identified as follows: ETOH 6% phase, ETOH 8% phase, and ETOH 10% phase. After that, baseline consumption was recorded: animals were given free access to two bottles, one containing ethanol (10%) and one containing filtered water for 10 days. Baseline consumption was considered as the average of the 4 days before the beginning of the deprivation phase (days in both male and female control groups with less than 20% variation from the average intake of ethanol 10%).

The next day after the baseline consumption phase, animals were submitted to the alcohol deprivation phase. Each deprivation period was comprised of 14 days of alcohol deprivation when only one bottle containing water was available followed by 7 days of free access to one bottle of 10% ethanol solution and one containing filtered water. The ADE was analyzed using the average consumption of four days after the end of each deprivation period. The animals were submitted to two deprivation periods. The experimental procedure indicating each ethanol exposure and deprivation phase is depicted in Figure 1.

Ethanol solutions (*v/v*) were prepared from absolute ethanol (Neon Comercial, São Paulo, SP, Brazil) diluted in filtered water and offered in plastic bottles fitted with stainless steel sipper tubes with ball-valve nipples in rubber stoppers. Ethanol and fluid intake was measured by weighing the bottles and reported as relative to body weight (g/kg). Liquid loss by leakage or evaporation was accessed using bottles placed in empty cages and was subtracted from the amount consumed by each animal.

Solutions were prepared fresh every day and presented to the animals at the same time. The position of the bottles was alternated every day to avoid side preference.

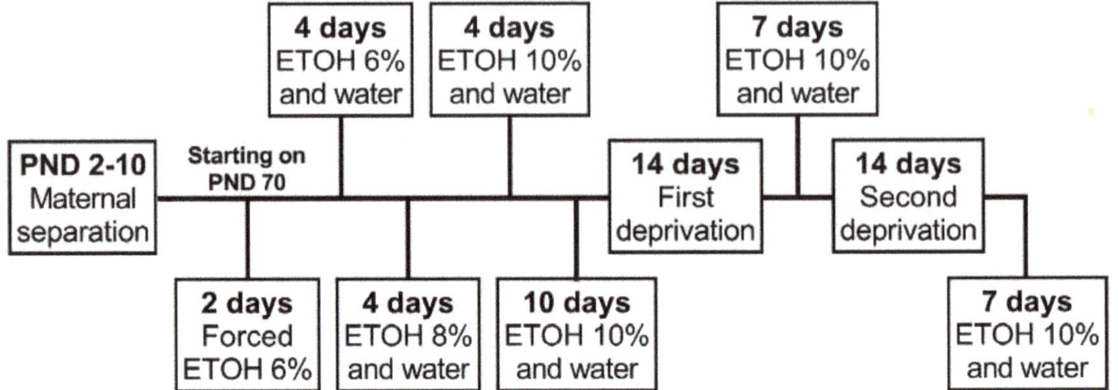

Figure 1. Experimental procedure timeline. Female and male rats were exposed to brief or prolonged maternal separation during early life and submitted to an alcohol deprivation effect protocol during adulthood. PND, postnatal day; ETOH, ethanol.

2.4. Statistics

Values were expressed as mean $\pm$ SEM and analyzed by repeated-measures ANOVA considering the independent factors sex and MS and the repeated measure phases. When ANOVA showed significant differences ($p \leq 0.05$), the Newman-Keuls *post hoc* test was performed.

Statistics were performed using Statistica 7.1 software (StatSoft, Inc., Tulsa, OK, USA) and graphs constructed using GraphPad Prism 7 software (GraphPad Software Inc., La Jolla, CA, USA).

3. Results

3.1. Ethanol Intake Prior to Deprivations

We found a significant difference in ethanol intake before deprivations, related to the sex and the phase of the experiment, but not MS (Figure 2). The repeated-measures ANOVA showed a significant effect for the factor sex ($F_{1,69} = 56.07$; $p < 0.001$), phase ($F_{3,207} = 13.14$; $p < 0.001$), and the interaction between sex and phase ($F_{3,207} = 3.50$; $p < 0.05$). The Newman-Keuls test revealed that females consumed more ethanol than males in all phases of the experiment before deprivations ($p < 0.001$). In males, ethanol intake during the baseline phase was increased when compared to ethanol consumption in ETOH 6% phase ($p < 0.001$). In females, ethanol consumption during the ETOH 10% phase was the greater prior deprivations ($p < 0.05$), while baseline ethanol intake was increased when compared to ethanol intake during ETOH 6% ($p < 0.01$) and ETOH 8% phases ($p < 0.05$).

3.2. Alcohol Deprivation Effect

Alcohol deprivation differentially affected the ethanol consumption in males and females, depending on the early-life experiences (brief or prolonged MS) (Figure 3A and B). The repeated-measures ANOVA showed a significant effect for MS ($F_{2,69} = 4.52$; $p < 0.05$), sex ($F_{1,69} = 17.59$; $p < 0.001$), phase ($F_{2,138} = 29.07$; $p < 0.001$), and for the interaction between MS and phase ($F_{4,138} = 3.22$; $p < 0.05$) and, MS, sex and phase ($F_{4,138} = 4.06$; $p < 0.01$).

Control males showed an increase in ethanol consumption only after the second deprivation ($p < 0.05$) compared to baseline. MS180 male group, similar to control males, showed a significant increase in ethanol consumption only after the second deprivation (emphp < 0.001) compared to baseline intake, while the MS15 male group did not show significant alterations in ethanol consumption after neither the first or second deprivation phases relative to baseline (Figure 3A).

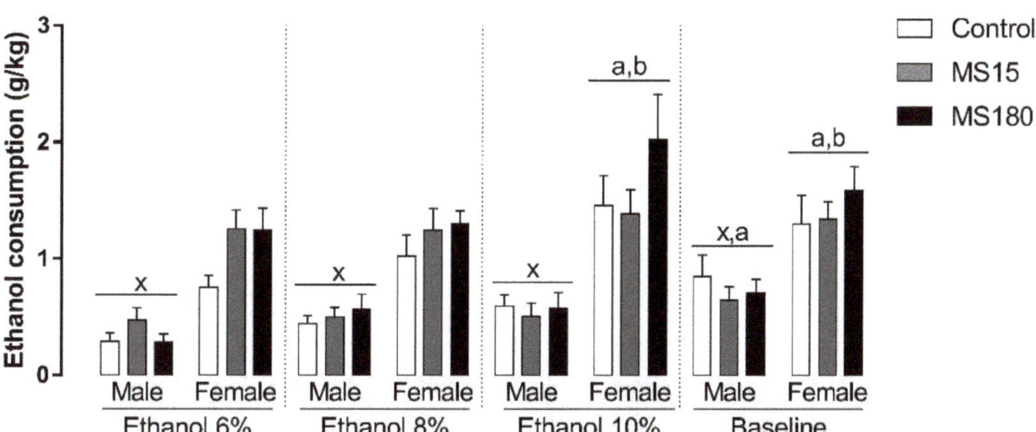

Figure 2. Ethanol intake before deprivations in female and male rats exposed to brief and prolonged maternal separation. Bars represent the mean ± SEM (n = 11–14 animals/group). Each concentration represents the average intake of the 4 days of exposition in each experimental phase of the procedure. x; $p < 0.05$ when compared to female groups of the same experimental phase, independent of manipulation; a, $p < 0.05$ when compared to the ETOH 6% phase of the same sex, independent of manipulation; b, $p < 0.05$ when compared to ETOH 8% phase of the same sex, independent of manipulation; c, $p < 0.05$ when compared to ETOH 10% phase of the same sex, independent of manipulation. ETOH, ethanol; MS15, brief maternal separation; MS180, prolonged maternal separation.

Control females showed an increase in ethanol consumption after either the first ($p < 0.01$) or second deprivation ($p < 0.001$) compared to baseline. Ethanol intake after the second deprivation was also greater than ethanol intake after the first deprivation ($p < 0.01$). Both MS15 and MS180 female groups did not show any significant alterations in ethanol consumption after neither the first or second deprivation phases. Additionally, ethanol consumption of the control female group was greater than the MS15 female group ($p < 0.05$), MS180 female group ($p < 0.05$) and control male group ($p < 0.05$) after the second deprivation (Figure 3B).

We also found significant alterations in ethanol preference (Figure 3C,D). The repeated-measures ANOVA showed a significant effect for MS ($F_{2,69} = 3.70$; $p < 0.05$), phase ($F_{2,138} = 37.68$; $p < 0.001$), and for the interaction between sex and phase ($F_{2,138} = 4.06$; $p < 0.05$) and, MS, sex and phase ($F_{4,138} = 2.41$; $p < 0.01$).

Control males showed an increase in ethanol preference only after the second deprivation compared to baseline ($p < 0.001$) and after the first deprivation ($p < 0.05$). MS180 male group, similar to control males, showed a significant increase in ethanol preference only after the second deprivation ($p > 0.001$) compared to baseline and after first deprivation ($p < 0.05$), while the MS15 male group did not show significant alterations in ethanol preference after neither the first or second deprivation phases relative to baseline (Figure 3C). Similar to control males, control females showed an increase in ethanol preference after the second deprivation compared to baseline ($p < 0.001$) and after the first deprivation ($p < 0.01$). MS15 and MS180 female groups did not show an increase in ethanol preference after neither the first or second deprivation phases. Ethanol preference of the control female group was greater than the MS180 female group ($p < 0.05$) after the second deprivation (Figure 3D).

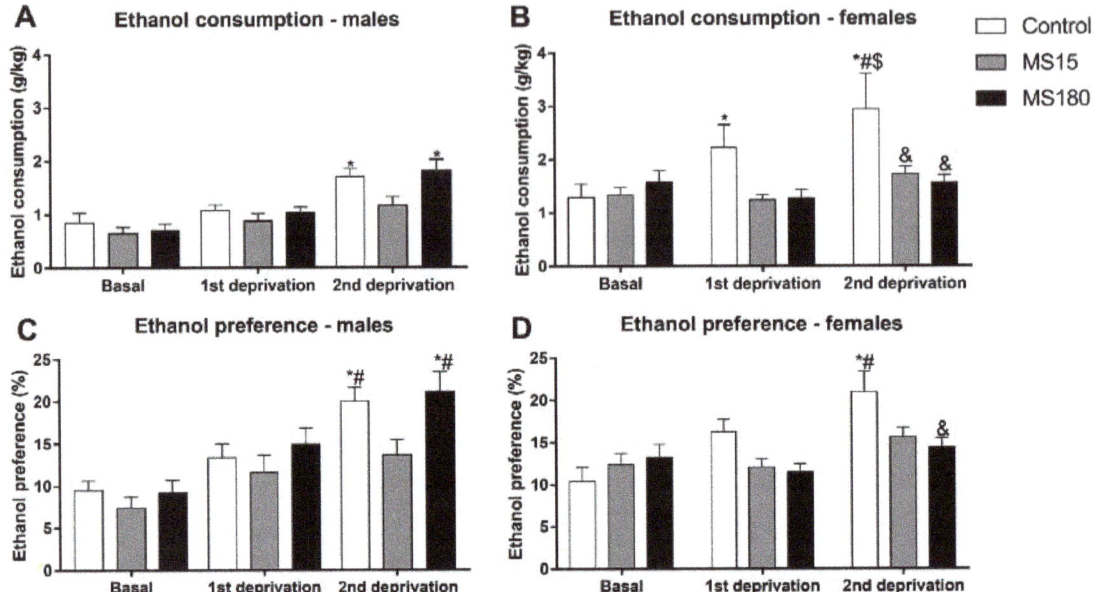

Figure 3. Alcohol deprivation effect (ADE) in female and male rats exposed to brief or prolonged maternal separation. The ADE was analyzed using the average ethanol intake or preference of four days after the end of each deprivation period compared to the last four days of the baseline period. The animals were submitted to two deprivation periods. Bars represent the mean ± SEM (n = 11–14 animals/group). (**A**), male ethanol consumption (g/kg); (**B**), female ethanol consumption (g/kg); (**C**), male ethanol preference; (**D**), female ethanol preference. *, $p < 0.05$ relative to the baseline ethanol consumption of the same gender control group; #, $p < 0.05$ relative to the ethanol consumption after the first deprivation of the same gender control group; &, $p < 0.05$ relative to the ethanol consumption after the second deprivation of the same gender control group; $, $p < 0.05$ relative to the ethanol consumption of the control male group after the second deprivation. MS15, brief maternal separation; MS180, prolonged maternal separation.

3.3. Fluid Intake

There was a significant effect of sex and phase, independent of the MS, on fluid intake through the experiment (Figure 4). The repeated measures ANOVA showed a significant effect for sex ($F_{1,69} = 201.83$; $p < 0.001$), phase ($F_{5,345} = 30.61$; $p < 0.001$) and for the interaction between sex and phase ($F_{5,345} = 2.60$; $p < 0.05$). Overall, female fluid intake was increased compared to males ($p < 0.001$). Moreover, fluid intake, independent of the sex or MS decreased through the experimental phases. In males, fluid intake in ETOH 10%, baseline, 1st re-exposure, and 2nd re-exposure phases was smaller than fluid intake in ETOH 6% and ETOH 8% phases, while fluid intake during 1st re-exposure was smaller than ETOH 10% phase. In females, baseline, 1st re-exposure, and 2nd re-exposure fluid intake were smaller than ETOH 6%, ETOH 8%, and ETOH 10% phases. The female fluid intake in ETOH 10% phase was smaller than fluid intake in ETOH 6% phase.

3.4. Body Weight

Body weight was evaluated through the different phases of the ADE protocol and reported as body weight gain in each period (Figure 5). First, we found a significant effect in initial body weight (Figure 5A) for the factor maternal separation ($F2,69 = 5.06$; $p < 0.01$) and sex ($F1,69 = 567.40$; $p < 0.001$). The post-hoc analysis revealed that males began the alcohol deprivation effect protocol heavier than females ($p < 0.001$), and both brief or prolonged maternal separation increased body weight in both males and females ($p < 0.05$).

On the other hand, total body weight (Figure 5B) through the experiment was only affected by sex (F1,69 = 55.57; $p < 0.001$), revealing that males gain more weight than females over time ($p < 0.001$). Next, we evaluated the body weight gain in each phase of our protocol (Figure 5C). Body weight gain during baseline phase was affected only by sex (F1,69 = 4.21; $p < 0.05$), revealing a greater body weight gain in males compared to females ($p < 0.05$). During the first deprivation period, the two-way ANOVA revealed a significant effect for maternal separation (F2,69 = 5.51; $p < 0.01$) and sex (F1,69 = 35.44; $p < 0.001$), showing that males gained more weight during this phase ($p < 0.001$) and maternal separation increased body weight gain during the ethanol deprivation phase ($p < 0.05$). We found similar results for the first re-exposure phase: a significant effect for maternal separation (F2,69 = 10.88; $p < 0.01$) and sex (F1,69 = 7.19; $p < 0.001$), showing that males gained more weight during this phase ($p < 0.001$) and maternal separation increased body weight gain during the ethanol deprivation phase ($p < 0.001$). We found a significant effect only for sex during the second deprivation phase (F1,69 = 7.90; $p < 0.01$), when males gained more weight than females ($p < 0.01$). Finally, we found a significant effect for the interaction between maternal separation and sex (F2,69 = 7.63; $p < 0.01$) in the second re-exposure phase, while the post-hoc test did not reveal any significant differences.

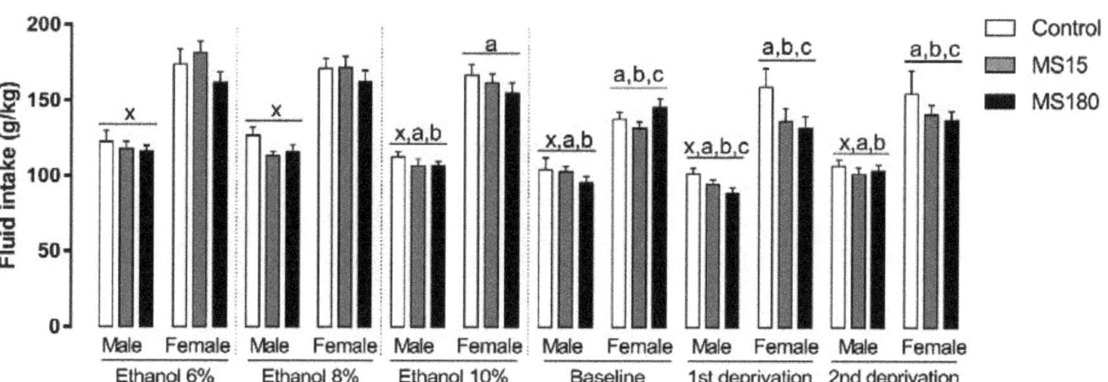

Figure 4. Fluid intake through the experiment in female and male rats exposed to brief and prolonged maternal separation. Bars represent the mean ± SEM (n = 11–14 animals/group). Each concentration represents the average intake of the 4 days of exposition in each experimental phase of the procedure. x; $p < 0.05$ when compared to female groups, independent of the manipulation or experimental phase; a, $p < 0.05$ when compared to the ETOH 6% phase, independent of sex or manipulation; b, $p < 0.05$ when compared to ETOH 8% phase, independent of sex or manipulation; c, $p < 0.05$ when compared to ETOH 10% phase, independent of sex or manipulation. ETOH, ethanol; MS15, brief maternal separation; MS180, prolonged maternal separation.

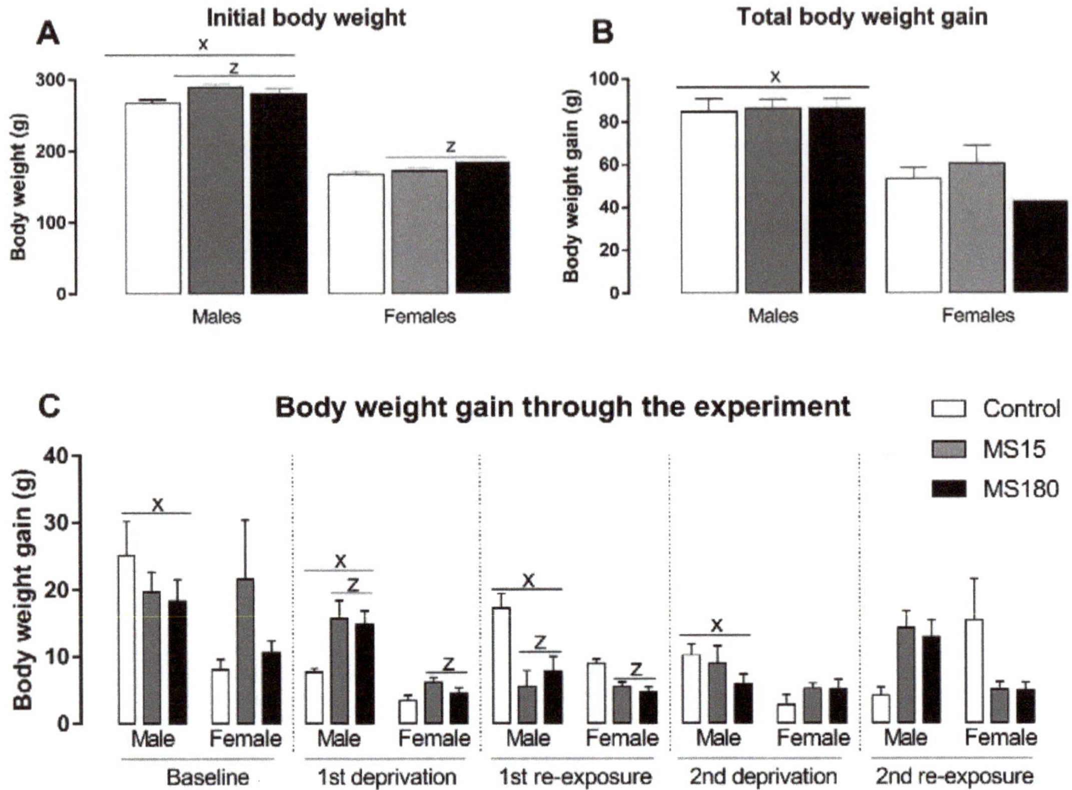

Figure 5. Initial body weight and body weight gain through the experiment in female and male rats exposed to brief and prolonged maternal separation. Bars represent the mean ± SEM (n = 11–14 animals/group). (**A**), initial body weight (g). (**B**), body weight gain during the whole period of alcohol deprivation effect (g/kg). (**C**), body weight gain in different phases of the alcohol deprivation effect protocol (g/kg). Body weight gain was calculated as the difference between the body weight on the last day of the respective phase and the first day of the same experimental phase. x; $p < 0.05$ when compared to female groups, independent of the manipulation; z, significant effect of maternal separation, regardless of sex. MS15, brief maternal separation; MS180, prolonged maternal separation.

4. Discussion

Here we showed that male and female rats are differentially affected by alcohol deprivation periods depending on the early-life experiences (brief or prolonged MS). First, females naturally drink more when compared to males. In males, the ADE was evident only after the 2nd ethanol re-exposure, and brief MS exposure was able to prevent it. In females, the ADE emerged earlier, in the 1st re-exposure period, while both brief and prolonged MS exposure prevented the ADE.

As already reported, we found remarkable differences in alcohol consumption between females and males. In both mice and rats, studies using several animal models of voluntary ethanol consumption showed that females consume more ethanol than males [16–20]. In a recent study from our lab, we showed that such differences are influenced by the concentration of the ethanol solution offered to the animals [16]. On the other hand, in humans, opposite results have been reported in epidemiological studies: men's ethanol intake and AUD prevalence are greater when compared to women, although women's

AUD prevalence has been growing recently [21]. Such results give strength to the view that alcohol drinking in humans is strongly influenced by cultural factors and that the underrepresentation of female individuals in biomedical research has a profound impact on substance use disorders research [41].

Our results also revealed that female Wistar rats are more susceptible to ADE since the increased alcohol intake in females showed up sooner than in males. As cited before, there is evidence suggesting that women who drink excessively are more prone to the development of medical problems compared to men [24], despite the greater incidence of AUD in men. In animal models, the results are highly dependent on the protocol used and the behavioral outcome tested. In this context, female rats showed enhanced sensitivity to the rewarding effects of ethanol [42] and increased ADE [19,43]. On the other hand, male rodents show increased behavioral signs of ethanol withdrawal [44] and greater withdrawal-induced ethanol-seeking behavior compared to females [45].

Studies in rodents, using mixed-sex analysis, showed that the prolonged disruption of pups-mother contact during early-life increases ethanol binge drinking and voluntary ethanol consumption in operant self-administration paradigms performed later in life [39,46–48]. Prolonged MS also increases ethanol consumption in male rats tested during adulthood in the 2-bottle [36,49,50] or 3-bottle choice paradigms [16]. On the other hand, brief MS during early life usually decreases ethanol intake in adult male rodents [36,50]. Both brief and prolonged periods of MS increase voluntary ethanol consumption in female rats submitted to a 3-bottle choice procedure [16], while there is no effect of MS when ethanol consumption is evaluated using the 2-bottle choice procedure [51,52]. Here, we found a more complex effect: while, in males, brief MS blocked the ADE, both brief and prolonged MS prevented the ADE in females. To date, few studies looked for the effects of MS on the ADE, especially using female individuals in the experimental procedure. One study found that neither brief nor prolonged MS affected the increased ethanol consumption after periods of intermittent ethanol exposure [53]. However, in this study, the authors did not use an animal facility hearing or handled control without separation periods, making difficult the comparison between the studies.

We found interesting results regarding body weight gain through different phases of our ADE protocol. As expected, females gained less body weight compared to males, regardless of the experimental phase, since male Wistar rats are larger and have an increased growth rate from puberty onwards than female Wistar rats [54]. Interestingly, animals exposed to MS, independent of MS length and sex, showed increased body weight. Moreover, maternally-separated rats showed increased body weight gain during periods of increased stress, that is, the first deprivation and first ethanol re-exposure phases. It should be related to an increased stress sensitivity in animals exposed to early-life stressors [35].

The variation in the period when brain maturation occurs could be related to the MS sex differences found in our study. Females' prefrontal cortex and amygdala undergo maturation faster when compared to males [13–15] and thus are more likely to be affected by the MS during the postnatal period that we performed our experiments. Prefrontal cortex dendritic morphology is more susceptible to the effects of MS in females than in males [55]. In this sense, the prefrontal cortex and the amygdala are involved in alcohol craving and relapse for alcohol drinking [56].

Two variables presented in our study could have impacted our results: the single-housing condition and the females' estrous cycle phase. First, social isolation is a known stressor for rodents and can affect voluntary ethanol intake [57,58]. This is an important limitation of our work since individual housing was necessary for the evaluation of ethanol intake. On the other hand, more recent works have been showing that social isolation does not impact consumption of ethanol solutions up to 10%, in both rats and mice [59,60]. Thus, the procedure used in our work should be taken into account when interpreting our results. Secondly, we did not evaluate the estrous cycle or synchronized females' cycle in our work. Literature reports have been showing that the estrous phase does not have a substantial impact on ethanol intake in rats [20], although it seems to influence ethanol consumption microstructure [61]. It could be explained by the fact that ethanol administration itself disrupts the estrous cycle in rats [62,63]. Moreover, human studies are inconsistent regarding the correlation between the menstrual cycle phase and alcohol intake [64].

In conclusion, brief MS exerted a protective effect against the ADE in both female and male rats. Despite the increased susceptibility to the ADE shown by females, prolonged MS also prevented the ADE in females, revealing sex-dependent effects of MS in rats.

Author Contributions: M.C., G.M.-S. and G.A.B.C. performed the experiments. G.M.-S. analyzed the data and drafted the manuscript. M.T.M. and V.C.S.A. designed the study, supervised the project, and revised the manuscript for intellectual content. All authors have read and agreed to the published version of the manuscript.

Funding: This research was funded by Fundação de Amparo a Pesquisa do Estado de Goias (FAPEG/CNPq Programa Primeiros Projetos [201610267001023]).

Institutional Review Board Statement: Experiments were carried out according to the principles and standards of the Brazilian National Council for the Control of Animal Experimentation (CONCEA), based on NIH Guidelines for the Care and Use of Laboratory Animals as approved by the Commission on Ethics in Animal Use (CEUA) of the State University of Goias (protocol number 008/2016).

Informed Consent Statement: Not applicable.

Data Availability Statement: The data presented in this study are available on request from the corresponding author.

Conflicts of Interest: The authors declare no conflict of interest.

References

1. Koob, G.F.; Volkow, N.D. Neurobiology of addiction: A neurocircuitry analysis. *Lancet Psychiatry* **2016**, *3*, 760–773. [CrossRef]
2. Feltenstein, M.W.; See, R.E.; Fuchs, R.A. Neural Substrates and Circuits of Drug Addiction. *Cold Spring Harb. Perspect. Med.* **2020**, *11*, a039628. [CrossRef] [PubMed]
3. Walters, H.; Kosten, T.A. Early life stress and the propensity to develop addictive behaviors. *Int. J. Dev. Neurosci.* **2019**, *78*, 156–169. [CrossRef] [PubMed]
4. Harrison, P.A.; Fulkerson, J.A.; Beebe, T.J. Multiple substance use among adolescent physical and sexual abuse victims. *Child Abus. Negl.* **1997**, *21*, 529–539. [CrossRef]
5. Dube, S.R.; Anda, R.F.; Whitfield, C.L.; Brown, D.W.; Felitti, V.J.; Dong, M.; Giles, W.H. Long-Term Consequences of Childhood Sexual Abuse by Gender of Victim. *Am. J. Prev. Med.* **2005**, *28*, 430–438. [CrossRef]
6. Felitti, V.J.; Anda, R.F.; Nordenberg, D.; Williamson, D.F.; Spitz, A.M.; Edwards, V.; Koss, M.P.; Marks, J.S. Relationship of Childhood Abuse and Household Dysfunction to Many of the Leading Causes of Death in Adults: The Adverse Childhood Experiences (ACE) Study. *Am. J. Prev. Med.* **1998**, *14*, 245–258. [CrossRef]
7. Pilowsky, D.J.; Keyes, K.M.; Hasin, D.S. Adverse Childhood Events and Lifetime Alcohol Dependence. *Am. J. Public Heal.* **2009**, *99*, 258–263. [CrossRef]
8. Green, J.; McLaughlin, K.; Berglund, P.A.; Gruber, M.J.; Sampson, N.A.; Zaslavsky, A.M.; Kessler, R.C. Childhood Adversities and Adult Psychiatric Disorders in the National Comorbidity Survey Replication I. *Arch. Gen. Psychiatry* **2010**, *67*, 113–123. [CrossRef]
9. McLaughlin, K.; Green, J.; Gruber, M.J.; Sampson, N.A.; Zaslavsky, A.M.; Kessler, R.C. Childhood Adversities and Adult Psychiatric Disorders in the National Comorbidity Survey Replication II. *Arch. Gen. Psychiatry* **2010**, *67*, 124–132. [CrossRef]
10. Bulik, C.M.; Prescott, C.A.; Kendler, K.S. Features of childhood sexual abuse and the development of psychiatric and substance use disorders. *Br. J. Psychiatry* **2001**, *179*, 444–449. [CrossRef]
11. Widom, C.S.; Ireland, T.; Glynn, P.J. Alcohol abuse in abused and neglected children followed-up: Are they at increased risk? *J. Stud. Alcohol* **1995**, *56*, 207–217. [CrossRef] [PubMed]
12. Young-Wolff, K.C.; Kendler, K.S.; Prescott, C.A. Interactive Effects of Childhood Maltreatment and Recent Stressful Life Events on Alcohol Consumption in Adulthood. *J. Stud. Alcohol Drugs* **2012**, *73*, 559–569. [CrossRef] [PubMed]
13. A Honeycutt, J.; Demaestri, C.; Peterzell, S.; Silveri, M.M.; Cai, X.; Kulkarni, P.; Cunningham, M.G.; Ferris, C.F.; Brenhouse, H.C. Altered corticolimbic connectivity reveals sex-specific adolescent outcomes in a rat model of early life adversity. *eLife* **2020**, *9*, 1–27. [CrossRef] [PubMed]
14. Mesquita, A.; Pêgo, J.M.; Summavielle, T.; Maciel, P.; Almeida, O.F.; Sousa, N. Neurodevelopment milestone abnormalities in rats exposed to stress in early life. *Neuroscience* **2007**, *147*, 1022–1033. [CrossRef]
15. Perry, C.J.; Campbell, E.J.; Drummond, K.D.; Lum, J.S.; Kim, J.H. Sex differences in the neurochemistry of frontal cortex: Impact of early life stress. *J. Neurochem.* **2020**, *157*, 963–981. [CrossRef]

16. Leichtweis, K.S.; Carvalho, M.; Morais-Silva, G.; Marin, M.T.; Amaral, V.C.S. Short and prolonged maternal separation impacts on ethanol-related behaviors in rats: Sex and age differences. *Stress* **2019**, *23*, 162–173. [CrossRef]

17. Blanchard, B.A.; Steindorf, S.; Wang, S.; Glick, S.D. Sex Differences in Ethanol-induced Dopamine Release in Nucleus Accumbens and in Ethanol Consumption in Rats. *Alcohol. Clin. Exp. Res.* **1993**, *17*, 968–973. [CrossRef]

18. Barker, J.M.; Torregrossa, M.M.; Arnold, A.P.; Taylor, J.R. Dissociation of Genetic and Hormonal Influences on Sex Differences in Alcoholism-Related Behaviors. *J. Neurosci.* **2010**, *30*, 9140–9144. [CrossRef]

19. Li, J.; Chen, P.; Han, X.; Zuo, W.; Mei, Q.; Bian, E.Y.; Umeugo, J.; Ye, J. Differences between male and female rats in alcohol drinking, negative affects and neuronal activity after acute and prolonged abstinence. *Int. J. Physiol. Pathophysiol. Pharmacol.* **2019**, *11*, 163–176.

20. Priddy, B.M.; Carmack, S.A.; Thomas, L.C.; Vendruscolo, J.C.; Koob, G.F.; Vendruscolo, L.F. Sex, strain, and estrous cycle influences on alcohol drinking in rats. *Pharmacol. Biochem. Behav.* **2016**, *152*, 61–67. [CrossRef]

21. WHO. *Global Status Report on Alcohol and Health*; World Health Organization: Geneva, Switzerland, 2018; Available online: https://www.who.int/substance_abuse/publications/global_alcohol_report/en/ (accessed on 10 January 2022).

22. Kerr-Corrêa, F.; Igami, T.Z.; Hiroce, V.; Tucci, A.M. Patterns of alcohol use between genders: A cross-cultural evaluation. *J. Affect. Disord.* **2007**, *102*, 265–275. [CrossRef] [PubMed]

23. Rahav, G.; Wilsnack, R.; Bloomfield, K.; Gmel, G.; Kuntsche, S. The influence of societal level factors on men's and women's alcohol consumption and alcohol problems. *Alcohol Alcohol.* **2006**, *41*, i47–i55. [CrossRef] [PubMed]

24. Erol, A.; Karpyak, V.M. Sex and gender-related differences in alcohol use and its consequences: Contemporary knowledge and future research considerations. *Drug Alcohol Depend.* **2015**, *156*, 1–13. [CrossRef] [PubMed]

25. Vengeliene, V.; Bilbao, A.; Spanagel, R. The alcohol deprivation effect model for studying relapse behavior: A comparison between rats and mice. *Alcohol* **2014**, *48*, 313–320. [CrossRef]

26. Martin-Fardon, R.; Weiss, F. Modeling relapse in animals. *Curr. Top. Behav. Neurosci.* **2012**, *13*, 403–432. [CrossRef]

27. Koob, G.F. Animal models of craving for ethanol. *Addiction* **2000**, *95* (Suppl. 2), S73–S81. Available online: http://www.ncbi.nlm.nih.gov/pubmed/11002904 (accessed on 10 January 2022). [CrossRef]

28. Heyser, C.J.; Moc, K.; Koob, G.F. Effects of Naltrexone Alone and In Combination With Acamprosate on the Alcohol Deprivation Effect in Rats. *Neuropsychopharmacology* **2003**, *28*, 1463–1471. [CrossRef]

29. Zhou, Y.; Crowley, R.; Prisinzano, T.; Kreek, M.J. Effects of mesyl salvinorin B alone and in combination with naltrexone on alcohol deprivation effect in male and female mice. *Neurosci. Lett.* **2018**, *673*, 19–23. [CrossRef]

30. Tomie, A.; Azogu, I.; Yu, L. Effects of naltrexone on post-abstinence alcohol drinking in C57BL/6NCRL and DBA/2J mice. *Prog. Neuro-Psychopharmacol. Biol. Psychiatry* **2013**, *44*, 240–247. [CrossRef]

31. Nylander, I.; Roman, E. Is the rodent maternal separation model a valid and effective model for studies on the early-life impact on ethanol consumption? *Psychopharmacology* **2013**, *229*, 555–569. [CrossRef]

32. Orso, R.; Creutzberg, K.C.; Wearick-Silva, L.E.; Viola, T.W.; Tractenberg, S.G.; Benetti, F.; Grassi-Oliveira, R. How Early Life Stress Impact Maternal Care: A Systematic Review of Rodent Studies. *Front. Behav. Neurosci.* **2019**, *13*. [CrossRef] [PubMed]

33. Raineki, C.; Lucion, A.B.; Weinberg, J. Neonatal handling: An overview of the positive and negative effects. *Dev. Psychobiol.* **2014**, *56*, 1613–1625. [CrossRef] [PubMed]

34. Newport, D.J.; Stowe, Z.N.; Nemeroff, C.B. Parental Depression: Animal Models of an Adverse Life Event. *Am. J. Psychiatry* **2002**, *159*, 1265–1283. [CrossRef] [PubMed]

35. Vetulani, J. Early maternal separation: A rodent model of depression and a prevailing human condition. *Pharmacol. Rep.* **2013**, *65*, 1451–1461. [CrossRef]

36. Ploj, K.; Roman, E.; Nylander, I. Long-term effects of maternal separation on ethanol intake and brain opioid and dopamine receptors in male wistar rats. *Neuroscience* **2003**, *121*, 787–799. [CrossRef]

37. Roman, E.; Nylander, I. The impact of emotional stress early in life on adult voluntary ethanol intake-results of maternal separation in rats. *Stress* **2005**, *8*, 157–174. [CrossRef]

38. Cruz, F.C.; Quadros, I.; Planeta, C.D.S.; Miczek, K.A. Maternal separation stress in male mice: Long-term increases in alcohol intake. *Psychopharmacology* **2008**, *201*, 459–468. [CrossRef]

39. Gondré-Lewis, M.C.; Warnock, K.T.; Wang, H.; June, H.L.; Bell, K.A.; Rabe, H.; Tiruveedhula, V.V.N.P.B.; Cook, J.; Lüddens, H.; Aurelian, L. Early life stress is a risk factor for excessive alcohol drinking and impulsivity in adults and is mediated via a CRF/GABAA mechanism. *Stress* **2016**, *19*, 235–247. [CrossRef]

40. Funk, D.; Vohra, S. Influence of stressors on the rewarding effects of alcohol in Wistar rats: Studies with alcohol deprivation and place conditioning. *Psychopharmacology* **2004**, *176*, 82–87. [CrossRef]

41. Hilderbrand, E.R.; Lasek, A.W. Studying Sex Differences in Animal Models of Addiction: An Emphasis on Alcohol-Related Behaviors. *ACS Chem. Neurosci.* **2017**, *9*, 1907–1916. [CrossRef]

42. Torres, O.V.; Walker, E.M.; Beas, B.S.; O'Dell, L.E. Female Rats Display Enhanced Rewarding Effects of Ethanol That Are Hormone Dependent. *Alcohol. Clin. Exp. Res.* **2013**, *38*, 108–115. [CrossRef] [PubMed]

43. García-Burgos, D.; Zuluaga, T.M.; Torre, M.G.; Reyes, F.G. Diferencias sexuales en el efecto de privación alcohólica en ratas. *Psicothema* **2010**, *22*, 887–892. [PubMed]

44. Becker, J.B.; Koob, G.F. Sex Differences in Animal Models: Focus on Addiction. *Pharmacol. Rev.* **2016**, *68*, 242–263. [CrossRef] [PubMed]

45. Randall, P.A.; Stewart, R.T.; Besheer, J. Sex differences in alcohol self-administration and relapse-like behavior in Long-Evans rats. *Pharmacol. Biochem. Behav.* **2017**, *156*, 1–9. [CrossRef] [PubMed]
46. Gondré-Lewis, M.C.; Darius, P.J.; Wang, H.; Allard, J.S. Stereological analyses of reward system nuclei in maternally deprived/separated alcohol drinking rats. *J. Chem. Neuroanat.* **2016**, *76*, 122–132. [CrossRef]
47. Bassey, R.; Gondré-Lewis, M.C. Combined early life stressors: Prenatal nicotine and maternal deprivation interact to influence affective and drug seeking behavioral phenotypes in rats. *Behav. Brain Res.* **2018**, *359*, 814–822. [CrossRef]
48. Bertagna, N.B.; Favoretto, C.A.; Rodolpho, B.T.; Palombo, P.; Yokoyama, T.S.; Righi, T.; Loss, C.M.; Leão, R.M.; Miguel, T.T.; Cruz, F.C. Maternal Separation Stress Affects Voluntary Ethanol Intake in a Sex Dependent Manner. *Front. Physiol.* **2021**, *12*, 2054. [CrossRef]
49. Amancio-Belmont, O.; Meléndez, A.L.B.; Ruiz-Contreras, A.E.; Méndez-Díaz, M.; Próspero-García, O. Opposed cannabinoid 1 receptor (CB1R) expression in the prefrontal cortex vs. nucleus accumbens is associated with alcohol consumption in male rats. *Brain Res.* **2019**, *1725*, 146485. [CrossRef]
50. Huot, R.L.; Thrivikraman, K.; Meaney, M.J.; Plotsky, P.M. Development of adult ethanol preference and anxiety as a consequence of neonatal maternal separation in Long Evans rats and reversal with antidepressant treatment. *Psychopharmacology* **2001**, *158*, 366–373. [CrossRef]
51. Gustafsson, L.; Ploj, K.; Nylander, I. Effects of maternal separation on voluntary ethanol intake and brain peptide systems in female Wistar rats. *Pharmacol. Biochem. Behav.* **2005**, *81*, 506–516. [CrossRef]
52. Berardo, L.R.; Fabio, M.C.; Pautassi, R.M. Post-weaning Environmental Enrichment, But Not Chronic Maternal Isolation, Enhanced Ethanol Intake during Periadolescence and Early Adulthood. *Front. Behav. Neurosci.* **2016**, *10*, 195. [CrossRef] [PubMed]
53. Lundberg, S.; Abelson, K.S.P.; Nylander, I.; Roman, E. Few long-term consequences after prolonged maternal separation in female Wistar rats. *PLoS ONE* **2017**, *12*, e0190042. [CrossRef] [PubMed]
54. Slob, A.; Bosch, J.J.V.D.W.T. Sex differences in body growth in the rat. *Physiol. Behav.* **1975**, *14*, 353–361. [CrossRef]
55. Farrell, M.; Holland, F.; Shansky, R.; Brenhouse, H. Sex-specific effects of early life stress on social interaction and prefrontal cortex dendritic morphology in young rats. *Behav. Brain Res.* **2016**, *310*, 119–125. [CrossRef] [PubMed]
56. Seo, D.; Sinha, R. The neurobiology of alcohol craving and relapse. In *Handbook of Clinical Neurology*; Elsevier: Amsterdam, The Netherlands, 2014; Volume 125, pp. 355–368. [CrossRef]
57. Parker, L.F.; Radow, B.L. Isolation stress and volitional ethanol consumption in the rat. *Physiol. Behav.* **1974**, *12*, 1–3. [CrossRef]
58. Mumtaz, F.; Khan, M.I.; Zubair, M.; Dehpour, A.R. Neurobiology and consequences of social isolation stress in animal model—A comprehensive review. *Biomed. Pharmacother.* **2018**, *105*, 1205–1222. [CrossRef]
59. Thorsell, A.; Slawecki, C.J.; Khoury, A.; Mathe, A.A.; Ehlers, C.L. Effect of social isolation on ethanol consumption and substance P/neurokinin expression in Wistar rats. *Alcohol* **2005**, *36*, 91–97. [CrossRef]
60. Evans, O.; Rodríguez-Borillo, O.; Font, L.; Currie, P.J.; Pastor, R. Alcohol Binge Drinking and Anxiety-Like Behavior in Socialized Versus Isolated C57BL/6J Mice. *Alcohol. Clin. Exp. Res.* **2019**, *44*, 244–254. [CrossRef]
61. Ford, M.M.; Eldridge, J.C.; Samson, H.H. Microanalysis of Ethanol Self-Administration: Estrous Cycle Phase-Related Changes in Consumption Patterns. *Alcohol. Clin. Exp. Res.* **2002**, *26*, 635–643. [CrossRef]
62. Emanuele, N.V.; LaPaglia, N.; Steiner, J.; Kirsteins, L.; Emanuele, M.A. Effect of Chronic Ethanol Exposure on Female Rat Reproductive Cyclicity and Hormone Secretion. *Alcohol. Clin. Exp. Res.* **2001**, *25*, 1025–1029. [CrossRef]
63. Sanchis, R.; Esquifino, A.; Guerri, C. Chronic ethanol intake modifies estrous cyclicity and alters prolactin and LH levels. *Pharmacol. Biochem. Behav.* **1985**, *23*, 221–224. [CrossRef]
64. Warren, J.G.; Fallon, V.M.; Goodwin, L.; Gage, S.H.; Rose, A.K. Menstrual Cycle Phase, Hormonal Contraception, and Alcohol Consumption in Premenopausal Females: A Systematic Review. *Front. Glob. Women's Heal.* **2021**, *2*. [CrossRef] [PubMed]

Article

Rethinking the Methods and Algorithms for Inner Speech Decoding and Making Them Reproducible

Foteini Simistira Liwicki *, Vibha Gupta, Rajkumar Saini, Kanjar De and Marcus Liwicki

Embedded Intelligent Systems LAB, Machine Learning, Department of Computer Science, Electrical and Space Engineering, Luleå University of Technology, 97187 Luleå, Sweden; vibha.gupta@ltu.se (V.G.); rajkumar.saini@ltu.se (R.S.); kanjar.de@ltu.se (K.D.); marcus.liwicki@ltu.se (M.L.)
* Correspondence: foteini.liwicki@ltu.se

Abstract: This study focuses on the automatic decoding of inner speech using noninvasive methods, such as Electroencephalography (EEG). While inner speech has been a research topic in philosophy and psychology for half a century, recent attempts have been made to decode nonvoiced spoken words by using various brain–computer interfaces. The main shortcomings of existing work are reproducibility and the availability of data and code. In this work, we investigate various methods (using Convolutional Neural Network (CNN), Gated Recurrent Unit (GRU), Long Short-Term Memory Networks (LSTM)) for the detection task of five vowels and six words on a publicly available EEG dataset. The main contributions of this work are (1) subject dependent vs. subject-independent approaches, (2) the effect of different preprocessing steps (Independent Component Analysis (ICA), down-sampling and filtering), and (3) word classification (where we achieve state-of-the-art performance on a publicly available dataset). Overall we achieve a performance accuracy of 35.20% and 29.21% when classifying five vowels and six words, respectively, in a publicly available dataset, using our tuned iSpeech-CNN architecture. All of our code and processed data are publicly available to ensure reproducibility. As such, this work contributes to a deeper understanding and reproducibility of experiments in the area of inner speech detection.

Keywords: brain–computer interface (BCI); inner speech; electroencephalography (EEG); deep learning; Convolutional Neural Network (CNN); independent component analysis; supervised learning

Citation: Simistira Liwicki, F.; Gupta, V.; Saini, R.; De, K.; Liwicki, M. Rethinking the Methods and Algorithms for Inner Speech Decoding and Making Them Reproducible. *NeuroSci* **2022**, *3*, 226–244. https://doi.org/10.3390/neurosci3020017

Academic Editor: Szczepan Paszkiel

Received: 1 March 2022
Accepted: 12 April 2022
Published: 19 April 2022

Publisher's Note: MDPI stays neutral with regard to jurisdictional claims in published maps and institutional affiliations.

1. Introduction

Thought is strongly related to inner speech [1,2], through a voice being inside the brain that does not actually speak. Inner speech, although not audible, occurs when reading, writing, and even when idle (i.e., "mind-wandering" [3]). Moreover, inner speech follows the same pattern, e.g., regional accents, as if the person is actually speaking aloud, for example [4]. This work focuses on inner speech decoding.

While inner speech has been a research topic in the philosophy of psychology since the second half of the 20th century [5], with results showing that the part of the brain responsible for the generation of inner speech is the frontal gyri, including Broca's area, the supplementary motor area, and the precentral gyrus, the automatic detection of inner speech has very recently become a popular research topic [6,7]; however, a core challenge of this research is to go beyond the closed vocabulary decoding of words and integrate other language domains (e.g., phonology and syntax) to reconstruct the entire speech stream.

In this work, we conducted extensive experiments using deep learning methods to decode five vowels and six words on a publicly available electroencephalography (EEG) dataset [8]. The backbone CNN architecture used in this work is based on the work of Cooney et al. [7].

The main contributions of this work are as follows: (i) providing code for reproducing the reported results, (ii) subject dependent vs. subject-independent approaches, (iii)

the effect of different preprocessing steps (ICA, down-sampling, and filtering), and (iv) achieving state-of-the-art performance on the six word classification task reporting a mean accuracy of 29.21% for all subjects on a publicly available dataset [8].

State-of-the-Art Literature

Research studies in inner speech decoding use data of invasive (e.g., Electrocorticography (ECoG) [9,10]) and non-invasive methods (e.g., Magnetoencephalography (MEG) [11,12], functional Magnetic Resonance Imaging (fMRI) [13], Functional Near-Infrared Spectroscopy (FNIRS) [14,15]) with EEG being the most dominate modality used so far [16]. Martin et al. [10] attempted to detect single words from inner speech using ECoG recordings from inner and outer speech. This study included six word pairs and achieved a binary classification accuracy of 58% using a Support Vector Machine (SVM). ECoG is not scalable as it is invasive but it advances our understanding and limit of decoding inner speech research. Recent methods used a CNN with the "MEG-as-an-image" [12] and "EEG-as-raw-data" [7,17] inputs.

The focus of this paper is on inner speech decoding in terms of the classification of the words and vowels. Classified words can be useful in many scenarios of human–computer communication, e.g., in smart homes or health-care devices, were the human wants to give simple commands via brain signals in a natural way. For human-to-human communication, the ultimate goal of inner speech decoding (in terms of representation learning) is often to synthesize speech [18,19]. In this related area, [18] uses a minimal invasive method called stereotactic EEG (sEEG) with one subject and 100 Dutch words in an open-loop stage for training the decoding models and close-loop stage to evaluate in real time the imagined and whispered speech. The attempt, although not yet intelligible, provides a proof of concept for tackling the close-loop synthesis of imagined speech in real time. Ref. [19] uses MEG data from seven subjects, using, as stimuli, five phrases (1. Do you understand me, 2. That's perfect, 3. How are you, 4. Good-bye, and 5. I need help.), and two words (yes/no). They follow a subject-dependent approach, where they train and tune a different model per subject. Using a bidirectional long short-term memory recurrent neural network, they achieve a correlation score of the reconstructed speech envelope of 0.41 for phrases and 0.77 for words.

Ref. [15] reported an average classification accuracy of 70.45 $\pm$ 19.19% for a binary word classification task using Regularized Linear Discriminant Analysis (RLDA) using FNIRS data. The EEGNet [20] is a CNN-based deep learning architecture for EEG signal analysis that includes a series of 2D convolutional layers, average pooling layers, and batch normalization layers with activations. Finally, there is a fully connected layer at the end of the network to classify the learned representations from the preceding layers. The EEGNet serves as the backbone network in our model; however, the proposed model extends the EEGNet in similar manner to [7].

There are two main approaches when it comes to brain data analysis: subject dependent and subject independent (see Table 1). In the subject-dependent approach, the analysis is taken for each subject individually and performance is reported per subject. Representative studies in the subject-dependent approach are detailed in the following paragraph. Ref. [8] reported a mean recognition rate of 22.32% in classifying five vowels and 18.58% in classifying six words using a Random Forest (RF) algorithm with a subject-dependent approach. Using the data from six subjects, Ref. [21] reported an average accuracy of 50.1% $\pm$ 3.5% for the three-word classification problem and 66.2% $\pm$ 4.8% for a binary classification problem (long vs. short words), following a subject-dependent approach using a Multi-Class Relevance Vector Machine (MRVM). In [12], MEG data from inner and outer speech was used; an average accuracy of 93% for the inner speech and 96% for the outer speech decoding of five phrases in a subject-dependent approach using a CNN was reported. Recently, Ref. [22] reported an average accuracy of 29.7% for a four-word classification task on a publicly available dataset of inner speech [23]. In the subject-independent approach, all subjects are taken into account and the performance is reported using the data of all subjects; therefore the generated decoding model can generalize the new subjects' data. The following studies use a subject-independent approach. In [6], the authors reported an overall accuracy of 90% on the binary classification of vowels compared with consonants using Deep-Belief Networks (DBN) and the combination of all modalities (inner and outer speech), in a subject-independent approach. In [7], the authors used a CNN with transfer learning to analyze inner speech on the EEG dataset of [8]. In these experiments, the CNN was trained on the raw EEG data of all subjects but one. A subset of the remaining subject's data was used to finely tune the CNN and the rest of the data were used to test the CNN model. They authors reported an overall accuracy of 35.68% (five-fold cross-validation) for the five-vowel classification task.

Table 1. Overview of inner speech studies (2015–2021). TL: Transfer learning.

Study	Technology	Number of Subjects	Number of Classes	Classifier	Results	Subject-Independent
2015—[6]	EEG, facial	6	2 phonemes	DBN	90%	yes
2017—[8]	EEG	15	5 vowels	RF	22.32%	no
2017—[8]	EEG	15	6 words	RF	18.58%	no
2017—[21]	EEG	6	3 words	MRVM	50.1% ± 3.5%	no
2017—[21]	EEG	6	2 words	MRVM	66.2% ± 4.8%	no
2018—[10]	ECoG	5	2 (6) words	SVM	58%	no
2019—[7]	EEG	15	5 vowels	CNN	35.68 (with TL), 32.75%	yes
2020—[24]	EEG	15	6 words	CNN	24.90%	no
2020—[24]	EEG	15	6 words	CNN	24.46%	yes
2019—[15]	FNIRS, EEG	11	2 words	RLDA	70.45% ± 19.19%	no
2020—[12]	MEG	8	5 phrases	CNN	93%	no
2021—[22]	EEG	8	4 words	CNN	29.7%	no

2. Materials and Methods

2.1. Dataset and Experimental Protocol

The current work used a publicly available EEG dataset as described in [8]. This dataset includes recordings from 15 subjects using their inner and outer speech to pronounce 5 vowels (/a/, /e/, /i /, /o/, /u/) and 6 words (arriba/up, abajo/down, derecha/right, izquierda/left, adelante/forward, and atr ás/backwards). A total of 3316 and 4025 imagined speech sample EEG recordings for vowels and words, respectively, are available in the dataset. An EEG with 6 electrodes was used in these recordings.

Figure 1 shows the experimental design followed in [8]. The experimental protocol consisted of a ready interval that was presented for 2 s, followed by the stimulus (vowel or word) presented for 2 s. The subjects were asked to use their inner or outer speech during the imagine interval to pronounce the stimulus. Finally, a rest interval of 4 s was presented, indicating that the subjects could move or blink their eyes before proceeding with the next stimulus. It is important to note that for the purpose of our study, only the inner speech part of the experiment was used.

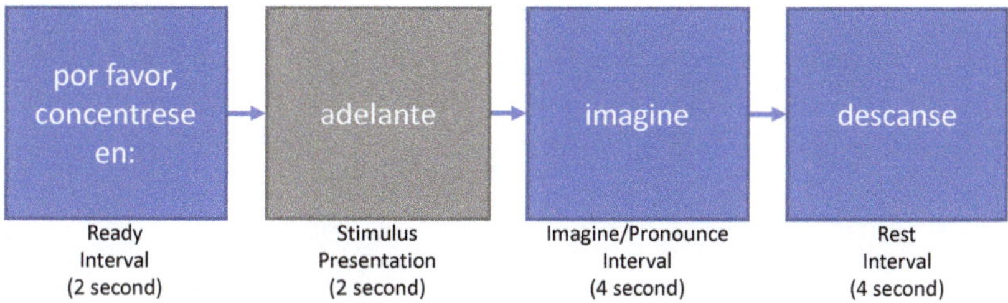

Figure 1. Experimental protocol used in [8]: Ready interval followed by a textual representation of the stimulus (vowel or word). The inner speech production took place during the stimulus interval for 4 s.

2.2. Methods

The proposed framework uses a deep CNN to extract representations from the input EEG signals. Before applying the proposed CNN, the signals are preprocessed and then the CNN network is trained on the preprocessed signals.

Figure 2 depicts the flow of the proposed work. Separate networks are trained for vowels and words following the architecture depicted in Figure 2. The proposed network is inspired by Cooney et al. [7]; they performed filtering, downsampling, and artifact removal before applying the CNN; however, we have noticed that downsampling degrades the recognition performance, see Section 4. As a result, we did not downsample the signals in our experiments. The downsampling block is represented by a cross in Figure 2 to indicate that this task is not included in our proposed system in comparison with [7]. The current work reports results on 3 different experimental approaches using preprocessed data and raw data. The 3 different approaches are discussed in detail in Sections 3.1.1 and 3.1.2. More information about the preproccessing techniques can be found in Section 2.3.

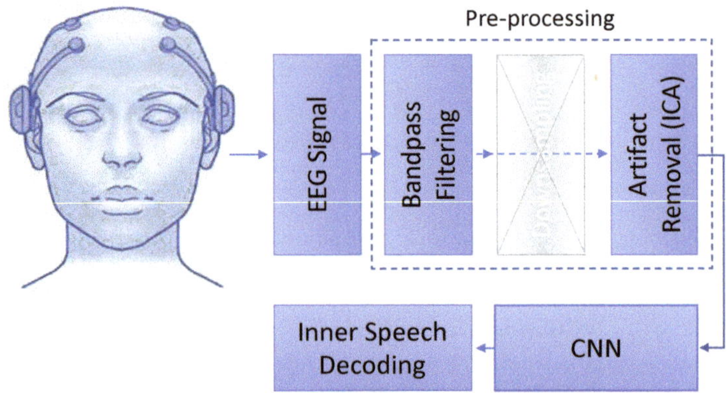

Figure 2. The figure illustrates the proposed workflow. The preprocessed EEG signals with or without downsampling are used to train a CNN model for inner speech decoding.

2.3. Preprocessing

In the current work, we apply the following preprocessing steps:

Filtering: A frequency between 2 Hz and 40 Hz is used for filtering [8].

Down-sampling: The filtered data are down-sampled to 128 HZ. The original frequency of the data is 1024 Hz.

Artifact removal: Independent component analysis (ICA) is known as a blind-source separation technique. When recording a multi-channel signal, the advantages of employing ICA become most obvious. ICA facilitates the extraction of independent components from mixed signals by transforming a multivariate random signal. Here, ICA applied to identify components in EEG signal that include artifacts such as eye blinks or eye movements. These are components then filtered out before the data are translated back from the source space to the sensor space. ICA effectively removes noise from the EEG data and is, therefore, an aid to classification. Given the small number of channels, we intact all the channels and instead use ICA [25] for artifact removal (https://github.com/pierreablin/picard/blob/master/matlab_octave/picard.m, accessed on 27 February 2022).

Figure A3 (see Appendix C) depicts the preprocessed signal after applying ICA. This figure shows the vowel *a* for two subjects. From this figure, it can be noted that the subject's model is not discriminative enough as overlapping is observed. The response from all electrodes' behavior for all vowels for Subject-02 can be seen in Figure A4 (see Appendix C). From this figure, it can be seen that all electrodes are adding information as they all differ in their characteristics.

2.4. iSpeech-CNN Architecture

In this section, we introduce the proposed CNN-based iSpeech architecture. After extensive experiments on the existing CNN architecture for inner speech classification tasks, we determined that downsampling the signal has an effect on the accuracy of the classification and thus removed it from the proposed architecture. The iSpeech-CNN architecture for imagined vowel and word recognition is shown in Figure 3. The same architecture is used in training for imagined vowels and words separately. The only difference is that the network for vowels has five classes; therefore, the softmax layer outputs five probability scores; one for each vowel. In the same manner, the network for words has six classes; therefore, the softmax layer outputs six probability scores; one for each word. Unlike [7], after extensive experimentation, we observed that the number of filters has an effect on the overall performance of the system; 40 filters are used in the first four layers of both networks. The next three layers have 100, 250, and 500 filters, respectively; however, the filter sizes are different. Filters of sizes (1×5), (6×1), (1×5), (1×3), (1×3), (1×3), and (1×3) are used in the first, second, third, fourth, fifth, sixth, and seventh layers, respectively.

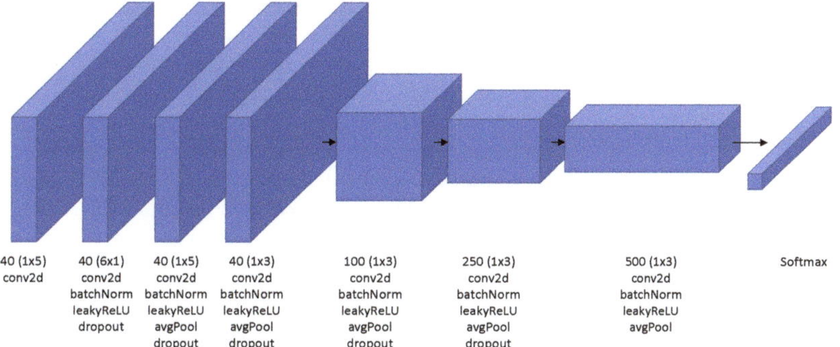

Figure 3. Proposed iSpeech-CNN architecture for imagined speech recognition based on the architecture described in [7]. This network is trained separately for vowels and words. Therefore, the difference lies in the last layer (softmax). The softmax layer for vowels have five outputs while for words has six outputs.

We used an Adam optimizer with a dropout of 0.0002 for the vowel classification and 0.0001 for the word classification. As the network is very small, dropping out more features will adversely affect the performance. The initial learning rate was fixed to 0.001 with a *piecewise* learning rate scheduler. Our network was trained for 60 epochs, and the best validation loss was chosen for the final network. The regularization was also fixed to a value of 0.001. Our proposed iSpeech-CNN architecture follows the same structure as [7] but with a different numbers of filters and training parameters and preprocessing.

3. Experimental Approaches and Performance Measures

This section describes the experimental approaches that have been utilized for the analysis of EEG data and the performance measures that quantify the obtained analysis.

3.1. Experimental Approaches

Three experimental approaches were used for analysis, and they are discussed in detail in the following subsections.

3.1.1. Subject-Dependent/Within-Subject Approach

Subject-dependent/within-subject classification is a baseline approach that is commonly used for the analysis of inner speech signals. In this approach, individual models are trained corresponding to each subject and for each subject, a separate model is created.

The training, validation, and testing sets all have data from the same subject. This approach essentially measures how much an individual subject's data changes (or varies) over time.

To divide the subject data into training, testing, and validation datasets, a ratio of 80-10-10 is used. The training, validation, and testing datasets contain all vowel/words category samples (five/six, respectively) in the mentioned ratio. To remove the bias towards the samples, five different trials are utilized. Furthermore, the mean accuracy and standard deviation are reported for all experimental approaches.

3.1.2. Subject Independent: Leave-One-Out Approach

The subject-dependent approach does not show generalization capability as it models one subject at a time (Testing data contain only samples of the subject that is being modeled). The leave-one-out approach is an independent approach where data of each subject are tested using models that are trained using the data of all other subjects but one, i.e., $n-1$ subjects out of total n will be used for the training model, and the rest will be used for testing. For example, Model-01 will be trained with data from subjects except Subject01, and will be tested with Subject01 (see Tables A3–A6).

This approach helps to obtain a deeper analysis when there are fewer subjects or entities and shows how each individual subject affects the overall estimate of the rest of the subjects. Hence, this approach may provide more generalizable remarks than subject-specific models that depend on individual models.

3.1.3. Mixed Approach

The mixed approach is a variation of subject-independent approach. Although leave-one-out is truly independent, we can see the mixed approach as less independent in nature as it includes data from all subjects in training, validation, and testing. As it contains the data of all subjects, we called it the mixed approach. This approach differs from the within-subject and leave-one-out approaches, where n models correspond to the total number of subjects in the data, are trained. In this approach, only one model will be trained for all subjects. Testing contains samples of all the subjects under all categories (vowels/words).

To run this experiment, 80% of the samples of all the subjects are included in the training set, 10% in the validation set, and the remaining in the test set. We also ensure class balancing, i.e., each class will have approximately the same number of samples of all vowel/word categories. The same experiment is repeated for five random trials, and the mean accuracy along with the standard deviation is reported.

3.2. Performance Measures

The mean and standard deviation are used to report the performance of all the approaches. For the final results, the F-scores are also given.

Mean: The mean is the average of a group of scores. The scores are totaled and then divided by the number of scores. The mean is sensitive to extreme scores when the population samples are small.

Standard deviation: In statistics, the standard deviation (SD) is a widely used measure of variability. It depicts the degree of deviation from the average (mean). A low SD implies that the data points are close to the mean, whereas a high SD suggests that the data span a wide range of values.

F-score: The F-score is a measure of a model's accuracy that is calculated by combining the precision and recall of the model. It is calculated by the following formula:

$$\text{F-score} = \frac{2 * \text{Precision} * \text{Recall}}{\text{Precision} + \text{Recall}} \tag{1}$$

where precision is the percentage of true positive examples among the positive examples classified by the model, and recall is the fraction of examples classified as positive, among the total number of positive examples.

4. Results and Discussion: Vowels (Five Classes)

The results estimated with the subject-specific approach are discussed first as this approach is common in most of the EEG-related papers. All code, raw data, and preprocessed data are provided on Github (https://github.com/LTU-Machine-Learning/Rethinking-Methods-Inner-Speech, accessed on 27 February 2022). Related approaches are discussed in later subsections.

4.1. Subject-Dependent/Within-Subject Classification

In this section, we report the results when applying the subject-dependent approach. Figures 4, 5 and A1 and Table A5 show the results of our proposed iSpeech-CNN architecture. Tables A1 and A2 show the results of the reference CNN architecture.

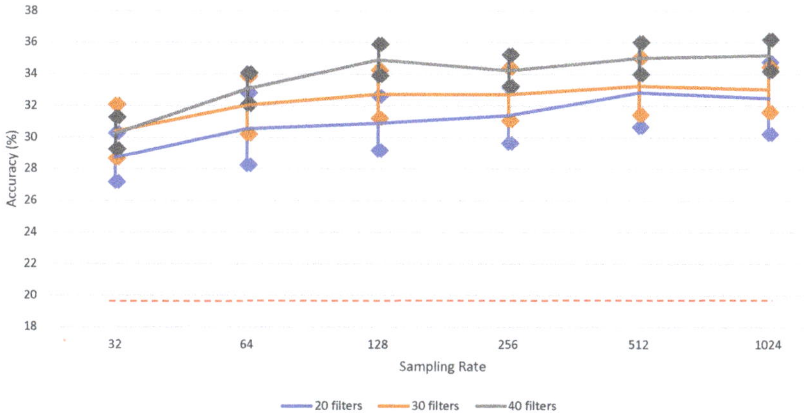

Figure 4. The impact of different sampling rates on vowel recognition performance of (iSpeech-CNN Architecture) with different filters in first three CNN layers. The bars indicate the standard error, sample size = 5. Theoretical chance accuracy = 20% (red-dotted line).

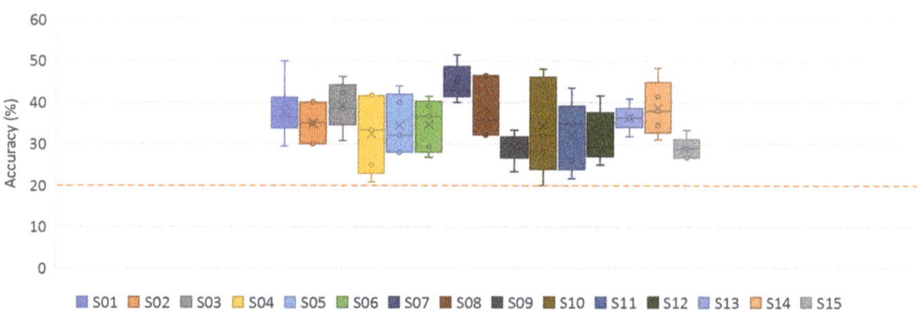

Figure 5. Subject dependent results for vowels without downsampling on preprocessed signals (iSpeech-CNN Architecture). Theoretical chance accuracy = 20% (red-dotted line).

4.1.1. Ablation Study—Influence of Downsampling

Table A1 shows the results with raw and downsampled data when used within the referenced CNN architecture framework.

It is clearly observed from Table A1 that downsampling the signals results in a loss of information. Figure 4 shows that there is a significant performance increase between 32 and 1024; however, some other differences (e.g., for 40 filters between 128 and 1024) are not significant. For clarity, the bars for standard error for each data point are added. The highest vowel recognition performance (35.20%) is observed at the highest sampling rate (1024), i.e., without downsampling.

In other words, the chosen sampling rate was not sufficient to retain the original information; therefore, further results will be reported for both raw data and downsampled data, in order to obtain a better insight into the preprocessing (i.e., filtering and ICA) stage.

4.1.2. Ablation Study—Influence of Preprocessing

Filtering and artifact removal plays an important role while analyzing the EEG signals. We applied both bandpass filtering (see Section 2), and picard (preconditioned ICA for real data) for artifact removal to obtain more informative signals. Table A2 shows the results of preprocessing when applied on the raw and downsampled data within the reference CNN architecture framework. The performance, i.e., the overall mean accuracy, decreased from 32.51% to 30.90%. The following points can be noted from Table A2: (1) Filtering and artifact removal highly influence the performance irrespective of raw and downsampled data. (2) The improved performance can also be observed with respect to each subject. A smaller standard deviation can also be seen. (3) The CNN framework generated higher performance than the handcrafted features and the GRU (see Table 2). We also performed experiments with the LSTM classifier and noticed the random behavior (theoretical chance accuracies); no significant difference as compared to GRU; therefore, iSpeech-CNN performs best among all classifiers.

Table 2. Average subject-dependent classification results on the [8] dataset.

Study	Classifier	Vowels	Words
2017—[8]	RF	22.32% ± 1.81%	18.58% ± 1.47%
2019, 2020—[7,24]	CNN	32.75% ± 3.23%	24.90% ± 0.93%
iSpeech-GRU	GRU	19.28% ± 2.15%	17.28% ± 1.45%
iSpeech-CNN (proposed)	CNN	35.20% ± 3.99%	29.21% ± 3.12%

4.1.3. Ablation Study—Influence of Architecture

Based on the CNN literature in the EEG paradigm [7,26], adding more layers to the reference CNN architecture does not help to obtain an improved performance; however, by changing the number of filters in the initial layers, some improvements can be observed. Based on the CNN literature for EEG signals, having a sufficient number of filters in the initial layers helps to obtain some improvement [7,27]. Here, we choose three initial layers, unlike in natural images, in speech, initial layers are more specific to the task rather than the last few layers. The results with a changing number of filters in the initial layer within the iSpeech-CNN architecture are shown in Table A5. In the reference CNN architecture, this filter number was 20 for the initial three layers; however, we have changed this number to 40 (decided based on experimentation) in the iSpeech-CNN architecture. Table A5 clearly shows that changing the filter parameter yields higher performance than with the number of filters (compare to the reference architecture results in Tables A1 and A2 in the Appendix A). This improvement is observed with and without downsampled data and with respect to the subject (see Figures A1 and 5). The standard deviation also decreases with these modifications (see Table A5).

4.2. Mixed Approach Results

This section discusses the results of the mixed approach. In this approach, data from all subjects are included in training, validation, and testing. Table A4 shows the results for the mixed approach with and without downsampling. These results were compiled with filtering and ICA in both reference and modified CNN architectures.

From these results, it is noted that the obtained accuracies are random in nature. The modified CNN architecture parameters do not help to obtain any improvements and show random accuracy behavior. In other words, it is difficult to achieve generalized performance with EEG signals. Based on the EEG literature, it has also been justified that models trained on data from one subject cannot be generalized to other subjects even though have been recorded using the same setup conditions.

Determining the optimal frequency sub-bands corresponding to each subject could be one possible direction that may be successful in such a scenario. We intent to explore such a direction in our future work.

4.3. Subject-Independent: Leave-One-Out Results

Having discussed the subject-specific and mixed results, in this section, the subject-independent results are discussed. The leave-one-out approach is a variation of the mixed approach; however, unlike the mixed approach, here, the data of the testing subject are not included in the training. For example, in Figure 6, except *Subject*01, all other subjects were used in the training of *Model*-01. Figure 6 and Table A6 show the results using the iSpeech-CNN architecture, while Table A3 shows the results using the reference architecture.

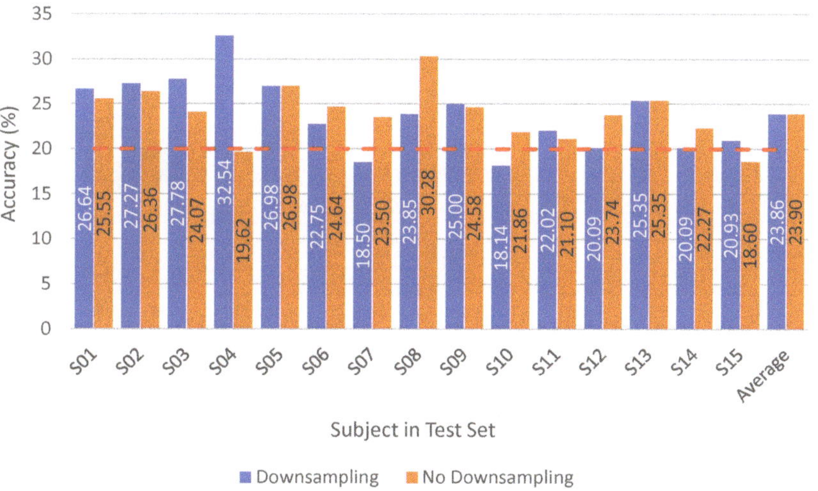

Figure 6. Leave-one-out results for vowels with and without downsampling on preprocessed signals (iSpeech-CNN Architecture). Theoretical chance accuracy = 20% (red-dotted line).

It can be noted that having fewer subjects in training (one less as compared to the mixed approach), shows slightly better behavior than the mixed approach, where all subjects were included in the training. Moreover, changing the reference CNN parameters to our proposed iSpeech-CNN architecture also shows improved performance (see Figure 6 and Table A6).

The mixed and leave-one-out approaches both showed that generalizing the performance over all subjects is difficult in the EEG scenario. Hence, there is a need for the preprocessing stage, which can make the data more discriminative.

5. Results and Discussion: Words (Six Classes)

Having discussed all the approaches for the category of vowels, we noticed that only the subject-specific approach showed performance that was not random in nature and hence makes sense; therefore, in this section, we only report results corresponding to the subject-specific approach for the word category.

This category contains six different classes (see Section 2.1). Table A8 and Figures A2 and 7 show the performance results for the classification of the six words, using the proposed iSpeech-CNN architecture. The performance results when using the reference architecture can be found in Appendix B. From these tables and figures, the same kind of behavior as vowels is observed. The change in the number of filters in the initial layers affected the performance as shown in Table A8. The downsampling of data also affects the overall performance. Figure 8 shows that the highest word recognition performance (29.12%) is observed at highest sampling rate (1024), i.e., without downsampling. For clarity, we added the bars for standard error for each data point. As opposed to vowel recognition, there is a steady increase in the performance when increasing the sampling rate (though again, not always significant among two neighboring values).

The iSpeech-CNN architecture shows better performance than handcrafted features such as real-time wavelet energy [8] and reference architecture (Appendix B).

Overall, we achieve a state-of-the-art performance of 29.21% when classifying the six words using our proposed iSpeech-CNN architecture and preprocessing methodology without downsampling.

The performance reported in this work is based on the CNN architecture of the reference network [7]. No other architecture was investigated. This is due to the reason that the goal of the proposed work is to reproduce the Cooney's results and making the network and codes available to the research community.

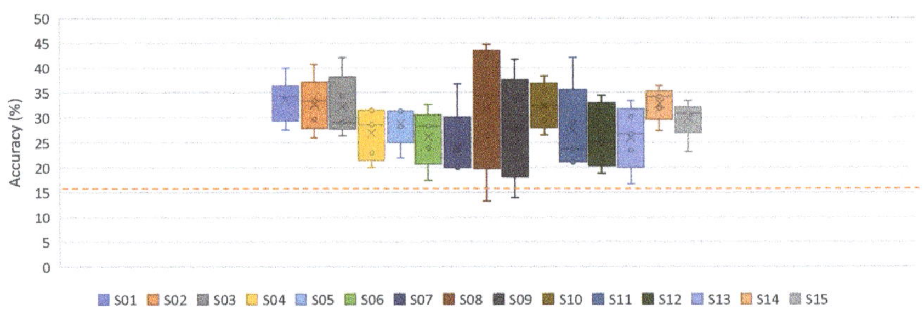

Figure 7. Subject-dependent results for words without downsampling on preprocessed signals (iSpeech-CNN Architecture). Theoretical chance accuracy = 16.66% (red-dotted line).

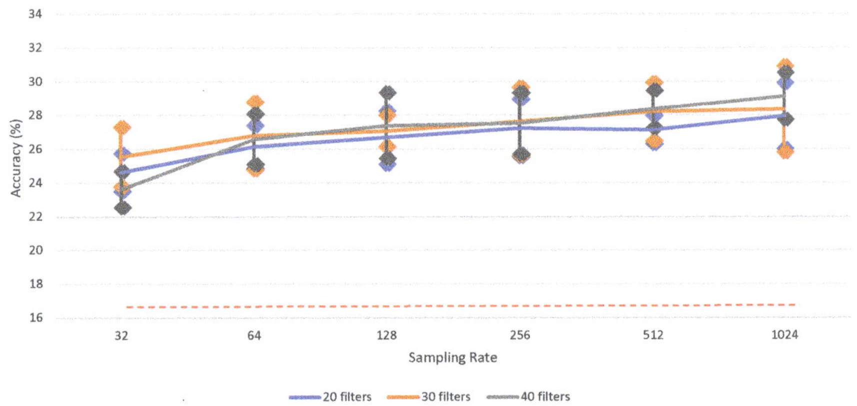

Figure 8. The impact of different sampling rates on word recognition performance of (iSpeech-CNN Architecture) with different filters in first three CNN layers. Performance increases with higher sampling rates. The bars indicate the standard error, sample size = 5. Theoretical chance accuracy = 16.66% (red-dotted line).

6. Performance Comparison and Related Discussion

In this section, we compare our results on the vowels and words dataset with existing work and discuss related findings. Based on the reported performances in the Table 2, it is clearly noted that the CNN performs better than the handcrafted features for both datasets.

The precision, weighted F-score, and F-score for our proposed iSpeech-CNN in comparison with the reported results of Cooney et al. [7] are shown in Table 3. From this table, we can note that our proposed system results in a higher precision; however, a lower F-score compared to the model in [7]. Hence, the reproducibility of the results reported in [7] is difficult.

Table 3. Precision and F-score (with respect to Tables A5–A8) for vowel and word classification (iSpeech-CNN Architecture).

	Vowel (iSpeech-CNN)		
	Precision	**Weighted F-Score**	**F-Score**
No Downsampling	34.85	41.12	28.45
Downsampling	34.62	38.99	30.02
Cooney et al. [7] (Downsampling)	33.00	-	33.17
	Words (iSpeech-CNN)		
	Precision	**Weighted F-Score**	**F-Score**
No Downsampling	29.04	36.18	21.84
Downsampling	26.84	31.94	21.50

Our proposed CNN architecture and preprocessing methodology outperform the existing work in word and vowel category when following subject-dependent approach, as shown in Table 2; however, it is worth to mention that for the vowel classification, unlike in [7], we do not downsampling the data. Furthermore, [7] when using transfer learning approach for the vowel classification task, they report an overall accuracy of 35.68%, which is slightly higher than our reported accuracy in the subject-dependent approach.

Based on the 1-tail paired *t*-test results, we found that there is statistical significant difference between iSpeech-CNN and the reference paper [7] for word classification and for vowel classification, if we compare to the work without transfer learning (which is the fair comparison, as transfer learning adds a new dimension). We also found that there is no significant difference between the best reported results with transfer learning [7,24] and iSpeech-CNN. Furthermore, when we run the 1-tail paired *t*-test results for iSpeech-CNN between downsampling and without downsampling, we found that these difference are significantly different for the words task ($p = 0.0005$), but not statistically significant for the vowels task. We are following 1-tail paired *t*-test and used 10% of the overall samples, i.e., 332 for vowels and 403 for words.

Hence, it is observed that the correct selection of preprocessing methods and the number of filters in the CNN, greatly add to the performance. The elaborated results for each category and with each approach have been added to Appendices A and B.

7. Conclusions

This study explores the effectiveness of preprocessing steps and the correct selection of filters in the initial layers of the CNN in the context of both vowel and word classification. The classification results are reported on a publicly available inner speech dataset of five vowels and six words [8]. Based on the obtained accuracies, it is found that such a direction of exploration truly adds to the performance. We report state-of-the-art classification performance for vowels and words with mean accuracies of 35.20% and 29.21%, respectively, without downsampling the original data. Mean accuracies of 34.88% and 27.38% have been reported for vowels and words, respectively, with downsampling. Furthermore, the proposed CNN code in this study is available to the public to ensure reproducibility of the research results and to promote open research. Our proposed iSpeech-CNN architecture and preprocessing methodology are the same for both datasets (vowels and words).

Evaluating our system in other publicly available datasets is part of our future work. Furthermore, we will address the issues related to the selection of the downsampling rate and the selection of the optimal frequency sub-band with respect to subjects.

Author Contributions: Conceptualization, F.S.L.; methodology, F.S.L. and V.G.; software, F.S.L. and V.G.; validation, F.S.L., V.G., R.S. and K.D.; formal analysis, F.S.L., V.G., R.S. and K.D.; investigation, F.S.L., V.G., R.S. and K.D.; writing—original draft preparation, F.S.L., V.G., R.S., K.D. and M.L.; writing—review and editing, F.S.L., V.G., R.S., K.D. and M.L.; visualization, F.S.L., V.G., R.S., and K.D.; supervision, F.S.L. and M.L.; project administration, F.S.L. and M.L.; funding acquisition, F.S.L. and M.L. All authors have read and agreed to the published version of the manuscript.

Funding: This research received funding by the Grants for excellent research projects proposals of SRT.ai 2022.

Institutional Review Board Statement: Not applicable.

Informed Consent Statement: Not applicable.

Data Availability Statement: The code used in this paper is publicly accessible on Github (https://github.com/LTU-Machine-Learning/Rethinking-Methods-Inner-Speech, accessed on 27 February 2022).

Conflicts of Interest: The authors declare no conflict of interest. The funders had no role in the design of the study; in the collection, analyses, or interpretation of data; in the writing of the manuscript, or in the decision to publish the results.

Appendix A. The Results on Vowels

Table A1. Subject-dependent results for the vowels using raw and downsampled data (Reference Architecture).

	Raw			Downsampled		
	Train	**Validation**	**Test**	**Train**	**Validation**	**Test**
S01	91.07	36.00	35.29	79.07	39.20	26.47
S02	92.44	29.00	17.00	72.00	24.00	16.00
S03	83.65	34.00	27.69	86.94	26.00	19.23
S04	89.94	38.00	33.33	89.33	30.00	31.67
S05	86.59	16.00	21.60	72.82	19.00	17.60
S06	91.73	27.00	25.85	80.00	29.00	27.80
S07	95.17	31.00	26.86	85.66	33.00	28.00
S08	88.12	29.00	21.43	83.06	24.00	21.43
S09	83.89	35.20	30.00	82.38	30.40	26.00
S10	88.35	29.00	20.80	63.88	26.00	19.20
S11	80.91	25.00	22.61	86.06	23.00	16.52
S12	88.57	29.00	26.67	96.23	37.00	30.83
S13	80.00	38.00	29.09	93.94	36.00	31.82
S14	79.66	27.20	22.76	73.49	27.20	23.45
S15	83.20	23.00	20.00	91.73	28.00	20.89
Average/Mean	**86.88**	**29.76**	**25.29**	**82.44**	**28.79**	**23.79**
Standard Deviation	**4.81**	**5.96**	**5.14**	**8.73**	**5.44**	**5.35**

Table A2. Subject-dependent results for vowels with and without downsampling on preprocessed data (Reference Architecture).

	Preprocessing (Filtering and Artifact Removal)					
	No Downsampling			Downsampling		
	Train	Validation	Test	Train	Validation	Test
S01	96.65	38.40	38.82	91.44	43.20	38.82
S02	92.78	40.00	25.00	99.33	32.00	27.00
S03	99.88	38.00	34.62	99.41	36.00	32.31
S04	99.15	43.00	32.50	97.45	42.00	30.83
S05	89.29	37.00	26.40	93.41	34.00	35.20
S06	95.47	38.00	37.56	98.80	38.00	32.20
S07	98.62	36.00	27.43	81.52	28.00	28.00
S08	97.06	45.00	38.57	98.59	39.00	32.14
S09	97.41	36.00	35.33	97.41	32.00	30.00
S10	96.24	35.00	37.60	98.47	35.00	28.80
S11	88.69	30.00	31.30	91.66	35.00	33.91
S12	86.86	35.00	30.83	91.31	28.00	24.17
S13	99.54	40.00	33.64	95.89	46.00	31.82
S14	87.43	39.20	35.86	84.11	40.00	33.79
S15	91.87	33.00	22.22	99.73	30.00	24.44
Average/Mean	**94.46**	**37.57**	**32.51**	**94.57**	**35.88**	**30.9**
Standard Deviation	**4.44**	**3.62**	**5.05**	**5.48**	**5.28**	**3.83**

Table A3. Leave-one-out results for vowels with and without downsampling on preprocessed data (Reference Architecture). Subject in test set.

	Preprocessing (Filtering and Artifact Removal)			
	No Downsampling		Downsampling	
	Validation	Test	Validation	Test
S01	71.43	21.17	23.08	18.61
S02	88.99	20.91	40.41	21.36
S03	86.00	25.93	57.48	19.91
S04	87.77	24.40	55.20	22.01
S05	69.33	19.53	58.72	24.65
S06	55.36	14.69	60.84	23.70
S07	60.30	26.00	36.36	30.50
S08	83.99	28.44	41.51	21.10
S09	45.45	26.25	44.05	24.58
S10	73.72	20.00	41.99	17.67
S11	51.45	25.23	68.88	24.31
S12	91.90	22.37	55.34	20.55
S13	63.63	20.28	61.47	22.12
S14	78.62	23.14	51.31	21.83
S15	84.42	25.58	24.28	22.33
Average/Mean	**72.82**	**22.93**	**48.06**	**22.35**
Standard Deviation	**14.84**	**3.55**	**13.10**	**2.95**

Table A4. Mixed-approach results for vowels with and without downsampling on preprocessed data (Filtering and Artifact Removal) (Reference Architecture; iSpeech-CNN Architecture).

	With Preprocessing; Reference Architecture Parameters					
	No Downsampling			Downsampling		
	Train	Validation	Test	Train	Validation	Test
Trial 1	72.45	20.95	22.27	62.61	20.32	17.63
Trial 2	76.73	22.22	19.03	58.75	20.63	24.36
Trial 3	64.82	22.54	19.26	50.58	20.32	20.65
Trial 4	67.55	18.10	19.95	57.04	18.41	20.19
Trial 5	60.78	20.32	22.04	47.78	22.86	23.90
Mean/Average	68.47	20.83	20.51	55.35	20.51	21.35
Standard Deviation	5.61	1.59	1.38	5.43	1.42	2.50
	With Preprocessing; iSpeech-CNN Architecture Parameters					
	No Downsampling			Downsampling		
	Train	Validation	Test	Train	Validation	Test
Trial 1	57.98	20.63	20.42	55.10	21.90	18.33
Trial 2	68.79	21.27	19.72	46.46	23.17	22.04
Trial 3	54.63	21.59	21.11	44.44	18.73	22.27
Trial 4	37.70	17.46	20.42	24.36	21.27	21.35
Trial 5	86.15	24.13	20.19	57.20	22.86	20.42
Average/Mean	61.05	21.02	20.37	45.51	21.59	20.88
Standard Deviation	16.04	2.14	0.45	11.64	1.58	1.43

Table A5. Subject-dependent results for vowels with and without downsampling on preprocessed signals (iSpeech-CNN Architecture).

	Preprocessing (Filtering and Artifact Removal)					
	No Downsampling			Downsampling		
	Train	Validation	Test	Train	Validation	Test
S01	86.33	48.80	37.65	81.86	33.60	37.65
S02	98.00	38.00	35.00	84.33	31.00	30.00
S03	85.76	38.00	39.23	98.00	37.00	36.15
S04	97.09	46.00	32.50	97.45	41.00	35.83
S05	91.41	44.00	34.40	94.59	38.00	36.00
S06	89.87	41.00	34.63	95.33	38.00	31.71
S07	98.34	39.00	44.57	98.21	39.00	41.14
S08	80.24	41.00	38.57	96.82	38.00	38.57
S09	96.65	32.80	29.33	92.22	35.20	30.00
S10	97.41	44.00	34.40	87.53	37.00	40.80
S11	95.43	41.00	32.17	98.06	31.00	30.43
S12	91.54	40.00	31.67	83.43	34.00	35.00
S13	82.74	46.00	36.36	85.60	46.00	37.27
S14	80.23	43.20	38.62	78.06	36.80	33.79
S15	89.60	31.00	28.89	90.80	38.00	28.89
Average/Mean	90.71	40.92	35.20	90.82	36.91	34.88
Standard Deviation	6.24	4.65	3.99	6.60	3.66	3.83

Table A6. Leave-one-out results for vowels with and without downsampling on preprocessed signals (iSpeech-CNN Architecture).

	With Preprocessing (Filtering and Artifact Removal)			
	No Downsampling		Downsampling	
	Validation	Test	Validation	Test
S01	46.65	26.64	38.23	25.55
S02	83.11	27.27	42.02	26.36
S03	74.32	27.78	43.16	24.07
S04	22.27	32.54	56.16	19.62
S05	64.79	26.98	56.24	26.98
S06	77.13	22.75	47.73	24.64
S07	47.21	18.50	40.76	23.50
S08	39.86	23.85	50.16	30.28
S09	82.38	25.00	46.36	24.58
S10	45.82	18.14	48.02	21.86
S11	52.42	22.02	54.23	21.10
S12	34.26	20.09	67.03	23.74
S13	46.05	25.35	50.08	25.35
S14	46.94	20.09	41.85	22.27
S15	51.69	20.93	46.31	18.60
Average/Mean	**54.33**	**23.86**	**48.56**	**23.90**
Standard Deviation	**18.13**	**4.02**	**7.51**	**2.97**

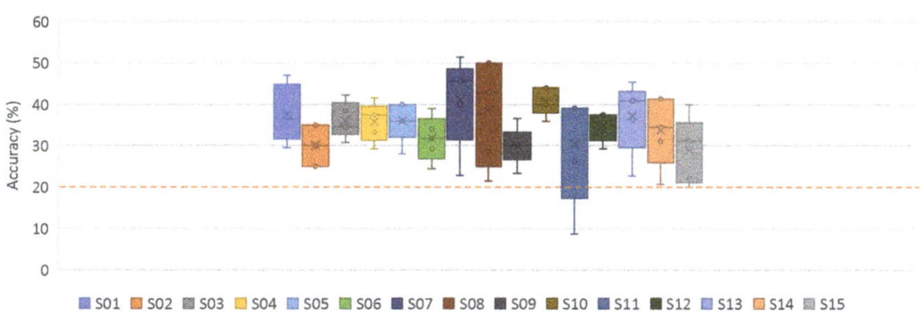

Figure A1. Subject-dependent results for vowels with downsampling on preprocessed signals (iSpeech-CNN Architecture). Chance accuracy 20%.

Appendix B. The Results on Words

Table A7. Subject-dependent results for words with and without downsampling on preprocessed data (Reference Architecture).

| | Preprocessing (Filtering and Artifact Removal) | | | | | |
| | No Downsampling | | | With Downsampling | | |
	Train	Validation	Test	Train	Validation	Test
S01	90.60	31.33	30.00	83.16	32.67	28.50
S02	96.57	30.83	23.70	79.62	30.00	22.22
S03	90.19	32.00	31.58	67.33	30.00	29.47
S04	92.75	36.67	25.14	75.49	33.33	25.71
S05	87.53	41.67	28.13	69.89	30.83	31.25
S06	83.12	39.17	25.65	71.72	29.17	23.04
S07	92.22	26.67	25.33	79.39	35.00	21.33
S08	85.37	38.33	30.53	86.48	34.17	23.16
S09	92.86	32.50	25.56	82.57	30.00	28.89
S10	89.05	31.67	29.41	92.00	30.00	30.59
S11	93.14	31.33	27.37	95.24	32.00	22.11
S12	88.00	33.33	25.62	82.00	34.67	32.50
S13	96.76	28.33	33.33	86.10	39.17	25.33
S14	86.10	35.33	28.64	68.95	36.00	28.18
S15	96.98	27.50	29.23	91.77	28.33	27.69
Average/Mean	**90.75**	**33.11**	**27.95**	**80.78**	**32.36**	**26.66**
Standard Deviation	**4.27**	**4.36**	**2.76**	**8.48**	**2.91**	**3.53**

Table A8. Subject-dependent results for words with and without downsampling on preprocessed signals (iSpeech-CNN Architecture).

| | Preprocessing (Filtering and Artifact Removal) | | | | | |
| | No Downsampling | | | Downsampling | | |
	Train	Validation	Test	Train	Validation	Test
S01	97.95	34.67	33.50	77.35	30.00	25.50
S02	92.86	34.17	32.59	89.05	28.33	23.70
S03	93.05	32.00	32.11	76.67	28.67	34.74
S04	94.51	35.00	26.86	95.59	29.17	31.43
S05	95.38	34.17	28.75	82.47	26.67	31.25
S06	96.99	38.33	26.09	67.20	30.83	26.09
S07	87.47	33.33	24.00	80.71	30.00	16.00
S08	98.33	35.00	32.11	83.70	30.00	28.95
S09	98.86	37.50	27.78	70.00	27.50	28.89
S10	92.38	32.50	32.35	93.52	34.17	27.06
S11	97.43	36.67	27.37	87.90	35.33	26.32
S12	92.86	37.33	26.25	69.05	32.00	25.62
S13	84.57	31.67	26.00	68.29	31.67	33.33
S14	96.38	37.33	32.73	66.38	30.67	27.27
S15	94.69	33.33	29.74	86.67	31.67	24.62
Average/Mean	**94.25**	**34.87**	**29.21**	**79.64**	**30.45**	**27.38**
Standard Deviation	**4.00**	**2.14**	**3.12**	**9.51**	**2.25**	**4.37**

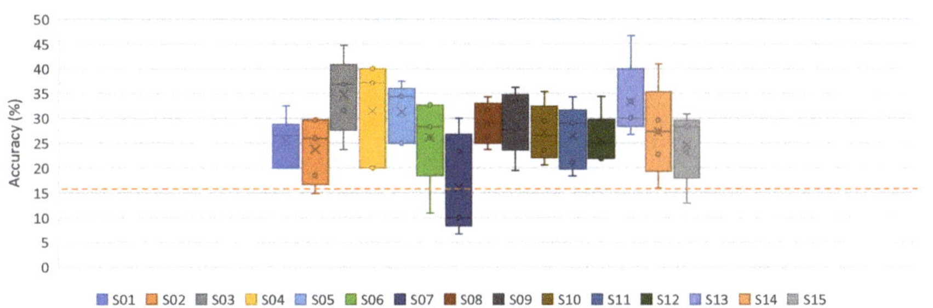

Figure A2. Subject-dependent results for words with downsampling on preprocessed signals (iSpeech-CNN Architecture). Chance accuracy 16.66%.

Appendix C. Dataset Samples

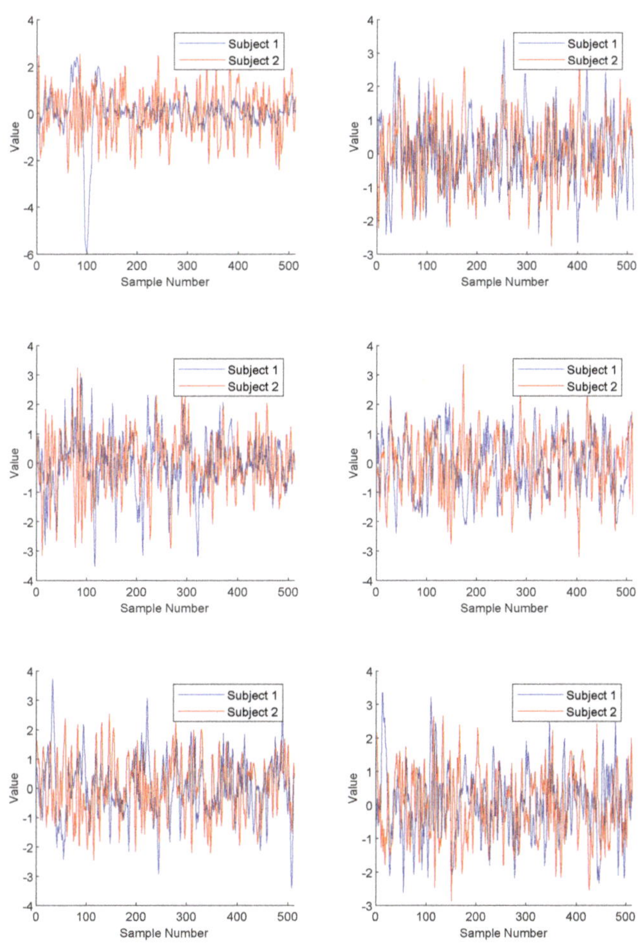

Figure A3. Example of preprocessed signals for all electrodes (after ICA) for the vowel /a/ for Subject01 and Subject02.

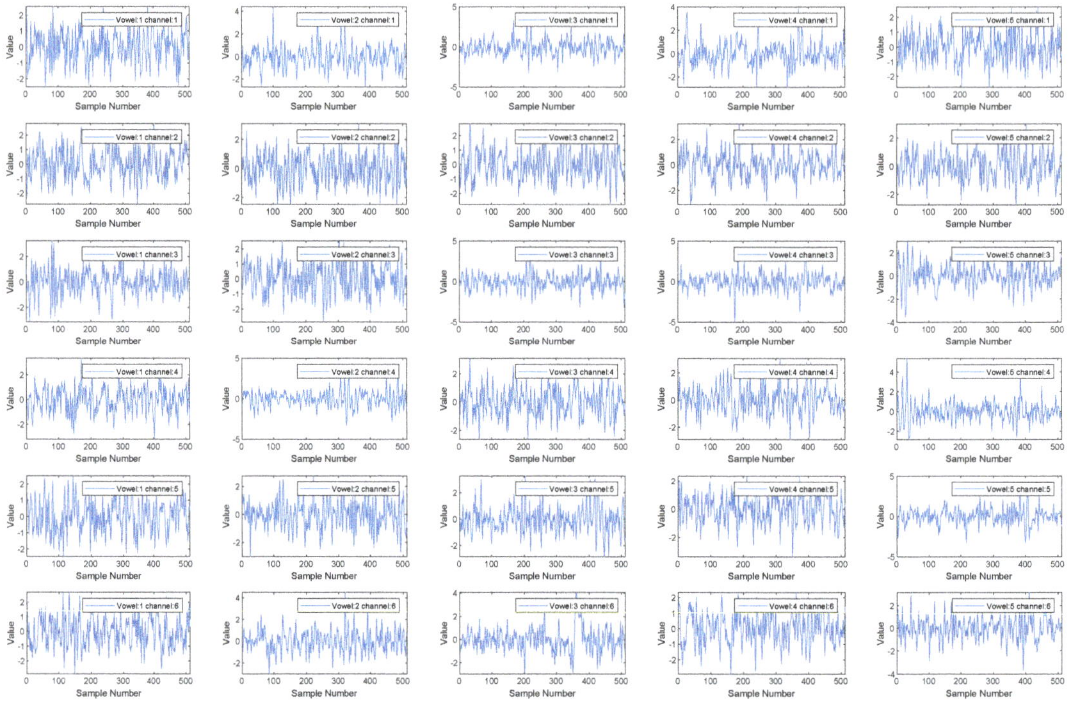

Figure A4. Example of preprocessed signals (after ICA) for all vowels and all electrodes for Subject02.

References

1. Alderson-Day, B.; Fernyhough, C. Inner speech: Development, cognitive functions, phenomenology, and neurobiology. *Psychol. Bull.* **2015**, *141*, 931. [CrossRef] [PubMed]
2. Whitford, T.J.; Jack, B.N.; Pearson, D.; Griffiths, O.; Luque, D.; Harris, A.W.; Spencer, K.M.; Le Pelley, M.E. Neurophysiological evidence of efference copies to inner speech. *Elife* **2017**, *6*, e28197. [CrossRef] [PubMed]
3. Smallwood, J.; Schooler, J.W. The science of mind wandering: Empirically navigating the stream of consciousness. *Annu. Rev. Psychol.* **2015**, *66*, 487–518. [CrossRef] [PubMed]
4. Filik, R.; Barber, E. Inner speech during silent reading reflects the reader's regional accent. *PLoS ONE* **2011**, *6*, e25782. [CrossRef]
5. Langland-Hassan, P.; Vicente, A. *Inner Speech: New Voices*; Oxford University Press: New York, NY, USA, 2018.
6. Zhao, S.; Rudzicz, F. Classifying phonological categories in imagined and articulated speech. In Proceedings of the 2015 IEEE International Conference on Acoustics, Speech and Signal Processing (ICASSP), South Brisbane, QLD, Australia, 19–24 April 2015; pp. 992–996.
7. Cooney, C.; Folli, R.; Coyle, D. Optimizing layers improves CNN generalization and transfer learning for imagined speech decoding from EEG. In Proceedings of the 2019 IEEE International Conference on Systems, Man and Cybernetics (SMC), Bari, Italy, 6–9 October 2019; pp. 1311–1316.
8. Coretto, G.A.P.; Gareis, I.E.; Rufiner, H.L. Open access database of EEG signals recorded during imagined speech. In Proceedings of the 12th International Symposium on Medical Information Processing and Analysis, Tandil, Argentina, 5–7 December 2017; Volume 10160, p. 1016002.
9. Herff, C.; Heger, D.; De Pesters, A.; Telaar, D.; Brunner, P.; Schalk, G.; Schultz, T. Brain-to-text: Decoding spoken phrases from phone representations in the brain. *Front. Neurosci.* **2015**, *9*, 217. [CrossRef]
10. Martin, S.; Iturrate, I.; Millán, J.d.R.; Knight, R.T.; Pasley, B.N. Decoding inner speech using electrocorticography: Progress and challenges toward a speech prosthesis. *Front. Neurosci.* **2018**, *12*, 422. [CrossRef] [PubMed]
11. Dash, D.; Wisler, A.; Ferrari, P.; Davenport, E.M.; Maldjian, J.; Wang, J. MEG sensor selection for neural speech decoding. *IEEE Access* **2020**, *8*, 182320–182337. [CrossRef] [PubMed]
12. Dash, D.; Ferrari, P.; Wang, J. Decoding imagined and spoken phrases from non-invasive neural (MEG) signals. *Front. Neurosci.* **2020**, *14*, 290. [CrossRef] [PubMed]
13. Yoo, S.S.; Fairneny, T.; Chen, N.K.; Choo, S.E.; Panych, L.P.; Park, H.; Lee, S.Y.; Jolesz, F.A. Brain–computer interface using fMRI: Spatial navigation by thoughts. *Neuroreport* **2004**, *15*, 1591–1595. [PubMed]

14. Kamavuako, E.N.; Sheikh, U.A.; Gilani, S.O.; Jamil, M.; Niazi, I.K. Classification of overt and covert speech for near-infrared spectroscopy-based brain computer interface. *Sensors* **2018**, *18*, 2989. [CrossRef] [PubMed]
15. Rezazadeh Sereshkeh, A.; Yousefi, R.; Wong, A.T.; Rudzicz, F.; Chau, T. Development of a ternary hybrid fNIRS-EEG brain–computer interface based on imagined speech. *Brain-Comput. Interfaces* **2019**, *6*, 128–140. [CrossRef]
16. Panachakel, J.T.; Ramakrishnan, A.G. Decoding covert speech from EEG-A comprehensive review. *Front. Neurosci.* **2021**, *15*, 642251. [CrossRef] [PubMed]
17. Schirrmeister, R.T.; Springenberg, J.T.; Fiederer, L.D.J.; Glasstetter, M.; Eggensperger, K.; Tangermann, M.; Hutter, F.; Burgard, W.; Ball, T. Deep learning with convolutional neural networks for EEG decoding and visualization. *Hum. Brain Mapp.* **2017**, *38*, 5391–5420. [CrossRef] [PubMed]
18. Angrick, M.; Ottenhoff, M.C.; Diener, L.; Ivucic, D.; Ivucic, G.; Goulis, S.; Saal, J.; Colon, A.J.; Wagner, L.; Krusienski, D.J.; et al. Real-time synthesis of imagined speech processes from minimally invasive recordings of neural activity. *Commun. Biol.* **2021**, *4*, 1055. [CrossRef] [PubMed]
19. Dash, D.; Ferrari, P.; Berstis, K.; Wang, J. Imagined, Intended, and Spoken Speech Envelope Synthesis from Neuromagnetic Signals. In Proceedings of the International Conference on Speech and Computer, St. Petersburg, Russia, 27–30 September 2021; pp. 134–145.
20. Lawhern, V.J.; Solon, A.J.; Waytowich, N.R.; Gordon, S.M.; Hung, C.P.; Lance, B.J. EEGNet: A compact convolutional neural network for EEG-based brain–computer interfaces. *J. Neural Eng.* **2018**, *15*, 056013. [CrossRef] [PubMed]
21. Nguyen, C.H.; Karavas, G.K.; Artemiadis, P. Inferring imagined speech using EEG signals: A new approach using Riemannian manifold features. *J. Neural Eng.* **2017**, *15*, 016002. [CrossRef] [PubMed]
22. van den Berg, B.; van Donkelaar, S.; Alimardani, M. Inner Speech Classification using EEG Signals: A Deep Learning Approach. In Proceedings of the 2021 IEEE 2nd International Conference on Human-Machine Systems (ICHMS), Magdeburg, Germany, 8–10 September 2021; pp. 1–4.
23. Nieto, N.; Peterson, V.; Rufiner, H.L.; Kamienkowski, J.E.; Spies, R. Thinking out loud, an open-access EEG-based BCI dataset for inner speech recognition. *Sci. Data* **2022**, *9*, 52. [CrossRef] [PubMed]
24. Cooney, C.; Korik, A.; Folli, R.; Coyle, D. Evaluation of hyperparameter optimization in machine and deep learning methods for decoding imagined speech EEG. *Sensors* **2020**, *20*, 4629. [CrossRef] [PubMed]
25. Ablin, P.; Cardoso, J.F.; Gramfort, A. Faster independent component analysis by preconditioning with Hessian approximations. *IEEE Trans. Signal Process.* **2018**, *66*, 4040–4049. [CrossRef]
26. Cheng, J.; Zou, Q.; Zhao, Y. ECG signal classification based on deep CNN and BiLSTM. *BMC Med. Inform. Decis. Mak.* **2021**, *21*, 365. [CrossRef] [PubMed]
27. Goodfellow, I.; Bengio, Y.; Courville, A. *Deep Learning*; MIT Press: Cambridge, MA, USA, 2016.

Article

Long COVID and the Autonomic Nervous System: The Journey from Dysautonomia to Therapeutic Neuro-Modulation through the Retrospective Analysis of 152 Patients

Joseph Colombo [1,*], Michael I. Weintraub [2,*], Ramona Munoz [1], Ashish Verma [1], Ghufran Ahmad [1], Karolina Kaczmarski [1], Luis Santos [3] and Nicholas L. DePace [1]

[1]   Franklin Cardiovascular, Autonomic Dysfunction and POTS Center, Sicklerville, NJ 08081, USA; rmunoz@franklincardio.com (R.M.); ashish@ashishverma.com (A.V.); ghufran.kmc@gmail.com (G.A.); kikaczmarski@gmail.com (K.K.); dovetech@erols.com (N.L.D.)
[2]   Department Neurology and Medicine, New York Medical College, Valhalla, NY 10595, USA
[3]   New Jersey Heart, Sicklerville, NJ 08081, USA; drlou214@icloud.com
*   Correspondence: jcolombo@physiops.com (J.C.); miwneuro@gmail.com (M.I.W.)

Abstract: Introduction. The severity and prevalence of Post-Acute COVID-19 Sequela (PACS) or long-COVID syndrome (long COVID) should not be a surprise. Long-COVID symptoms may be explained by oxidative stress and parasympathetic and sympathetic (P&S) dysfunction. This is a retrospective, hypothesis generating, outcomes study. Methods. From two suburban practices in northeastern United States, 152 long COVID patients were exposed to the following practices: (1) first, they were P&S tested (P&S Monitor 4.0; Physio PS, Inc., Atlanta, GA, USA) prior to being infected with COVID-19 due to other causes of autonomic dysfunction; (2) received a pre-COVID-19 follow-up P&S test after autonomic therapy; (3) then, they were infected with COVID-19; (4) P&S tested within three months of surviving the COVID-19 infection with long-COVID symptoms; and, finally, (5) post-COVID-19, follow-up P&S tested, again, after autonomic therapy. All the patients completed autonomic questionnaires with each test. This cohort included 88 females (57.8%), with an average age of 47.0 years (ranging from 14 to 79 years), and an average BMI of 26.9 #/in^2. Results. More pre-COVID-19 patients presented with sympathetic withdrawal than parasympathetic excess. Post-COVID-19, these patients presented with this ratio reversed and, on average, 49.9% more autonomic symptoms than they did pre-COVID-19. Discussion. Both parasympathetic excess and sympathetic withdrawal are separate and treatable autonomic dysfunctions and autonomic treatment significantly reduces the prevalence of autonomic symptoms. Conclusion. SARS-CoV-2, via its oxidative stress, can lead to P&S dysfunction, which, in turn, affects the control and coordination of all systems throughout the whole body and may explain all of the symptoms of long-COVID syndrome. Autonomic therapy leads to positive outcomes and patient quality of life may be restored.

Keywords: long COVID; parasympathetic; sympathetic; autonomic dysfunction; autonomic therapy; outcomes

Citation: Colombo, J.; Weintraub, M.I.; Munoz, R.; Verma, A.; Ahmad, G.; Kaczmarski, K.; Santos, L.; DePace, N.L. Long COVID and the Autonomic Nervous System: The Journey from Dysautonomia to Therapeutic Neuro-Modulation through the Retrospective Analysis of 152 Patients. *NeuroSci* **2022**, *3*, 300–310. https://doi.org/10.3390/neurosci3020021

Academic Editor: Masato Nakafuku

Received: 4 March 2022
Accepted: 12 April 2022
Published: 23 May 2022

Publisher's Note: MDPI stays neutral with regard to jurisdictional claims in published maps and institutional affiliations.

1. Introduction

The severity and prevalence of Post-Acute COVID-19 Sequela (PACS) or long-COVID syndrome (long COVID) should not be a surprise. SARS-CoV-2 targets diverse organs and tissues after entry into the human body [1]. Long-COVID syndrome is defined as persistent symptoms beyond 12 weeks after acute COVID-19 infection [2,3]. Viruses, by inducing an inflammatory state, can damage tissue. At a cellular level, the mitochondria are susceptible to the effects of inflammation and oxidative stress [4]. Given that nerve cells, including brain cells, and heart muscle cells contain significantly more mitochondria than other cells in the body, it is to be expected that they will be the most affected by oxidative stress. The results of mitochondrial dysfunction includes primarily autonomic dysfunction (including both parasympathetic and sympathetic (P&S)) and cardiovascular dysfunction [5]. Arguably, the first symptom of P&S dysfunction is orthostatic dysfunction [5,6]. Orthostatic dysfunction is a significant contributor to poor cardiac and cerebral perfusion (and, of course, all structures around and above the heart). Autonomic dysfunction is also induced as a result of the severity of the infection [7].

Furthermore, COVID-19 injures the lungs, reducing their ability to exchange oxygen, exacerbating the poor perfusion and resulting dysfunctions [8]. The initial respiratory compromise, due to the COVID-19 virus, on the medullary respiratory control centers (including the pre-Bötzinger complex) [9–11] may be so dramatic that P&S symptoms and signs are often overlooked or misunderstood. Respiratory pacing from the pre-Bötzinger complex involves (1) vagus nerve afferents, among other brainstem structures; (2) feedback from the COVID-19-damaged lung; (3) aortic and carotid chemo-, baro-, and vagal receptors; and (4) medullary chemoreceptors. All involving P&S nerves [9,12]. Brainstem cardiorespiratory centers (e.g., the Nucleus Tractus Solitarius, Dorsal Vagal Motor Nucleus, and Nucleus Ambiguus, all of which are autonomic nuclei) are also implicated in COVID-19 infection [13]. Furthermore, sympathetic involvement in cytokine storms [14–17] and the angiotensin system [18,19], and parasympathetic involvement in immune function [20–22], provides further evidence of P&S compromise in COVID-19 infections. Any resulting damage to these nerves further implicates P&S dysfunction in long-COVID syndrome.

Long-COVID symptoms [23] may be explained by a pro-inflammatory state with oxidative stress and P&S dysfunction [24]. This study presents the data obtained from autonomic dysfunction patients who were P&S tested and treated prior to COVID-19 infection due to other causes of autonomic dysfunction. Then, they were P&S tested and treated after surviving COVID-19 infection.

Long-COVID symptoms may be explained by a pro-inflammatory state with oxidative stress and P&S dysfunction. This is hypothesis generating.

Long COVID is characterized by parasympathetic excess and alpha-sympathetic withdrawal.

Anti-cholinergic therapy may relieve post-COVID-19 symptoms associated with parasympathetic excess. This is hypothesis generating and further trials are needed.

2. Methods

The data presented are from 2 suburban practices in northeastern United States (Sicklerville, NJ, USA and Valhalla, NY, USA), a cardiovascular and autonomic dysfunction clinic and a neurology clinic (respectively). From these two practices, 152 long-COVID patients from around the world who (1) had been under medical therapy for autonomic dysfunction, had been evaluated and underwent P&S testing prior to being infected with COVID-19; (2) a follow-up P&S test was conducted after autonomic therapy; (3) patients were infected with COVID-19; (4) patients were P&S tested within three months of surviving COVID-19 infection with long-COVID symptoms, typically more than the pre-COVID condition, with continued autonomic therapy adjusted to individual patients' needs; and, finally, (5) patients were follow-up P&S tested, again, after autonomic therapy. This cohort includes 88 females (57.8%). The average patient age is 47.0 years (ranging from 14 to 79 years), with an average BMI of 26.9 #/in^2. All the patients were tested with P&S monitoring (P&S Monitor 4.0; Physio PS, Inc., Atlanta, GA, USA) and completed a 28-symptom questionnaire (Table 1). This is a retrospective, observational, hypothesis-generating, outcomes study. All the patients permitted their data to be included in this large population study and patient data were maintained according to the HIPPA guidelines.

P&S monitoring collects EKG, respiratory activity, and BP during four challenges: (1) rest (baseline), (2) deep breathing (0.1 Hz, a parasympathetic challenge), (3) short Valsalva maneuvers (<15 s, as a sympathetic challenge), and (4) head-up postural change (stand, which is equivalent to tilt [25]). Stand is both an orthostatic challenge and a measure of the coordination between the P&S branches. With spectral analyses, these data are analyzed and independent, and simultaneous P&S activity is measured throughout the clinical study [5]. Normal and abnormal P&S response plots are depicted in Figures 1–5, including, in order, (1) a resting baseline response (Figure 1) depicting normal and abnormal ranges; (2) a normal stand or upright posture response (Figure 2); (3) an abnormal stand response depicting alpha-sympathetic withdrawal (upon standing) which indicates orthostatic dysfunction (Figure 3); (4) an abnormal stand response depicting parasympathetic excess (upon standing) which indicates vagal excess (Figure 4); and (5) an abnormal stand response depicting parasympathetic excess with a hyperadrenergic response (upon standing) which indicates vasovagal syncope (Figure 5) [5].

Table 1. 28-symptom autonomic dysfunction questionnaire.

1. Lightheaded	2. Fatigue	3.Chest Pain, Palpitations	4. Short of Breath	5. Fainting and Near Fainting	6. Difficulty Standing	7. Sweat Too Much, Too Little
8. Brain fog or mental cloudiness	9. Difficulty finding words	10. Short-term memory loss	11. Insomnia, sleep difficulty	12. Depression, anxiety	13. Tension headaches	14. Migraine, headache
15. Chronic pain	16. Coat hanger pain in neck and shoulders	17. Pins and needs in arms/legs	18. Numbness in hands and feet	19. Hypermobility of joints, joints pop out	20. Nausea, vomiting	21. Diarrhea, constipation
22. Sensory: hypersensitive to light, sound, motion, touch	23. Sensory deficits: vision, hearing, taste, smell	24. Cold hands or feet	25. Ringing in ears	26. Does hot or cold weather bother you?	27. Hands or feet turn different colors (red, white, or blue) in cold temperatures	28. Salivate too little, dry mouth

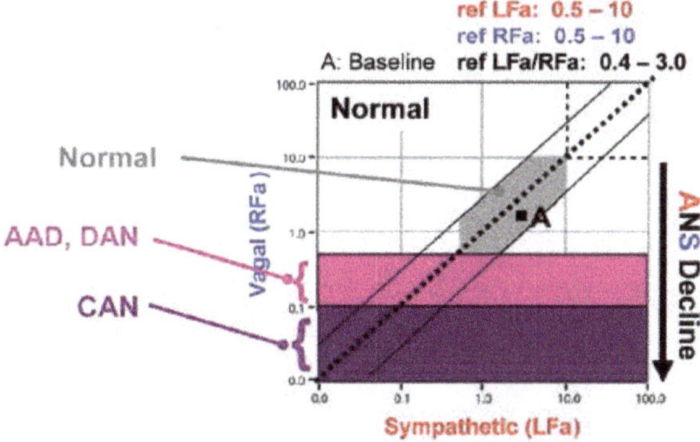

Figure 1. "Normal at Rest". A resting (baseline) P&S response plot depicting normal and abnormal ranges. The gray area depicts the normal response region. The purple highlighted areas depict the definitions of Advanced Autonomic Dysfunction (AAD, light purple) or Diabetic Autonomic Neuropathy (DAN, also light purple), and Cardiovascular Autonomic Neuropathy (CAN, dark purple). AAD and DAN indicate an increased morbidity risk and CAN indicates an increased mortality risk. Risk is stratified by sympathovagal balance ("LFa/RFa" = S/P). The space between the two outer diagonal lines defines a normal sympathovagal balance, regardless of the resting autonomic state. A normal sympathovagal balance normalizes the morbidity and mortality risks. Above and to the left of the upper diagonal line indicates a low sympathovagal balance, which is a resting parasympathetic excess. Below and to the right of the lower diagonal line indicate a high sympathovagal balance, which is a resting sympathetic excess.

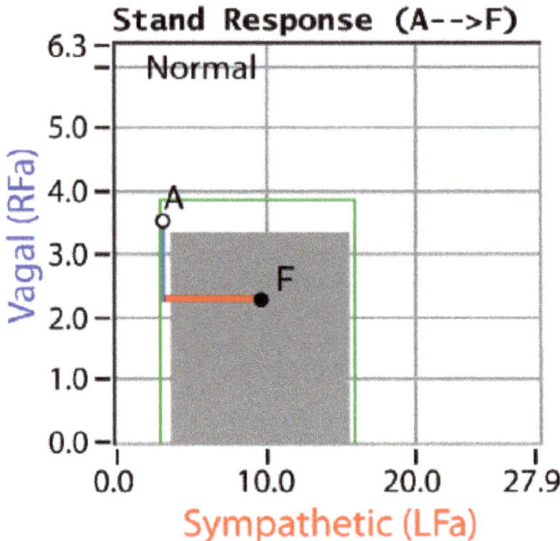

Figure 2. "Normal upon Standing". An example normal stand P&S response plot. Active standing is equivalent to a positive, head-up, tilt [25]. Point "A" is the patient's resting, baseline response and point "F" is the patient's stand response. In the normal stand response, the parasympathetics (the blue line) first decrease and then the (alpha-)sympathetics (the red line) increase [5].

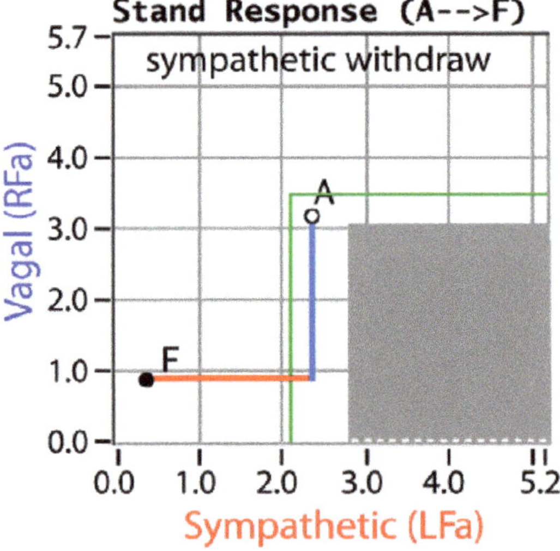

Figure 3. "Sympathetic Withdrawal". An example of an abnormal stand P&S response plot depicting alpha-sympathetic withdrawal. Point "A" is the patient's resting, baseline response and point "F" is the patient's stand response. Here, the parasympathetic response is normal (see Figure 2), but the sympathetic response decreases abnormally, indicating orthostatic dysfunction, possibly leading to all forms of POTS, orthostatic intolerance, orthostatic hypotension, neurogenic orthostatic hypotension, etc. [5].

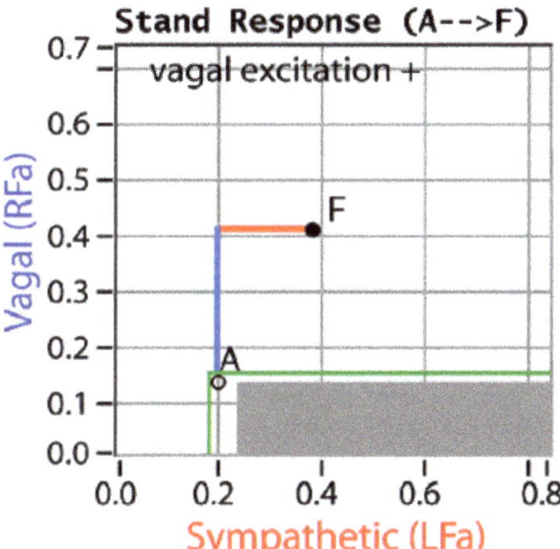

Figure 4. "Vagal Excitation". An example of an abnormal stand P&S response plot depicting parasympathetic excess. Point "A" is the patient's resting, baseline response and point "F" is the patient's stand response. Here, the sympathetic response is normal (see Figure 2), but the parasympathetic response increases abnormally, indicating vagal or parasympathetic excess, associated with difficult-to-control BP, blood glucose, hormone levels, or weight; difficult-to-describe pain syndromes (including CRPS); unexplained arrhythmia (palpitations) or seizures; temperature dysregulation (both the response to heat or cold and sweat responses); symptoms of depression or anxiety, ADD/ADHD, fatigue, exercise intolerance, sex dysfunction, sleep or GI disturbance, lightheadedness, cognitive dysfunction or "brain fog"; and frequent headaches or migraines. Parasympathetic excess and sympathetic withdrawal may concurrently occur, including the fact that parasympathetic excess may mask sympathetic withdrawal. This masking is indicated by an abnormal BP response to stand as compared with resting BP [5].

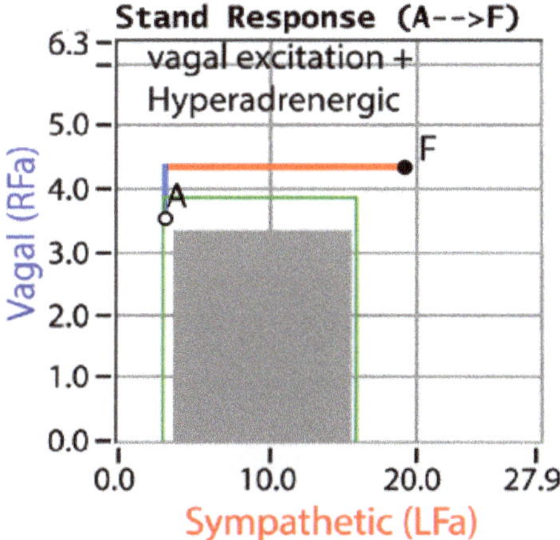

Figure 5. "Vagal Excitation + Hyperadrenergic". An example of an abnormal stand P&S response plot depicting parasympathetic excess with sympathetic excess. Point "A" is the patient's resting, baseline response and point "F" is the patient's stand response. Here, the parasympathetic response is abnormal (see Figure 2), as is the sympathetic response which increases too significantly, exceeding the normal area. The combination indicates vasovagal syncope. The parasympathetic excess is the vagal component, and the sympathetic excess (hyperadrenergic response) indicates the nervous system's response to syncope and the accompanying poor cerebral perfusion [5].

P&S monitoring differs from the other autonomic monitors, in that it is uniquely capable of independently and simultaneously measuring the two individual autonomic branches without assumption and approximation [26]. P&S monitoring permits follow-up testing, and includes indications for peripheral autonomic neuropathy (including Small C-Fiber Disease) [5], as well as P&S dysfunctions (including autonomic neuropathies) not detected by typical autonomic monitors, including sympathetic withdrawal (an alpha-adrenergic insufficiency upon assuming a head-up posture, associated with orthostatic dysfunction) [5,27] and parasympathetic excess (an excessive cholinergic response to a stress, as modeled by the Valsalva challenge or upon assuming a head-up posture, associated with Vagal over-reactions) [5,28].

Sympathetic withdrawal (see Figure 3) and parasympathetic excess (see Figure 4) are two of the P&S dysfunctions typically demonstrated by long-COVID patients. The others include (1) sympathetic excess with up-right posture (a beta-adrenergic response associated with syncope and pre-syncopal symptoms, see Figure 5); (2) low and (3) high sympatho-vagal balance (a measure of the ratio of sympathetic-to-parasympathetic activity at rest, see Figure 1); (4) low resting sympathetic or parasympathetic activity associated with advanced autonomic dysfunction or diabetic autonomic neuropathy if diabetic (see Figure 1); and (5) very-low resting parasympathetic activity at rest, associated with cardiovascular autonomic neuropathy (see Figure 1).

Based on their P&S test results, the patients were prescribed therapy, typically for both sympathetic withdrawal (and associated orthostatic dysfunction) and for parasympathetic excess. Therapy for sympathetic withdrawal (after ruling out vascular causes) typically included: (1) 2.5 mg, tid, of Midodrine (ProAmatine, an alpha-adrenergic antagonist); and (2) up to 600 mg, tid, of Alpha-Lipoic Acid (an antioxidant selective for nerves [29,30]). Therapy for parasympathetic excess included: (1) 10 mg, qd, of Nortriptyline (as a low-dose anti-cholinergic), and (2) up to 40 minutes of low-and-slow exercise [31]. The Pearson correlation and Student's *t*-test statistics are based on SPSS v. 20.

3. Results

In general, the patients reported poor health. Patients first presented (pre-COVID-19) with lightheadedness (100%), due to (1) pre-syncope (28.3%) or syncope (2.6%); (2) orthostatic dysfunction, including Postural Orthostatic Tachycardia Syndrome (POTS, 8.6%) and orthostatic intolerance or orthostatic hypotension (36.8%); or (3) excessive vagal symptoms (27.0%). Approximately, a quarter (25.7%) of the cohort first presented with anxiety-like symptoms, including palpitations and shortness of breath. Over a third (36.9%) of the cohort reported fatigue, nearly half (46.9%) reported generalized pain, including headaches and migraines, and 25.7% of the patients were diagnosed with Ehlers-Danlos Syndrome—Hypermobility (see Table 2). The prevalence of the autonomic dysfunctions are listed in Table 3. Sympathetic withdrawal is the most prevalent autonomic dysfunction pre-COVID-19, and parasympathetic excess is the most prevalent post-COVID-19.

Table 2. Patient demographics upon the first presentation.

Cohort	No.	No. Female	Ave. Age	Ave. BMI	LH	Fatigue	Anxiety	Headache, Migraine	EDSh
# (%)	152	88 (57.8)	47.0 yrs	26.9 lbs/ft^2	152 (100)	56 (36.9)	39 (25.7)	71 (46.9)	39 (25.7)

Ave.: average; BMI: body mass index; EDSh: Ehlers-Danlos Syndrome/Hypermobility; LH: lightheadedness; and No.: number.

From the last column of Table 3, upon the initial presentation (pre-COVID-19), these autonomic dysfunction patients presented with an average of 2.34 of the 7 P&S dysfunctions listed in the first seven (7) columns of Table 3. With less than 9 months of therapy, the pre-COVID-19 patients were found with an average of 0.95 of the 7 P&S dysfunctions ($p < 0.001$). Post-COVID-19, these patients demonstrated an average of 3.67 of the 7 P&S

dysfunctions ($p = 0.004$). With less than 6 months of continued and, as needed, additional autonomic therapy, the post-COVID-19 patients were found with an average of 1.63 of the 7 P&S dysfunctions ($p = 0.003$).

Table 3. Percentage prevalence of the autonomic dysfunctions in long-COVID patients. See the text for details and abbreviations.

N = 152 # (%)	SW	PE	SE	Low SB	Hi SB	AAD	CAN	Ave. Sx	Ave. ADs
Pre-COVID-19	69 (45.4)	41 (27.0)	19 (12.5)	23 (15.1)	41 (27.0)	26 (17.1)	8 (5.3)	9.74	2.34
Pre-COVID-19 Follow-up	41 (27.2)	25 (16.2)	11 (7.5)	14 (9.1)	25 (16.2)	6 (3.9)	0 (0)	6.25	0.95
p-value	0.023	0.011	0.009	0.001	0.002	<0.001	<0.001	0.009	<0.001
Post-COVID-19	55 (36.2)	71 (46.7)	44 (28.9)	58 (38.2)	69 (45.4)	31 (20.4)	10 (6.6)	14.6	3.67
Post-COVID-19 Follow-up	33 (21.7)	43 (28.0)	26 (17.4)	29 (19.1)	31 (20.4)	6 (3.9)	2 (1.3)	7.44	1.63
p-value	0.041	0.024	0.016	0.010	0.002	<0.001	0.001	0.004	0.003

AD: average number of autonomic dysfunctions based on the seven possible dysfunctions listed as the first seven column headers of this table; AAD: Advanced Autonomic Dysfunction, an indication of morbidity risk; CAN: Cardiovascular Autonomic Neuropathy, an indication of mortality risk; PE: parasympathetic excess, an abnormal parasympathetic response to a sympathetic challenge or stress; SB: sympathovagal balance, the ratio of resting sympathetic-to-resting parasympathetic activity; SE: sympathetic excess, a beta-adrenergic response to challenge; SW: sympathetic withdrawal, an alpha-adrenergic response to positive, head-up postural change (e.g., stand); and Sx: average number of autonomic symptoms from the 28-question survey in Table 1.

From the second to last column of Table 3, at the pre-COVID-19 baseline, these patients complained of an average of 9.74 of the 28 symptoms. Upon the pre-COVID-19 follow-up, the patients' complaints were reduced, on average, to 6.25 symptoms ($p = 0.009$). Post-COVID-19, the patients complained of an average of 14.6 of the 28 symptoms ($p < 0.001$). Upon the post-COVID-19 follow-up, the patients complained of an average of 7.44 symptoms ($p = 0.004$).

COVID-19 infection returned and added to the number of (56.8% more) autonomic dysfunctions demonstrated by these 152 patients. Also, COVID-19 infection returned and increased the number of (49.9% more) associated symptoms reported by these patients. Upon follow-up testing, both pre- and post-COVID infection, all the patients reported improved outcomes, which was evidenced by the fewer P&S dysfunctions and fewer symptoms reported upon follow-up.

From Table 3, acute COVID-19 infection also reversed the order of the top two autonomic dysfunctions from sympathetic withdrawal being more predominant pre-COVID-19 to parasympathetic excess being more predominant post-COVID-19. An abnormal sympathovagal balance also become more significant. Those who also demonstrated a low sympathovagal balance (resting vagal excess) also reported more significant symptoms of depression/anxiety and fatigue. Those who demonstrated a high sympathovagal balance (resting sympathetic excess) also reported more significant symptoms of pain and hypertension. However, the high sympathovagal balance results may be biased by the number of Ehlers-Danlos Syndrome-Hypermobility patients.

4. Discussion

COVID-19 is documented to adversely affect the autonomic nervous system [32]. In many patients, the lingering effect on the autonomic nervous system results in what has been termed long COVID [33]. Long COVID is well documented to involve the autonomic nervous system [34–36]. Autonomic dysfunctions may be peripheral or central. In central cases, autonomic dysfunctions may be related to microglial hyperactivation inside the brainstem autonomic centers [37]. Microglial hyperactivation is associated with PE [38]. Autonomic dysfunctions may also be highly influenced by psychological factors.

In our findings, long COVID is largely characterized by parasympathetic excess and sympathetic withdrawal. Both potentially contributing to hypoperfusion of the brain and all structures above and around the heart. Pre-COVID-19 infection, patients presented to the clinics with more sympathetic withdrawal (45.7%) than parasympathetic excess (27.0%). Post-COVID-19 infection, these patients presented with that ratio reversed (36.2% and 46.7%, respectively). The etiology of this is not well known; however, parasympathetic excess may be more prominent post-COVID-19, due to an over-active immune system, which the parasympathetics help to control and coordinate and leads to parasympathetic excess.

Given that the parasympathetic nervous system controls and coordinates the immune system, severe infections lead to excessive and prolonged parasympathetic activation in response to challenges or stressors (known as parasympathetic excess) [7], which exacerbates autonomic and cardiovascular dysfunctions. A common, and perhaps first cause of autonomic dysfunction, due to mitochondrial dysfunction and associated oxidative stress, is orthostatic dysfunction [6], resulting in poor cardiac and cerebral perfusions (and, of course, all the structures around and above the heart). Orthostatic dysfunction is caused by poor vasoconstriction due to alpha-adrenergic (sympathetic) dysfunction, known as sympathetic withdrawal [5]. Poor perfusion and dysfunction are exacerbated by the effect of COVID-19 on the lungs.

Both parasympathetic excess and sympathetic withdrawal are separate and treatable dysfunctions. As in this study, parasympathetic excess was treated, pharmaceutically, with anti-cholinergics (e.g., Nortriptyline, see the Methods Section) [31] and sympathetic withdrawal was treated, pharmaceutically, with oral vasoactives (e.g., Midodrine, see the Methods Section) [30].

Our findings demonstrate an initial worsening of autonomic dysfunction and symptoms associated with COVID-19 infection, and then, with autonomic treatment, these dysfunctions and symptoms may again be relieved. Traditionally, upon COVID-19 infection, there is a marked increase in the resting sympathetic activity and a decrease in anti-inflammatory resting parasympathetic activity [16], causing a high (resting) sympathovagal balance in all patients. However, in post-COVID-19 syndrome patients, after 12 weeks or more, our data shows that there is a significant percentage of patients that develop a parasympathetic dominance as indicated by the low (resting) sympathovagal balance. This is also indicative of increasing and prolonged parasympathetic activity. Parasympathetic activation is meant to be protective; including, since the parasympathetics are anti-inflammatory. However, prolonged and increased parasympathetic activity, especially in response to stressors, seems to exaggerate sympathetic inflammatory activity. Within this cohort, and anecdotally with the vast majority of our patients, anti-cholinergic therapy relieves parasympathetic excess. Further studies are required to elaborate whether anti-cholinergic therapy may relieve post-COVID-19 symptoms.

All symptoms of long COVID may be explained by oxidative stress and P&S dysfunction. For example, P&S dysfunction leading to orthostatic dysfunction underlies poor cerebral (including all structures above the heart) perfusion, which causes fatigue, brain-fog, cognitive and memory difficulties, sleep difficulties, and other depression-like symptoms, including "coat-hanger" pain, headaches and migraines; cranial nerve dysfunctions, including visual and auditory effects (including tinnitus), taste and smell deficits, and facial sensations due to trigeminal nerve dysfunction. P&S dysfunction may also increase BP (and may eventually lead to hypertension) as a compensatory mechanism to promote cerebral perfusion. Further decreases in cerebral perfusion may lead to "adrenaline storms", which cycle anxiety-like symptoms, including shortness of breath and palpitations which may cause chest pressure or chest pain. The effects of sympathetic withdrawal and orthostatic dysfunction are exacerbated by parasympathetic excess, which may limit or decrease the heart rate and blood pressure, reducing cerebral perfusion. The decrease in BP is also associated with excessive vasodilation from parasympathetic excess.

If the parasympathetics increase in response to a stress (known as parasympathetic excess), the result is a secondary sympathetic excess [5]. Our findings of prolonged parasympathetic excess in long-COVID patients appears to prolong sympathetic excess responses causing more and chronic symptoms, suggesting that this may be a mechanism contributing to long-COVID syndrome.

Pharmaceutical therapy for P&S dysfunction (anti-cholinergics for parasympathetic excess [28] and oral vasoactives for sympathetic withdrawal [39]) needs to be very low to prevent additional symptoms, thereby exacerbating P&S dysfunction. From Table 3, COVID-19 significantly increases autonomic dysfunctions and the associated symptoms, and autonomic therapy significantly reduces autonomic dysfunctions and the associated symptoms. Further studies are needed, including blinded, controlled studies.

5. Conclusions

The current body of evidence suggests that SARS-CoV-2 can affect the nervous system in previously unexpected ways [40]. Mitochondrial damage causing oxidative stress leads to P&S dysfunction. In turn, oxidative stress and P&S dysfunction affects the control and coordination of systems throughout the body and may explain the clinical symptoms recognized as long-COVID syndrome. Autonomic therapy has been shown to provide positive outcomes and improvement in patients quality of life.

The association of anxiety, Postural Orthostatic Tachycardia Syndrome (POTS), and Chronic Fatigue Syndrome/Myalgic Encephalomyelitis (CFS/ME) with long COVID is interesting, because all are also characterized by P&S dysfunction. However, to diagnose these conditions, including long COVID, independent and simultaneous (direct) measures of P&S activity are required; the assumptions and approximations required by other autonomic tests are not typically appropriate for these patients. These direct measures are needed because both autonomic branches are involved in long COVID in different ways and must be treated separately, and may be treated simultaneously. Therapy is low and slow, and patient expectations must be properly established for optimum compliance. Follow-up testing is needed to help with compliance and ensure that therapy is properly titrated to the individual patient. Based on all of this, positive outcomes are realized and patient quality of life may be restored. While this study serially followed patients with underlying autonomic dysfunctions pre- and post-COVID-19, future studies should assess the effects of autonomic functions on normal subjects pre- and post-COVID-19.

Author Contributions: Conceptualization and writing, J.C. and N.L.D.; formal analysis, N.L.D.; investigation, J.C.; data collection, R.M. and L.S.; data mining, A.V., G.A. and K.K.; supervision, M.I.W. All authors have read and agreed to the published version of the manuscript.

Funding: This research received no external funding.

Institutional Review Board Statement: All patients permitted their data to be included in this large population study and patient data were maintained according to HIPPA guidelines.

Informed Consent Statement: Patient consent was waived due to all patients attending these clinics are informed that their data may be used for large population clinical studies, unless the patient objects. None of the 152 patients included in this study objected.

Data Availability Statement: All data are HIPAA protected and available upon request.

Conflicts of Interest: Only Colombo has a conflict of interest as Co-Founder and Sr. Medical Director of Physio PS, Inc. No other individual associated with this manuscript has a conflict of interest.

References

1. DePace, N.L.; Colombo, J. Long-covid syndrome: A multi-organ disorder. *Cardio. Open* **2022**, *7*, 212–223.
2. Alwan, N.A.; Johnson, L. Defining long COVID: Going back to the start. *Med* **2021**, *2*, 501–504. [CrossRef] [PubMed]
3. DePace, N.L.; Colombo, J. Long-COVID syndrome: A review of what we have learned clinically to date. 2022, *Submitted*.

4. DePace, N.L.; Colombo, J. *Autonomic and Mitochondrial Dysfunction in Clinical Diseases: Diagnostic, Prevention, and Therapy*; Springer Science + Business Media: New York, NY, USA, 2019.

5. Colombo, J.; Arora, R.R.; DePace, N.L.; Vinik, A.I. *Clinical Autonomic Dysfunction: Measurement, Indications, Therapies, and Outcomes*; Springer Science + Business Media: New York, NY, USA, 2014.

6. Vinik, A.I.; Maser, R.E.; Nakave, A.A. Diabetic cardiovascular autonomic nerve dysfunction. *US Endocr. Dis.* **2007**, *2*, 2–9. [CrossRef]

7. Felton, D.L. Neural influence on immune responses: Underlying suppositions and basic principles of neural-immune signaling. *Prog. Brain Res.* **2000**, *122*, 381–389.

8. Dixit, N.M.; Churchill, A.; Nsair, A.; Hsu, J.J. Post-acute COVID-19 syndrome and the cardiovascular system: What is known? *Am. Heart J. Plus.* **2021**, *5*, 100025. [CrossRef] [PubMed]

9. Baig, A.M. Computing the effects of SARS-CoV-2 on respiration regulatory mechanisms in COVID-19. *ACS Chem. Neurosci.* **2020**, *11*, 2416–2421. [CrossRef]

10. Li, Z.; Liu, T.; Yang, N.; Han, D.; Mi, X.; Li, Y.; Liu, K.; Vuylsteke, A.; Xiang, H.; Guo, X. Neurological manifestations of patients with COVID-19: Potential routes of SARS-CoV-2 neuroinvasion from the periphery to the brain. *Front Med.* **2020**, *14*, 533–541. [CrossRef]

11. Román, G.C.; Spencer, P.S.; Reis, J.; Buguet, A.; Faris, M.E.A.; Katrak, S.M.; Láinez, M.; Medina, M.T.; Meshram, C.; Mizusawa, H.; et al. The neurology of COVID-19 revisited: A proposal from the environmental neurology specialty group of the world federation of neurology to implement international neurological registries. *J. Neurol. Sci.* **2020**, *414*, 116884. [CrossRef]

12. Baig, A.M.; Khaleeq, A.; Ali, U.; Syeda, H. Evidence of the COVID-19 virus targeting the CNS: Tissue distribution, host-virus interaction, and proposed neurotropic mechanisms. *ACS Chem. Neurosci.* **2020**, *11*, 995–998. [CrossRef]

13. Pergolizzi, J.V., Jr.; Raffa, R.B.; Varrassi, G.; Magnusson, P.; LeQuang, J.A.; Paladini, A.; Taylor, R.; Wollmuth, C.; Breve, F.; Chopra, M.; et al. Potential neurological manifestations of COVID-19: A narra-tive review. *Postgrad. Med.* **2021**, *11*, 1–11. [CrossRef]

14. Fischer, L.; Barop, H.; Ludin, S.M.; Schaible, H.G. Regulation of acute reflectory hyperinflammation in viral and other diseases by means of stellate ganglion block. A conceptual view with a focus on COVID-19. *Auton. Neurosci.* **2022**, *237*, 102903. [CrossRef] [PubMed]

15. Shah, R.M.; Shah, M.; Shah, S.; Li, A.; Jauhar, S. Takotsubo syndrome and COVID-19: Associations and implications. *Curr. Probl. Cardiol.* **2021**, *46*, 100763. [CrossRef]

16. Al-Kuraishy, H.M.; Al-Gareeb, A.I.; Qusti, S.; Alshammari, E.M.; Gyebi, G.A.; Batiha, G.E. COVID-19-induced dysautonomia: A menace of sympathetic storm. *ASN Neuro.* **2021**, *13*, 17590914211057635. [CrossRef] [PubMed]

17. Woiciechowsky, C.; Asadullah, K.; Nestler, D.; Eberhardt, B.; Platzer, C.; Schöning, B.; Glöckner, F.; Lanksch, W.R.; Volk, H.D.; Döcke, W.D. Sympathetic activation triggers systemic interleukin-10 release in immunode-pression induced by brain injury. *Nat. Med.* **1998**, *4*, 808–813. [CrossRef]

18. Miller, A.J.; Arnold, A.C. The renin-angiotensin system in cardiovascular autonomic control: Recent developments and clinical implications. *Clin. Auton. Res.* **2019**, *29*, 231–243. [CrossRef] [PubMed]

19. Porzionato, A.; Emmi, A.; Barbon, S.; Boscolo-Berto, R.; Stecco, C.; Stocco, E.; Macchi, V.; De Caro, R. Sympathetic activation: A potential link between comorbidities and COVID-19. *FEBS J.* **2020**, *287*, 3681–3688. [CrossRef]

20. Forsythe, P. The nervous system as a critical regulator of immune responses underlying allergy. *Curr. Pharm. Des.* **2012**, *18*, 2290–2304. [CrossRef]

21. Pavlov, V.A.; Tracey, K.J. The vagus nerve and the inflammatory reflex—Linking immunity and metabolism. *Nat. Rev. Endocrinol.* **2012**, *8*, 743–754. [CrossRef]

22. Breit, S.; Kupferberg, A.; Rogler, G.; Hasler, G. Vagus nerve as modulator of the brain-gut axis in psychiatric and inflammatory disorders. *Front Psychiatry.* **2018**, *9*, 44. [CrossRef]

23. Lambert, N.J. ; *Survivor Corps. COVID-19 "Long Hauler" Symptoms Survey Report*; Indiana University School of Medicine: Indianapolis, IN, USA, 2020. Available online: https://dig.abclocal.go.com/wls/documents/2020/072720-wls-covid-symptom-study-doc.pdf (accessed on 20 August 2021).

24. Bisaccia, G.; Ricci, F.; Recce, V.; Serio, A.; Iannetti, G.; Chahal, A.A.; Ståhlberg, M.; Khanji, M.Y.; Fedorowski, A.; Gallina, S. Post-acute sequelae of COVID-19 and cardiovascular autonomic dysfunction: What do we know? *J. Cardiovasc. Dev. Dis.* **2021**, *8*, 156. [CrossRef]

25. Bloomfield, D.M.; Kaufman, E.S.; Bigger, J.T., Jr.; Fleiss, J.; Rolnitzky, L.; Steinman, R. Passive head-up tilt and actively standing up produce similar overall changes in autonomic balance. *Am. Heart J.* **1997**, *134*, 316–320. [CrossRef]

26. Akselrod, S. Spectral analysis of fluctuations in cardiovascular parameters: A quantitative tool for the investigation of autonomic control. *Trends Pharmacol. Sci.* **1988**, *9*, 6–9. [CrossRef]

27. Arora, R.R.; Bulgarelli, R.J.; Ghosh-Dastidar, S.; Colombo, J. Autonomic mechanisms and therapeutic implications of postural diabetic cardiovascular abnormalities. *J. Diabetes Sci. Technol.* **2008**, *2*, 568–571. [CrossRef] [PubMed]

28. Tobias, H.; Vinitsky, A.; Bulgarelli, R.J.; Ghosh-Dastidar, S.; Colombo, J. Autonomic nervous system monitoring of patients with excess parasympathetic responses to sympathetic challenges—Clinical observations. *US Neurol.* **2010**, *5*, 62–66. [CrossRef]

29. Prendergast, J.J. Diabetic autonomic neuropathy: Part 2. Treatment. *Pract. Diabetol.* **2001**, *18*, 30–36.

30. Ziegler, D.; Low, P.A.; Litchy, W.J.; Boulton, A.J.M.; Vinik, A.I.; Freeman, R.; The NATHAN 1 Trial Group. Efficacy and safety of antioxidant treatment with α-lipoic acid over 4 years in diabetic polyneuropathy. *Diabetes Care* **2011**, *34*, 2054–2060. [CrossRef] [PubMed]
31. Fu, Q.; Levine, B.D. Exercise and the autonomic nervous system. *Handb. Clin. Neurol.* **2013**, *117*, 147–160. [CrossRef]
32. Goldstein, D.S. The extended autonomic system, dyshomeostasis, and COVID-19. *Clin. Auton. Res.* **2020**, *30*, 299–315. [CrossRef]
33. Larsen, N.W.; Stiles, L.E.; Miglis, M.G. Preparing for the long-haul: Autonomic complications of COVID-19. *Auton. Neurosci.* **2021**, *235*, 102841. [CrossRef]
34. Dani, M.; Dirksen, A.; Taraborrelli, P.; Torocastro, M.; Panagopoulos, D.; Sutton, R.; Lim, P.B. Autonomic dys-function in 'long COVID': Rationale, physiology and management strategies. *Clin. Med.* **2021**, *21*, e63–e67. [CrossRef]
35. Raj, S.R.; Arnold, A.C.; Barboi, A.; Claydon, V.E.; Limberg, J.K.; Lucci, V.M.; Numan, M.; Peltier, A.; Snapper, H.; Vernino, S.; et al. Long-COVID postural tachycardia syndrome: An American Auto-nomic Society statement. *Clin. Auton. Res.* **2021**, *31*, 365–368. [CrossRef] [PubMed]
36. Shouman, K.; Vanichkachorn, G.; Cheshire, W.P.; Suarez, M.D.; Shelly, S.; Lamotte, G.J.; Sandroni, P.; Benarroch, E.E.; Berini, S.E.; Cutsforth-Gregory, J.K.; et al. Autonomic dysfunction following COVID-19 infection: An early experience. *Clin. Auton. Res.* **2021**, *31*, 385–394. [CrossRef] [PubMed]
37. Poloni, T.E.; Medici, V.; Moretti, M.; Visonà, S.D.; Cirrincione, A.; Carlos, A.F.; Davin, A.; Gagliardi, S.; Pansarasa, O.; Cereda, C.; et al. COVID-19-related neuropathology and microglial activation in elderly with and without dementia. *Brain Pathol.* **2021**, *31*, e12997. [CrossRef] [PubMed]
38. Wirtz, M.R.; Moekotte, J.; Balvers, K.; Admiraal, M.M.; Pittet, J.F.; Colombo, J.; Wagener, B.M.; Goslings, J.C.; Juffermans, N. Autonomic nervous system activity and the risk of nosocomial infection in critically ill patients with brain injury. *Intensive Care Med. Exp.* **2020**, *8*, 69. [CrossRef]
39. DePace, N.L.; Vinik, A.I.; Acosta, C.R.; Eisen, H.J.; Colombo, J. Oral vasoactive medications: A summary of midodrine and droxidopa as applied to orthostatic dysfunction. *Cardio. Open* **2022**, *7*, 224–240.
40. Abboud, H.; Abboud, F.Z.; Kharbouch, H.; Arkha, Y.; El Abbadi, N.; El Ouahabi, A. COVID-19 and SARS-COV-2 infection: Pathophysiology and clinical effects on the nervous system. *World Neurosurg.* **2020**, *140*, 49–53. [CrossRef]

Case Report

Bilateral Facial Palsy as the Onset of Neurosarcoidosis: A Case Report and a Revision of Literature

Chiara Gallo [1]**, Letizia Mazzini** [1]**, Claudia Varrasi** [1]**, Domizia Vecchio** [1]**, Eleonora Virgilio** [1,2,*] **and Roberto Cantello** [1]

[1] Department of Neurology, University of Eastern Piedmont, Maggiore della Carità Hospital, 28100 Novara, Italy; 20032512@studenti.uniupo.it (C.G.); mazzini.letizia5@gmail.com (L.M.); claudia.varrasi@libero.it (C.V.); domizia.vecchio@gmail.com (D.V.); roberto.cantello@med.uniupo.it (R.C.)

[2] Ph.D. Program in Medical Sciences and Biotechnologies, Department of Translational Medicine, University of Eastern Piedmont, 28100 Novara, Italy

* Correspondence: virgilioeleonora88@gmail.com

Abstract: Unilateral facial nerve palsy (FNP) is one of the most common cranial mononeuropathies. Among rare etiologies, neurosarcoidosis (NS) can cause bilateral involvement (both recurring and simultaneous) only in 15% to 25% of cases. The rarity of this systemic disease and its clinical heterogeneity, due to granulomatous inflammation that may affect many anatomic substrates, frequently make the diagnosis a real challenge for the clinician. Based on laboratory and instrumental tests, a careful diagnostic algorithm must be adopted to avoid misdiagnosis and delay in treatment. We present a 52-year-old woman with an acute onset of unilateral right FNP, rapidly developing contralateral involvement (simultaneous bilateral FNP). Lung findings pointed towards a systemic disease, and then lymph node biopsy confirmed NS. Corticosteroid therapy was started. After three years of follow-up, the patient is still in remission with a low prednisone dose. We discuss the differential diagnosis of bilateral FNP, focusing on clinical presentation, diagnosis, and treatment of NS. We have performed a literature revision, confirming bilateral FNP, outside Heerfordt syndrome, to be rare and sometimes represent the only neurological manifestation of NS onset.

Keywords: neurosarcoidosis; sarcoidosis; cranial neuropathy; facial diplegia; lymph node biopsy

Citation: Gallo, C.; Mazzini, L.; Varrasi, C.; Vecchio, D.; Virgilio, E.; Cantello, R. Bilateral Facial Palsy as the Onset of Neurosarcoidosis: A Case Report and a Revision of Literature. *NeuroSci* **2022**, *3*, 321–331. https://doi.org/10.3390/neurosci3020023

Academic Editors: Lucilla Parnetti, Federico Paolini Paoletti and Xavier Gallart-Palau

Received: 21 April 2022
Accepted: 25 May 2022
Published: 29 May 2022

Publisher's Note: MDPI stays neutral with regard to jurisdictional claims in published maps and institutional affiliations.

1. Introduction

Facial nerve palsy (FNP) is a frequent cranial mononeuropathy [1]. Overall, 70% of cases are idiopathic, not recognizing a specific etiology, best known as Bell's palsy. This condition may leave a residual facial weakness, but it is not a life-threatening condition. Furthermore, many different traumatic, iatrogenic, infectious, and systemic diseases can manifest as FNP more rarely than Bell's palsy [1–3]. Additionally, bilateral FNP is even more uncommon, ranging from 0.3% to 2% of all FNP cases, and recognizes fewer identifiable causes [4], which could be amenable to prompt medical management or surgical approach. The ruling out of all life-threatening diseases, such as leukemia or Guillain-Barre syndrome (GBS), is mandatory and prioritizes workups [3–5]. The diagnostic algorithm includes targeted laboratory and instrumental findings (such as brain MRI), depending upon then history, which may lead to identifying even the rarest causes of simultaneous presentation throughout a real diagnostic challenge. Cranial neuropathy is the most common manifestation of neurosarcoidosis (NS), with cranial nerves II, VII, and VIII being the most frequently affected [6] and a high frequency of subclinical leptomeningeal involvement. About one-third of FNP are bilateral and could be recurrent (RBFP) or simultaneous (SBFP). SBFP is defined as the involvement of the opposite side within 30 days of the onset of the first side [6,7]. Facial diplegia in NS is also described as SBFP/RBFP, associated with fever and the involvement of parotid glands and uveitis (also known as "Heerfordt's syndrome") [8–12]. Since NS is a rare disease, the recognition of bilateral FNP may represent a red flag for the clinician to direct early diagnosis and intervention. Thus, we present the case of a 52-year-old Caucasian woman with SBFP and a review of the existing literature regarding bilateral facial palsy/plegia as a manifestation of NS.

2. Case Report

Our patient presented to the Emergency Department with House–Brackmann scale grade V (corresponding to severe dysfunction, barely perceptible motion, and asymmetry at rest) right peripheral FNP. Her past medical history included situational depression with anxiety and mild tension-type headache occurring since adolescence. She was on no medications, and she denied any tobacco, alcohol, or drug abuse, exposition to toxic agents, or recent infection. Vital parameters were normal. Lungs auscultation and abdomen evaluation were unremarkable. Herpes Zoster infection was first excluded by searching for vesicles or scabbing. Then, a complete neurological examination showed right tongue deviation, impaired sensation to the right side of the face, and an alteration to taste sensation. A brain CT scan ruled out a cerebrovascular accident of posterior circulation. Subsequent brain MRI with contrast was normal. A complete blood count and a comprehensive metabolic panel showed no abnormality. A routine chest X-ray showed multiple bilateral lymphadenopathies confirmed by chest CT (Figure 1a). We excluded occult neoplasia, performed a mammography and breast ultrasound (BI-RADS 1 bilaterally), and planned a total body PET scan (Figure 1b).

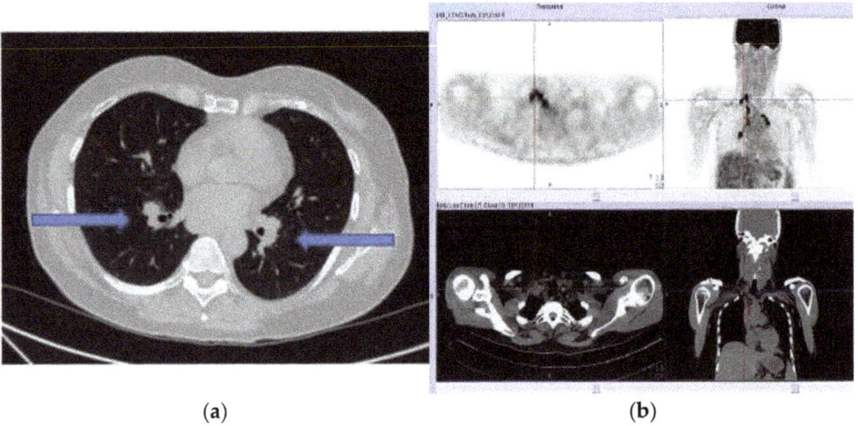

(a) (b)

Figure 1. (a) Pulmonary computed tomography (CT) scan showing multiple bilateral lymphadenopathies. (b) Positron emission tomography (PET) showing multiple adenopathies with a high metabolic component.

After a week, she presented with left peripheral FNP, bilateral hearing loss, limitation in left gaze, expression of left sixth cranial nerve palsy, and liquid dysphagia. The examination confirmed the presence of cranial multineuritis. Following readmission, a lumbar puncture was performed with cerebrospinal fluid (CSF) analysis showing a slight increase in cell count (17 cells/mm^3—monocytes), normal protein (31 mg/dL), and glucose level (70 mg/dL); malignant cells were not seen on cytology. The PET scan showed multiple supraclavicular, paratracheal, epiaortic, paraesophageal, and parahilar adenopathies with a high metabolic component. Serological tests, including ANCA, ANA, HIV antibody test, tuberculosis, and B.Borrelia serology, were negative, but the sedimentation rate of the erythrocytes (ESR) was 24 mm/h. Flow cytometry was not performed. Based on clinical presentation and exam results, the suspect of a granulomatous disorder was made, and corticosteroid therapy (1 mg/kg of prednisone–55 mg/day) was started with a partial clinical benefit over four days. A needle biopsy was performed via bronchoscopy and showed epithelioid macrophages in granulomatous aggregation to characterize the lymph node alterations (Figure 2).

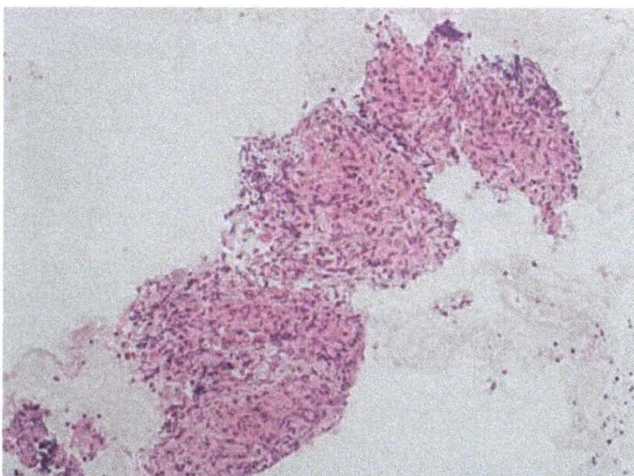

Figure 2. Parahilar lymph node biopsy showing epithelioid macrophages in granulomatous aggregation (courtesy of Professor R. Boldorini Division of Pathology, Department of Health Science, University of Piemonte Orientale (UPO), Novara, Italy). Scale bar: 100 μm (40× magnification).

The histological finding was compatible with sarcoidosis. The serum ACE value was increased (67.7 U/L—normal range 8.0–52.0), and the CSF-ACE level was not performed. The clinical presentation, the evidence of central nervous system (CNS) inflammation (CSF findings), typical histological findings of granulomatous inflammation of tissue biopsy in at least one extraneural organ, and the exclusion of other causes confirmed a diagnosis of probable neurosarcoidosis in pulmonary sarcoidosis grade I. She improved on this regimen with partial resolution of her facial palsies after six weeks of steroid therapy (right paralysis: House–Brackmann grade III; left paralysis: grade II) and complete hearing loss regression. Her steroids were slowly tapered after eight weeks of therapy, and she continues to do well with 12.5 mg of prednisone after three years of follow-up. After a pneumological examination, immunosuppressant treatment with methotrexate was recommended to reduce and discontinue steroid therapy.

3. Literature Review
Search Strategy and Selection Process

According to the guidance on narrative reviews, we conducted a literature search in PubMed, up to 17 March 2022, combining the main terms "facial diplegia neurosarcoidosis", (("bi-lateral facial paralysis") OR ("bilateral facial palsy") OR ("bilateral seventh nerve palsy")) AND (("neurosarcoidosis") OR ("sarcoidosis")). Clinical studies, case reports, and case series reporting NS patients with acute bilateral recurrent or simultaneous FNP published in English, Dutch, Spanish, and French between 1970 and March 2022 were considered for inclusion. Two authors (CG and EV) independently screened seventy titles, selecting sixty abstracts and retaining fifty-one full texts of all relevant articles. Three articles were excluded, published between 1970 and 1980, due to difficulty finding the full-text and only partial clinical information. Finally, twenty-one articles were selected [8,13–30]. Two articles conducted in the same center had only partial information regarding patients with BFP and, therefore, could not be included in the final results [31,32]. Data on study characteristics, demographic features, clinical manifestations, treatment, and outcome are reported below in Table 1, Table 2 and in Table S1 (reported in Supplementary Materials).

Table 1. Demographic and clinical properties of patients with bilateral facial palsy/plegia as the onset of NS (Heerfordt's syndrome cases not included).

N°	Age/Gender/Diagnosis *	Onset/HB (R-L)/Other	Systemic Involvement /Instrumental Findings	Serum //CFS Lab	Serum ACE Onset//FU	Ref
1	48Y/F/Probable NS in P.S I	SBFP/-/ bilat VIII c.n	Lungs/diffused LNP+ -CT: hilar LNP+	N//-	ACE↑//-	[13]
2	25Y/M/Possible NS in P.S II	RBFP recurrences /-/ TIAs, ischemic stroke	Lungs/Kidney/Hematological -CT: LNP+; multiple rounded lungs opacities; BS: nodules	↑Ca^{2+}; ↑γ-glob//NA	ACE↑//-	[14]
3	40Y/F/Possible Paradoxical NS TNFα-related in P.S I	SBFD/V-V/ lymphocytic meningitis, anosmia	Lung/Eye/Kidney/Weight loss -Ophthalmological test: papilledema, anterior uveitis -18F-PET-CT: + hilar LNP, parotid	Ca^{2+} ↑; proteinemia↑ /proteinorachia ↑	ACE↑//-	[15]
4	28Y/F/Probable NS in P.S I	Headache, dizziness/ SBFD/-/ bilateral VI c.n/ diffuse leptomeningitis	Lung/Skin/URT -67Ga scintigraphy: hilar LNP+, salivary glands, nasal sinus and cubital fossa	N//proteinorachia ↑ (125 mg/dl); ↑ CSF ACE 2.8 IU F.U.:normal	ACE N//-	[16]
5	33Y/M/Probable NS in P.S II	RBFP recurrences + VI c.n/-/ III c.n/ O.N	Lung/Eye/Skin/Kidney -67Ga scintigraphy: hilar LNP+, salivary glands (uptake +); frontal meningities, lungs,	ESR ↑// NA	ACE↑//-	[17]
6	34Y/F/Probable NS in P:S Grade I	SBFP/V-III/ headache, dysgeusia	Lung/Weight loss -CT: hilar, mediastinal LNP	N// proteinorachia ↑ (125 mg/dl);	ACE N//-	[18]
7	35Y/M/ Probable Isolated NS	SBFP/VI-III/	PSN -EMG: sensorimotor demyelinating PNP -EMG VII nerve: axonal loss, partial denervation	N// lymphocytes ↑; CFS-ACE increased ↑	-//-	[19]
8	24Y/F/NS	SBFP/-/ parotid swelling	Parotid gland	-	-	[20]
9	-/-/NS	SBFP/-/	-	-	-	[21]
10–11	-/-/NS	SBFP/-/	-	-	-	[22]
12	62Y/F/Probable NS in P.S II	SBFP/II-IV/	Lung	N//N	ACE N//-	[23]
13–14	-/-/NS	RBFP/-/	-	-	-	[24] **
15	-/-/NS	SBFD/VI-VI/	-	-	-	[25] **
16	60Y/F/Definite NS in P.S I	SBFD/VI-VI/ aseptic meningitis/ dysarthria, distal limb dysesthesia	Lung/Eye/PNS -Ophthalmological test: bilateral papilledema -67Ga scintigraphy: mediastimal LNP+ (uptake +) -EMG:mixed PNP	N //Aseptic meningitis //CSF ACE-test NA	-//-	[26]
17	43/F/Possible N.S. in P.S Grade I	SBFP/-/-/ Bilateral V c.n Hyposmia, Dysphagia, dysgeusia Hypoesthesia (C8-T12) level (Multiple postganglionic neuropathy—sarcoid polyradiculopathy)	Lung/Eye/PNS -Ophthalmological test: bilateral uveitis -67Ga scintigraphy: mediastimal LNP+ (uptake +) -EMG: F-wave frequency ↓; ↓ bilateral ulnar, median nerve SNAP	N//lymphocytosis, proteinorrachia ↑// CSF ACE-test-	ACE↑// improved	[27]

Abbreviations: ↑: augmented, above the upper limit of normal; ↓ reduced, under the lower limit of normal BS: bronchoscopy; BP: brain biopsy; /-/: unknown; CN: cranial nerve; CT: computed tomography; ESR: erythrocyte sedimentation rate; F: female; F.U.: follow-up; HB Scale: House–Brackmann Scale; LNP+: lymphadenopathy; LNP: lymphonodes; M: male; N: normal; NR: not recovered; NS: neurosarcoidosis; ORL: otorhinolaryngology; PDL: prednisolone; PDN: prednisone; PNP: polyneuropathy; PR: partial recovery; PS: pulmonary sarcoidosis; SBFD: simultaneous bilateral facial di-plegia; R: remission; SBFP/RBFP: simultaneous/recurrent bilateral facial palsy; SCN: subcutaneous nodule;TB: transbronchial; URT: Upper respiratory tract; VP: ventriculoperitoneal; * according to NCCG' criteria; ** Review article.

Table 2. Demographic and clinical properties of patients with bilateral facial palsy/plegia as the onset of NS (Heerfordt's syndrome cases not included).

N°	Brain Imaging	Biopsy	Treatment Onset//FU	Prognosis/HB (R-L)	Ref
1	Focal leptomenigeal thickening (Gd+)	Soft tissue LNP biopsy +	PDN 60 mg/day //PDN (20 mg/day) + MTX	R/II	[13]
2	Ischemic cerebral stroke	Forearm Kveim test biopsy +	-	R/I-I	[14]
3	Left VII, V, VIII nerves, Gasser's ganglia (Gd+); FU: Normal	TB biopsy—(1M after therapy	Steroid bolus, PDL (1 mg/kg/day) //PDN + MTX (25 mg/week)	NR/-	[15]
4	Left VII, VIII nerves diffuse leptomeningeal, cerebellar cortex (Gd+); severe hydrocephalus -FU: Left VII, VIII nerves (Gd-)	SCN biopsy+	VP shunt, PDL 60 mg/day //PDN 15 mg/day	NA	[16]
5	Temporal leptomeninges (Gd+)	SCN biopsy	PDN// PDN 20 mg/day	NA	[17]
6	N	TB biopsy +	M-PDL(1g/day for 5 days) -> PDN 60 mg //no therapy	R/I-I	[18]
7	Sub/cortical white matter hemispheres lesions	-	IvIgG//-	PR/IV-I * * 3 months after	[19]
8	-	-	-	-	[20]
9	-	-	-	-	[21]
10–11	-	-	-	-	[22]
12	N	TB biopsy+	PDN 30 mg/die * //tapering * low dose due to diabetes	PR/-/* * 1 month after	[23]
13–14	-	-	-	-	[24] **
15	-	-	-	-	[25] **
16	-	Skin and sural nerve biopsy+	-	NA	[26]
17	N	-	M-PDL iv// oral PSL (30 mg/day)	PR/-/	[27]

Abbreviations: ↑: augmented, above the upper limit of normal; ↓ reduced, under the lower limit of normal BS: bronchoscopy; BP: brain biopsy; /-/: unknown; CN: cranial nerve; CT: computed tomography; ESR: erythrocyte sedimentation rate; F: female; F.U.: follow-up; HB Scale: House–Brackmann Scale; LNP+: lymphadenopathy; LNP: lymphonodes; M: male; N: normal; NR: not recovered; NS: neurosarcoidosis; ORL: otorhinolaryngology; PDL: prednisolone; PDN: prednisone; PNP: polyneuropathy; PR: partial recovery; PS: pulmonary sarcoidosis; SBFD: simultaneous bilateral facial di-plegia; R: remission; SBFP/RBFP: simultaneous/recurrent bilateral facial palsy; SCN: subcutaneous nodule;TB: transbronchial; URT: Upper respiratory tract; VP: ventriculoperitoneal; * according to NCCG' criteria; ** Review article.

When analyzing the literature, the diagnosis of pulmonary sarcoidosis was defined according to the known radiological criteria by Siltzbach (from "Grade 0—no radiological findings" to "Grade 4: pulmonary fibrosis") [33]. Diagnostic criteria for NS (as by Zajicek et al.; Marangoni et al.), used by various authors before 2018, have been revised. NS cases were re-classified as possible, probable, and definite according to revised The Neurosarcoidosis Consortium Consensus Group (NCCG) diagnostic criteria (2018) [34]. The pathologic confirmation of systemic granulomatous disease with biopsy, consistent with sarcoidosis, was also analyzed. NS diagnosis, based on the Forearm Kveim test biopsy [14], was redefined as "possible": this test is no longer used in clinical practice and is not explicitly included in the latest diagnostic update criteria [34]. In one case, biopsy resulted negative, and we redefined the patients as "possible NS" [15]. In this patient, to note, the tissue sampling was performed one month after the steroid therapy, possibly affecting the result. In addition, two cases of NS were excluded [35,36], respectively, for lacking diagnostic data on NS according to NCGG diagnostic criteria [34]. Therapy was classified as first-line, second-line, or third-line therapy. First-line therapy corresponds to corticosteroid treatment, second-line treatment consists of immunosuppressive therapy with methotrexate, azathioprine, mycophenolate mofetil, and hydroxychloroquine, and third-line therapy either consists of tumor necrosis factor-alpha inhibitors (principally infliximab) and B-cell targeted therapy [37], according to the last ESR clinical practice guidelines [38]. The prognosis was established using the House–Brackmann Scale (H-B Scale), and remission was defined as a complete improvement without residual symptoms.

4. Results

Overall, excluding our patient, 17 adults with bilateral FNP as the onset of NS (not showing Heerfordt's syndrome signs) have been reported in the literature from January 1970 to March 2022. Tables 1 and 2 shows the demographic and clinical characteristics of these patients. Kidd reported 12 bilateral and simultaneous FNP patients in a population of 166 (7%) "highly probable" NS patients, according to the WASOG sarcoidosis organ assessment instrument criteria. No other data for specifically these 12 patients were available in the papers; therefore, they were not considered in Tables 1 and 2 [31,32]. Among the NS patients, the majority were females (eight; 73%), and three were males (gender was not reported in six cases). The median age at presentation was 39 years (ranging from 24 to 62 years old). Fourteen patients (82%) presented with SBFP involvement: simultaneous diplegia (SBFD) was detected in five cases. RBFP were only observed in three patients. According to the NCCG diagnostic criteria, the diagnosis was "probable NS" in six cases. Only in one patient in which a sural nerve biopsy was performed was it possible to define a diagnosis of "definite NS" [26]. Three "possible" diagnoses were found. In seven patients only described as "NS," the revision of diagnosis was impossible due to lacking information. We will proceed by showing the characteristics of "definite," "probable," and "possible" NS. A patient presented with a "probable isolated NS"; systemic sarcoidosis was detected in nine patients, all with pulmonary sarcoidosis (grade I was observed in 80% of the cases; grade II in 20%). Multiorgan involvement was found in nine patients (53%) with systemic (weight loss and soft tissue lymphadenopathy—three), renal (two), ocular (two), upper respiratory tract (URT) (two), skin (two), and hematological manifestations (one). In all cases, bilateral facial nerve involvement did not appear in an already diagnosed sarcoidosis and was one of the first clinical red flags finally leading to the diagnosis.

Only a patient with self-limited RBFP during adolescence was initially misdiagnosed as recurrent Bell's palsy; after an ischemic stroke which led the patient to the emergency department, his diagnosis was revised to NS [14]. In another patient, RBFP initially self-resolved with steroid therapy. After two years, NS manifested through other symptoms [17]. Multiple cranial neuropathies were presented in five cases (29%), with a prevalence of acoustic nerve (II (1), III (1), V (2), VI (1), VIII (3) cranial nerves); also clinical, laboratoristic/instrumental signs of meningeal involvement were described in six patients. Leptomeningeal involvement, presented in three patients, often represents a more severe disorder with a risk of hydrocephalus. A patient developed ventricular obstructions by an inflammatory or granulomatous process, making it necessary to perform an emergency ventriculoperitoneal shunt [16]. CSF revealed a protein elevation (>1 g/L) and increased CSF-ACE level, respectively, in 67% and 50% of cases where (CFS data were available on 4/6 and 3/6). Serum ACE levels were normal in more than half of the cases when dosed. Brain imaging was available in nine patients; MRI acquired in the acute phase showed: cranial nerve contrast enhancement (22%); leptomeningeal involvement (33%). Multiple non-enhancing white matter lesions were found in isolated NS patients. No intraparenchymal granulomatous lesions, pituitary gland, or spinal cord involvement were described. Two patients presented with peripheral nervous system (PNS) involvement: EMG demonstrated demyelinating polyneuropathy and a mixed (axonal-demyelinating) pattern [19–26]. No dysautonomic symptoms (orthostatic hypotension, palpitations, hyperhidrosis, gastrointestinal dysmotility, or bowel/bladder dysfunction) or small-fiber neuropathy were described. Tissue biopsies were available for eight patients (47%), significant for sarcoidosis (except in a case in which it was performed one month after steroid therapy) [15]. Transbronchial biopsy was the most frequently performed. No brain biopsies were performed. First-line treatment with corticosteroids was started in all patients and then tapered, except for an isolated NS with demyelinating PNP, treated immediately with intravenous immunoglobulins [19].

Immunosuppressive therapy with methotrexate (MTX) was started only in two patients (12%). Third-line treatment was not described. Follow-up information was available for eight patients: three patients had a complete recovery (remission, R), four patients had a partial improvement without a full recovery (partial recovery, PR), and one patient remained stable (not recovered, NR). No patients died.

Including a literature review by Chappity et al. [8], eight adults with Heerfordt's syndrome and bilateral facial palsy/plegia have been reported in the literature from January 1970 to March 2022, (Table S1 in Supplementary Materials) summarizes their demographic and clinical characteristics. No prevalence of sex was documented (gender was not reported in four cases). At presentation, the median age was 38 years (ranging from 26 to 52 years old). Four patients (50%) presented with SBF involvement. No details were provided on the chronology of the presentation of bilateral facial deficit in patients reported in the literature review [8]. Lung involvement was prevalent, and it was found in three cases where systemic manifestation data were available. The treatment strategy and follow-up information were available for only half patients: all underwent first-line therapy; an immunosuppressant was started in one patient. Recovery was achieved in all patients (50% of them in R).

5. Discussion

Unilateral facial palsy (UFP) is frequent in NS, while BFP only occurs in 15% to 25% of cases [1,4,34]. We confirm that multiple cranial neuropathies are among the most frequent manifestations in NS, presenting in 55% of patients with neurological involvement [34] and more globally in 5–6% of systemic sarcoidosis [9]. Moreover, approximately in half of the suspected NS cases, neurological involvement is the first manifestation that leads to the identification of the disease [32,34]. All of the cranial nerves can be affected, with II, VII, and VIII being the most commonly involved [10], underlying an epineural/perineural granulomatous inflammation or a direct compression by granulomatous inflammation of the leptomeningeal compartment. Our revision confirmed that in 29% of patients, multiple cranial nerve involvement and UFP occur in 20% of the cases among cranial neuropathies in NS, with bilateral involvement reported in 30% of these patients. Our literature revision confirmed that BFP occurs as the first presentation of NS, mostly in young females, and more frequently as a simultaneous presentation (SBFD) than recurrent (RBFP). In 53% of cases, a multiorgan involvement was reported, where BFP was always the first manifestation, representing the first red flag for NS diagnosis.

The prognosis of NS UFP is similar to Bell's palsy, with complete recovery in about 90% of patients under corticosteroids. However, the resolution of BFP is more often non-simultaneous [10]. Many disorders may cause a BFP, and NS is one of the rarer [2]. Teller and Murphy's [5] review shows that Lyme disease is responsible for 36% of the cases of BFP. Guillain- Barre syndrome (5%), trauma (4%), sarcoidosis (0.9%), and AIDS (0.9%) are other rarer causes. Diagnosis workup includes serological laboratory tests such as complete blood count, fluorescent treponemal antibody test, HIV test, fasting glucose, erythrocyte sedimentation rate (ESR), Lyme titer, and an antinuclear antibody level measurement. A lumbar puncture could help diagnose an inflammatory process (>50% of the reported patients showed elevated CSF proteins), but a lumbar puncture is also mandatory to rule out other alternative diagnoses. Brain MRI with gadolinium may also reveal cranial nerve enhancement, meningeal involvement, and neoplastic processes, including Internal Acoustic Channel (IAC) and cerebellopontine angle. Frequently, facial diplegia warrants investigations to unveil signs of systemic disorder, such as NS, especially through chest CT and CT-total body PET in selected cases. If clinically indicated, other laboratory investigations may be conducted, such as serum and CSF ACE assay. However, serum and CSF elevated ACE levels are only sometimes found in NS patients, but several studies demonstrated low sensitivity and specificity of this test [34], and it must be stressed that normal ACE results do not exclude the possibility of NS [31,39,40]. Our case showed elevated serum-ACE. However, our literature search confirmed that most patients showed unremarkable serum ACE levels (C SF-ACE was dosed only in six patients with a 50% positivity). NS may manifest in many different ways, making the diagnosis difficult without histologic evidence. Because of the high risks of CNS biopsy, its use is limited to patients with a radiologically confirmed focal lesion in an accessible location or seriously ill patients with rapidly progressive disease. This finding was also confirmed in our literature revision. In patients with less severe disease, it is often more appropriate to look for systemic disease elsewhere in other tissues, as we undertook in the pulmonary district [5,6], even if labial salivary gland [6] and conjunctival [7] could lead to the right diagnosis [11,12].

In the literature, facial diplegia in NS is also described as RBFP, associated with Heerfordt's syndrome" [8–12]. However, SBFP, defined as the involvement of the opposite side within 30 days of the onset of the first side [18], is only described in 17 adults affected by sarcoidosis without uveoparotitis fever association (Tables 1 and 2). The clinical presentation of NS can be extremely heterogeneous. The reviewed literature allowed us to recognize that the most frequent clinical scenario of NS' SBFP is represented by a new onset that develops more frequently in young adult females without a known history of sarcoidosis. Subsequent clinical medical evaluations revealed multiorgan involvement in 81% of cases. If present, SBFP represents one of the first and more frequently recognized manifestations of systemic disease. Neuroimaging, particularly MRI, is usually used to detect and localize neurologic lesions, and it plays a strategical role, especially in the absence of other systemic manifestations of sarcoidosis. Unfortunately, the present case did not show an enhancement to the seventh left cranial nerves, as described previously in NS (Tables 1 and 2) [14,35,37]. Finally, in our case, acute onset unilateral facial palsy evolving into SBFP, radiologic and serological findings, the response to steroid therapy strongly support the diagnosis of NS as the first manifestation of systemic sarcoidosis [33]; transbronchial lymph node biopsy confirmed the "probable NS" diagnosis by NCCG criteria. In our literature search, only seven patients (41%) fulfilled the criteria for at least probable NS. [32].

As described in the literature, steroids are the most used first-line treatment, then tapered to a minimum maintenance dose. These agents typically suppress inflammation and may relieve acute symptoms. They are recommended in most cases of NS as spontaneous recovery cannot be predicted. An initial dose of prednisone or prednisolone of 20 mg once a day is typically recommended; followed by 5–10 mg once a day to once every other day [38], but in severe cases, high doses of IV methylprednisolone may be used for a few days to achieve a good clinical response in severely ill or deteriorating patients (Tables 1 and 2) [22,36]. Second-line therapy is generally initiated after steroid tapering and is associated with severe neurological and systemic manifestations [34]. Perhaps reflecting the publication date of some included papers and the lack of long reported follow-up, our search found only two patients treated with methotrexate.

Facial palsy/plegia also represents a neurological manifestation of "Heerfordt's syndrome" (HS), an acute subtype of sarcoidosis seen in 0.3–1.2% of the cases of sarcoidosis [37]. HS is characterized by facial palsy, parotid gland enlargement, and uveitis associated with low-grade fever. If only two of the three characteristic symptoms are present, an "incomplete Heerfordt's syndrome" can be defined. HS can be rarely associated with other cranial nerve involvement, particularly affecting the trigeminal nerve [41]. Sarcoid uveitis has a favorable visual outcome since most patients experience mild or no visual impairment [42,43]. However, 2.4 to 10% of patients with sarcoid uveitis develop severe visual impairment [42–45]. Therefore, ophthalmologic screening is recommended for all patients with newly diagnosed sarcoidosis, even in the absence of symptomatic ocular sarcoidosis. When bilateral VII cranial nerves are involved, ophthalmologic screening and salivatory palpation are mandatory to exclude this specific type of sarcoidosis.

6. Conclusions

Neurosarcoidosis is a great mimicker. A high degree of clinical suspicion and investigations should always be part of the extensive diagnostic workup when multiple cranial nerves are involved. We reported a case of NS that debuted with bilateral facial palsy and reviewed published literature. The biggest limitations of this paper are the small sample size of the historical cases and the lack of complete clinical information and follow-up. More recent NS cohorts often do not clearly specify the unilateral or bilateral facial nerve involvement and therefore make it difficult to draw clear conclusions. However, we believe that even with those limitations, the present review highlights that SBFP/RBFP as the onset of NS is rarely described, representing a diagnostic challenge for clinicians. The prompt recognition of NS and the initiation of appropriate steroid therapy could partially or entirely reverse neurologic sequelae, thus changing the natural progression of the disease. So that when bilateral facial palsy, especially in young adult females, occurs, it is essential to consider the neurological onset of systemic sarcoidosis. Larger studies and revisions are needed to improve knowledge in the field.

Supplementary Materials: The following supporting information can be downloaded at: https:// www.mdpi.com/article/10.3390/neurosci3020023/s1. Table S1: Demographic and clinical properties of patients with bilateral facial palsy/diplegia in complete Heerfordt's syndrome.

Author Contributions: Conceptualization, C.G., D.V. and E.V.; methodology, C.G. and E.V.; software, E.V.; data curation, D.V., C.V., L.M. and R.C.; writing—original draft preparation, C.G. and E.V.; writing—review and editing, E.V., D.V., L.M. and R.C.; visualization, E.V. and R.C.; supervision, C.V., L.M. and R.C. All authors have read and agreed to the published version of the manuscript.

Funding: This research received no external funding.

Institutional Review Board Statement: The study was conducted in accordance with the Declaration of Helsinki.

Informed Consent Statement: Informed consent was obtained from the patient involved in the study.

Data Availability Statement: Not applicable.

Acknowledgments: R. Boldorini Division of Pathology, Department of Health Science, University of Piemonte Orientale (UPO), Novara, Italy.

Conflicts of Interest: The authors declare no conflict of interest.

References

1. Gilchrist, J.M. Seventh cranial neuropathy. *Semin. Neurol.* **2009**, *29*, 5–13. [CrossRef] [PubMed]
2. Steenerson, R.L. Bilateral facial paralysis. *Am. J. Otol.* **1986**, *7*, 99–103. [PubMed]
3. Gevers, G.; Lemkens, P. Bilateral simultaneous facial paralysis–differential diagnosis and treatment options. A case report and review of literature. *Acta Otorhinolaryngol. Belg.* **2003**, *57*, 139–146. [PubMed]
4. Domeshek, L.F.; Zuker, R.M.; Borschel, G.H. Management of Bilateral Facial Palsy. *Otolaryngol. Clin. N. Am.* **2018**, *51*, 1213–1226. [CrossRef]
5. Teller, D.C.C.; Murphy, T.P. Bilateral facial paralysis: A case presentation and literature review. *J. Otolaryngol.* **1992**, *21*, 44–47.
6. Carlson, M.L.; White, J.R., Jr.; Espahbodi, M.; Haynes, D.S.; Driscoll, C.L.; Aksamit, A.J.; Pawate, S.; Lane, J.I.; Link, M.J. Cranial base manifestations of neurosarcoidosis: A review of 305 patients. *Otol. Neurotol.* **2015**, *36*, 156–166. [CrossRef]
7. Stern, B.J.; Krumholz, A.; Johns, C.; Scott, P.; Nissim, J. Sarcoidosis and its neurological manifestations. *Arch. Neurol.* **1985**, *42*, 909–917. [CrossRef]
8. Chappity, P.; Kumar, R.; Sahoo, A.K. Heerfordt's Syndrome Presenting with Recurrent Facial Nerve Palsy: Case report and 10-year literature review. *Sultan Qaboos Univ. Med. J.* **2015**, *15*, e124–e128.
9. Burns, T.M. Neurosarcoidosis. *Arch. Neurol.* **2003**, *60*, 1166–1168. [CrossRef]
10. Sève, P.; Pacheco, Y.; Durupt, F.; Jamilloux, Y.; Gerfaud-Valentin, M.; Isaac, S.; Boussel, L.; Calender, A.; Androdias, G.; Valeyre, D.; et al. Sarcoidosis: A Clinical Overview from Symptoms to Diagnosis. *Cells* **2021**, *10*, 766. [CrossRef]

11. Krainik, A.; Casselman, J.W. IDKD Springer Series Imaging Evaluation of Patients with Cranial Nerve Disorders. In *Diseases of the Brain, Head and Neck, Spine 2020–2023: Diagnostic Imaging*; Hodler, J., Kubik-Huch, R.A., von Schulthess, G.K.K., Eds.; Springer: Berlin/Heidelberg, Germany, 2020; pp. 143–161.

12. Hingwala, D.; Chatterjee, S.; Kesavadas, C.; Thomas, B.; Kapilamoorthy, T.R. Applications of 3D CISS sequence for problem solving in neuroimaging. *Indian J. Radiol. Imaging* **2011**, *21*, 90–97. [CrossRef] [PubMed]

13. Medhat, B.M.; Behiry, M.E.; Fateen, M.; El-Ghobashy, N.; Fouda, R.; Embaby, A.; Seif, E.M.; Taha, M.M.; Hasswa, M.K.; Sobhy, D.; et al. Sarcoidosis beyond pulmonary involvement: A case series of unusual presentations. *Respir. Med. Case Rep.* **2021**, *34*, 101495. [CrossRef] [PubMed]

14. Brown, M.M.; Thompson, A.J.; Wedzicha, J.A.; Swash, M. Sarcoidosis presenting with stroke. *Stroke* **1989**, *20*, 400–405. [CrossRef] [PubMed]

15. Durel, C.A.; Feurer, E.; Pialat, J.B.; Berthoux, E.; Chapurlat, R.D.; Confavreux, C.B. Etanercept may induce neurosarcoidosis in a patient treated for rheumatoid arthritis. *BMC Neurol.* **2013**, *13*, 212. [CrossRef]

16. Sugita, M.; Sano, M.; Uchigata, M.; Aruga, T.; Matsuoka, R. Facial nerve enhancement on gadolinium-DTPA in a case with neurosarcoidosis. *Intern. Med.* **1997**, *36*, 825–828. [CrossRef]

17. Estrada-Correa, G.; Molina-Carrión, L.E.; Ysita-Morales, A. Neurosarcoidosis. Report of one case in Mexico. *Rev. Med. Inst. Mex. Seguro Soc.* **2006**, *44*, 469–472.

18. Jain, V.; Deshmukh, A.; Gollomp, S. Bilateral facial paralysis: Case presentation and discussion of differential diagnosis. *J. Gen. Intern. Med.* **2006**, *21*, C7–C10. [CrossRef]

19. Varol, S.; Ozdemir, H.H.; Akil, E.; Arslan, D.; Aluclu, M.U.; Demir, C.F.F.; Yucel, Y. Facial diplegia: Etiology, clinical manifestations, and diagnostic evaluation. *Arq. Neuropsiquiatr.* **2015**, *73*, 998–1001. [CrossRef]

20. Magliocca, K.R.; Leung, E.M.; Desmond, J.S. Parotid swelling and facial nerve palsy: An uncommon presentation of sarcoidosis. *Gen. Dent.* **2009**, *57*, 180–182.

21. George, M.K.; Pahor, A.L. Sarcoidosis: A cause for bilateral facial palsy. *Ear Nose Throat J.* **1991**, *70*, 492–493.

22. Sharma, S.K.; Soneja, M.; Sharma, A.; Sharma, M.C.; Hari, S. Rare manifestations of sarcoidosis in modern era of new diagnostic tools. *Indian J. Med. Res.* **2012**, *135*, 621–629. [PubMed]

23. Haydar, A.A.A.; Hujairi, N.M.; Tawil, A.; Sawaya, R.A. Bilateral facial paralysis: What's the cause? *Med. J. Aust.* **2003**, *179*, 553. [CrossRef] [PubMed]

24. Chweya, C.M.; Anzalone, C.L.; Driscoll, C.L.W.; Lane, J.I.; Carlson, M.L. For Whom the Bell's Toll: Recurrent Facial Nerve Paralysis, A Retrospective Study and Systematic Review of the Literature. *Otol. Neurotol.* **2019**, *40*, 517–528. [CrossRef] [PubMed]

25. Keane, J.R. Bilateral seventh nerve palsy: Analysis of 43 cases and review of the literature. *Neurology* **1994**, *44*, 1198–1202. [CrossRef]

26. Campello-Morer, I.; López-Gastón, J.I.; Giménez-Mas, J.A.; Gracia-Naya, M. Sarcoidosis: The presentation of a case with clinical features limited to the nervous system. *Rev. Neurol.* **1997**, *25*, 1079–1081. [PubMed]

27. Uzawa, A.; Kojima, S.; Yonezu, T.; Kanesaka, T. Truncal polyradiculopathy due to sarcoidosis. *J. Neurol. Sci.* **2009**, *281*, 108–109. [CrossRef]

28. Srirangaramasamy, J.; Kathirvelu, S. A Rare Case of Heerfordt's Syndrome with Bilateral Facial Palsy. *Indian J. Otolaryngol. Head Neck Surg.* **2019**, *71*, 1027–1029. [CrossRef]

29. King, C.S.; Zapor, M. Bilateral parotitis and facial nerve palsy. *Cleve. Clin. J. Med.* **2007**, *74*, 512–513. [CrossRef]

30. Glocker, F.X.; Seifert, C.; Lücking, C.H. Facial palsy in Heerfordt's syndrome: Electrophysiological localization of the lesion. *Muscle Nerve* **1999**, *22*, 1279–1282. [CrossRef]

31. Kidd, D.P. Sarcoidosis of the central nervous system: Clinical features, imaging, and CSF results. *J. Neurol.* **2018**, *265*, 1906–1915. [CrossRef]

32. Kidd, D.P. Sarcoidosis of the central nervous system: Safety and efficacy of treatment, and experience of biological therapies. *Clin. Neurol. Neurosurg.* **2020**, *194*, 105811. [CrossRef]

33. Criado, E.; Sánchez, M.; Ramírez, J.; Arguis, P.; de Caralt, T.M.; Perea, R.J.; Xaubet, A. Pulmonary sarcoidosis: Typical and atypical manifestations at high-resolution CT with pathologic correlation. *Radiographics* **2010**, *30*, 1567–1586. [CrossRef] [PubMed]

34. Stern, B.J.; Royal, W., III; Gelfand, J.M.; Clifford, D.B.; Tavee, J.; Pawate, S.; Berger, J.R.; Aksamit, A.J.; Krumholz, A.; Pardo, C.A.; et al. Definition and Consensus Diagnostic Criteria for Neurosarcoidosis: From the Neurosarcoidosis Consortium Consensus Group. *JAMA Neurol.* **2018**, *75*, 1546–1553. [CrossRef] [PubMed]

35. Jha, G.; Azhar, S.; Kuttuva, S.; Elahi, S.; Baseer, A. Bilateral Facial Palsy: A Case Study of an Exceedingly Rare and Difficult Diagnosis. *Cureus* **2021**, *13*, e18900. [CrossRef] [PubMed]

36. Sharma, S.K.; Mohan, A. Uncommon manifestations of sarcoidosis. *J. Assoc. Physicians India* **2004**, *52*, 210–214.

37. Nozaki, K.; Judson, M.A. Neurosarcoidosis. *Curr. Treat. Options Neurol.* **2013**, *15*, 492–504. [CrossRef]

38. Baughman, R.P.; Valeyre, D.; Korsten, P.; Mathioudakis, A.G.; Wuyts, W.A.; Wells, A.; Rottoli, P.; Nunes, H.; Lower, E.E.; Judson, M.A.; et al. ERS clinical practice guidelines on treatment of sarcoidosis. *Eur. Respir. J.* **2021**, *58*, 2004079. [CrossRef]

39. Fritz, D.; van de Beek, D.; Brouwer, M.C. Clinical features, treatment and outcome in neurosarcoidosis: Systematic review and meta-analysis. *BMC Neurol.* **2016**, *16*, 220. [CrossRef]

40. Zajicek, J.P.; Scolding, N.J.J.; Foster, O.; Rovaris, M.; Evanson, J.; Moseley, I.F.; Scadding, J.W.W.; Thompson, E.J.; Chamoun, V.; Miller, D.H.; et al. Central nervous system sarcoidosis–diagnosis and management. *QJM* **1999**, *92*, 103–117. [CrossRef]

41. Baumann, R.J.; Robertson, W.C., Jr. Neurosarcoid presents differently in children than in adults. *Pediatrics* **2003**, *112*, e480–e486. [CrossRef]
42. Rochepeau, C.; Jamilloux, Y.; Kerever, S.; Febvay, C.; Perard, L.; Broussolle, C.; Burillon, C.; Kodjikian, L.; Seve, P. Long-term visual and systemic prognoses of 83 cases of biopsy-proven sarcoid uveitis. *Br. J. Ophthalmol.* **2017**, *101*, 856–861. [CrossRef] [PubMed]
43. Sève, P.; Jamilloux, Y.; Tilikete, C.; Gerfaud-Valentin, M.; Kodjikian, L.; El Jammal, T. Ocular Sarcoidosis. *Semin. Respir. Crit. Care Med.* **2020**, *41*, 673–688. [CrossRef] [PubMed]
44. Miserocchi, E.; Modorati, G.; Di Matteo, F.; Galli, L.; Rama, P.; Bandello, F. Visual outcome in ocular sarcoidosis: Retrospective evaluation of risk factors. *Eur. J. Ophthalmol.* **2011**, *21*, 802–810. [CrossRef] [PubMed]
45. Dana, M.R.; Merayo-Lloves, J.; Schaumberg, D.A.; Foster, C.S. Prognosticators for visual outcome in sarcoid uveitis. *Ophthalmology* **1996**, *103*, 1846–1853. [CrossRef]

Article

Obesity and Neurocognitive Performance of Memory, Attention, and Executive Function

Antonio G. Lentoor

Department of Clinical Psychology, School of Medicine, Sefako Makgatho Health Sciences University, Ga-Rankuwa, Pretoria 0208, South Africa; dr.lentoor.antonio@gmail.com; Tel.: +27-(0)-125214767

Abstract: Background: Obesity has been linked to an increased risk of dementia in the future. Obesity is known to affect core neural structures, such as the hippocampus, and frontotemporal parts of the brain, and is linked to memory, attention, and executive function decline. The overwhelming majority of the data, however, comes from high-income countries. In undeveloped countries, there is little evidence of a link between obesity and neurocognition. The aim of this study was to investigate the effects of BMI on the key cognitive functioning tasks of attention, memory, and executive function in a South African cohort. Methods: A total of 175 females (NW: BMI = 18.5–24.9 kg/m^2 and OB: BMI > 30.0 kg/m^2) aged 18–59 years (M = 28, SD = 8.87 years) completed tasks on memory, attention, and executive functioning. Results: There was a statistically significant difference between the groups. The participants who had a BMI corresponding with obesity performed poorly on the tasks measuring memory ($p = 0.01$), attention ($p = 0.01$), and executive function ($p = 0.02$) compared to the normal-weight group. Conclusions: When compared to normal-weight participants, the findings confirm the existence of lowered cognitive performance in obese persons on tasks involving planning, decision making, self-control, and regulation. Further research into the potential underlying mechanism by which obesity impacts cognition is indicated.

Keywords: body mass index; brain function; cognition; developing context; neuropsychological tests; obesity

Citation: Lentoor, A.G. Obesity and Neurocognitive Performance of Memory, Attention, and Executive Function. *NeuroSci* **2022**, *3*, 376–386. https://doi.org/10.3390/neurosci3030027

Academic Editor: Ugo Nocentini

Received: 26 May 2022
Accepted: 23 June 2022
Published: 28 June 2022

Publisher's Note: MDPI stays neutral with regard to jurisdictional claims in published maps and institutional affiliations.

1. Introduction

Obesity is a significant public health problem that contributes to the overall burden of disease globally [1]. Obesity is an excess of fat mass caused by an imbalance in energy intake and energy expenditure [2], and a complex interplay between genes and the environment [3]. Obesity is a risk factor for multiple health issues, such as diabetes and hypertension, and a major cause of premature mortality. Body mass index (BMI) given by dividing the weight by height (kg/m^2), is a cost-effective marker of identifying a person as obese (BMI above 30 kg/m^2) and has been extensively used as a proxy measure for adiposity. The enormous health and financial burden associated with obesity makes it an important research topic [4]. While the physical health sequelae of obesity are well understood, recent empirical evidence suggests a significant impact on the brain [5]. Obesity is a risk factor for neurodegenerative changes and has a deleterious impact on brain function and structure [6]. Atrophy of the temporal brain region, hippocampal, and frontal structure have been found in obese individuals [7]. While the mechanism by which obesity is associated with cognitive function requires furthers explication, pathophysiological changes including oxidative stress, metabolic changes, neuroendocrine dysregulation, and systematic neuroinflammation have been suggested as key mechanisms in hippocampal and frontostriatal dysfunction in obesity [8–10]. Particularly, adipokines, including leptin, interleukin (IL-6), and tumour necrosis factor (TNF-α), have been linked to a weakened blood–brain barrier and obesity-related brain dysfunction [7,8]. Changes in the function and structures of the brain include cognitive problems in memory, attention, and executive function. Memory, which is the ability to store, maintain, and retrieve information, relies on the integrity of the hippocampal structure [11]. There is increasing evidence from obesity-cognition research to suggest that episodic memory has an important role in regulating consumption in humans [12]. For example, memory of recent meals showed to have an impact on long-term satiation of meals, while both executive function and episodic memory is crucial for the regulation of consumption (i.e., control of food intake) in obese individuals [13]. On the other hand, attention and executive function, which involve the skills of regulation, control, decision making, and appropriate behavioural responses, is largely mediated by the frontostriatal regions of the brain [14].

Evidence from neuropsychological studies suggests that early adult and middle-aged obese individuals compared to normal-weight people showed lower performance on higher-order cognitive function tasks [15]. Several studies found that obese individuals have a lowered performance on executive tasks of planning, problem solving, and cognitive flexibility when compared to normal-weight individuals [8]. Coinciding with these findings is the observation that obesity negatively impacts memory. Cheke and colleagues [12] found that a higher BMI was associated with significantly lower performance on episodic memory tests when compared with individuals with a normal BMI. Likewise, Nguyen and colleagues showed that obesity is associated with poor performance on a task measuring short-term memory that is essential for comprehension, learning, and planning [16].

Obesity [17] and cognitive disorders [18] are substantial contributions to the worldwide burden of non-communicable disease disability, especially in the sub-Saharan African (SSA) region. In South Africa, like in other SSA regions [19], the burden of obesity is disproportionately higher in women than males [17]. Because of the shifting socioeconomic situation in post-apartheid South Africa, women now have access to higher education, are financially secure enough to acquire and purchase high-calorie-density meals, and work in positions that limit physical activity [17]. Evidence also suggests that weight increase throughout reproductive years is a key cause of obesity in women [20]. It is worth noting that African women of reproductive age are more likely to use injectable hormonal contraceptives such as Depo-Provera, which has been linked to a considerable gain in body fat and eventually obesity [21]. Depo-Provera was linked to changes in menstruation pattern, bone mineral density loss, considerable weight gain, and higher BMI in Ethiopian women who took it compared to non-users, according to research done in Northwest Ethiopia [22]. Depo-Provera was related in another study to causing alterations in the hypothalamus appetite control centre [23], which led to weight gain in Depo-Provera users and was linked to increased food consumption associated with higher hunger and, as a result, more carbohydrate-rich diets.

Obesity has been associated with a number of physiological and metabolic changes [24] that could possibly act as pathways for underlying neurodegeneration and dementia risk in later life. Endocrine functional alteration activated by hormonal changes and adipokines due to increased BMI may have a role in the development of increased inflammation and cardiovascular changes that may result in alterations in brain structure, blood–brain barrier integrity, and atrophy which increases vulnerability to neurodegeneration [25].

While obesity and cognitive disorders have grown considerably more common in developing countries, the bulk of existing obesity-cognition neuroscience research is from developed countries, making direct comparisons to the developing setting challenging. In South Africa, there is a dearth of studies on obesity, including its correlates. If we can understand the relationship between obesity and cognitive vulnerability, it can help us take the necessary steps towards developing interventions that can prevent cognitive decline and reduce the risk of dementia in the future. Hence, the current study investigated the association between obesity and cognition, with a specific focus on memory, attention, and executive function, in a cohort of women in a developing context. Based on existing research findings [11,15], we expected to find differences in the neurocognitive performance of obese and normal-weight women. More specifically, we expected to find significantly lower neurocognitive performance scores on the domains of memory, attention, and executive function for obese than normal-weight women.

2. Materials and Methods

2.1. Study Design and Participants

In this cross-sectional, quantitative study, a total of 308 participants were screened for potential participation, while 133 individuals were excluded (Figure 1). A total of 175 females, aged 18–59 years (28 + 8.87 years) with normal (NW: BMI = 18.5–24.9 kg/m^2) or obese (OB: BMI $\geq$ 30.0 kg/m^2) weight were enrolled into the study through purposive sampling techniques from a local hospital, university, and community between January and December 2018 via poster advertisements and word of mouth, in the northern parts of Gauteng Province. The community from which the participants were recruited can be classified as peri-urban, with access to public transport, an academic hospital, schools, universities, malls, and other services.

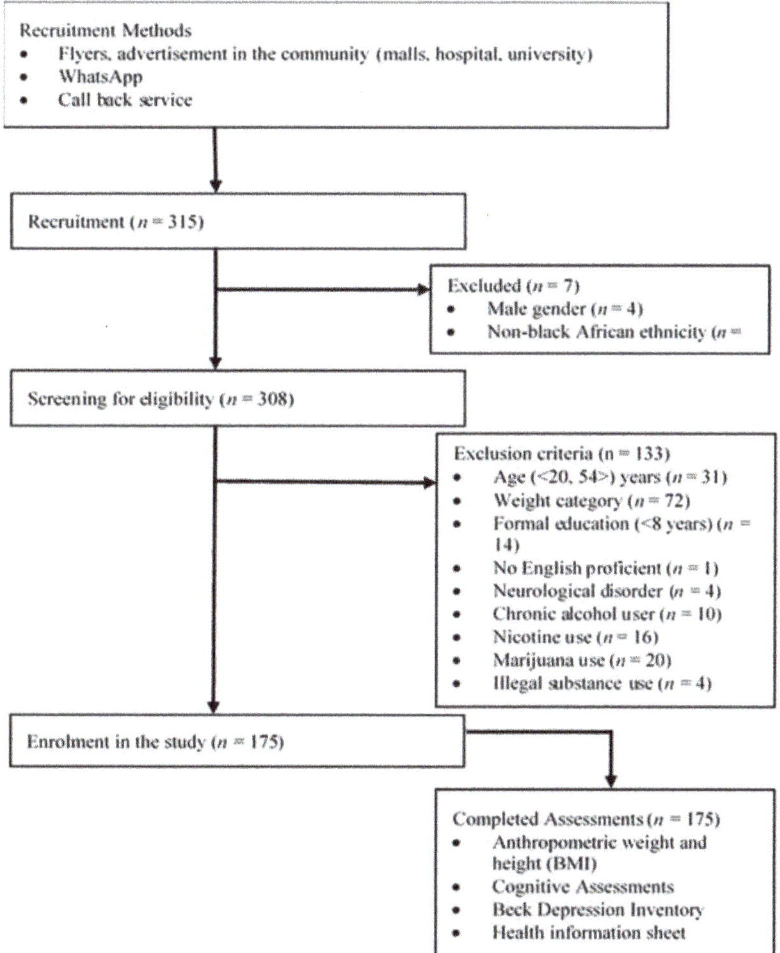

Figure 1. Flow chart of the study recruitment procedure.

2.2. Eligibility Criteria

Women, according to Nglazi and Ataguba [17], bear a disproportionate burden of obesity as compared to males. Based on a report from the South African health ministry, over 41% of women and 11% of males aged 15 and above were obese in 2019. Therefore, this study recruited females aged 18 to 59 years. Volunteers who responded were initially screened for eligibility, with those eligible for inclusion reporting no psychiatric and neurological conditions and no use of medication/substances known to alter mood and cognitive capacity that could impact on neurocognitive performance; a proficiency in English to be able to complete the neurocognitive assessment; not being younger than 18 or older than 59 years; and no visual, hearing, or motor coordination problems that would restrict them from completing the assessments.

2.3. Data Collection

After completing a screening, all participants meeting the inclusion criteria completed an anthropometric assessment, followed by a neurocognitive functioning assessment that was conducted by a research assistant who was a trained graduate-level student psychologist, and confirmed by the author who is a licenced clinical psychologist trained in neuropsychology. Written informed consent was obtained from all the participants prior to the completion of the assessments. All tests were conducted in a quiet and private space to avoid any distractions and noise interferences.

2.4. Measures

A trained graduate-level student psychologist, under the supervision of a licenced clinical psychologist, administered all the assessments. The participants completed domain-specific neurocognitive assessments [26–28].

2.4.1. Sociodemographic and Health Information Sheet

The participants completed an information sheet on age, gender, ethnicity, and health comorbidities (i.e., diabetes, hypertension). The 21-item Beck Depression Inventory-III (BDI-III) [29,30] was administered to screen for depression. A score of 0–9 indicates no depression, while a score of 10 and above is indicative of depression. The BDI-III is a reliable and valid tool that has been use cross-culturally.

2.4.2. Anthropometry

BMI was calculated as the ratio of weight (in kilograms) to height (in metres) squared. Height was measured to the nearest 0.1 cm with a measuring tape. Weight was measured to the nearest 0.1 kg (kilogram) on a digital scale. A BMI of 18.5–24.9 kg/m^2 was considered normal weight, and a BMI $\geq$ 30.0 kg/m^2 obese.

2.4.3. Memory

The research assistant read out five words with unrelated meaning. Two learning trials of the five words were administered [31]. To assess short-term memory, the participants recalled the words immediately. After approximately 10 min, the participants recalled the five words. The total score of five for delayed memory was assigned. Points were given for uncued recall only.

2.4.4. Attention

The Digit Span subtest has two tasks in which the participant recites five digits forward and recites three digits in a reverse order [31]. The participant's score was the sum of the two tasks, with a total score of two being obtainable if performed correctly.

2.4.5. Executive Function

Multiple aspects of executive functions were assessed using the alternating task adapted from the trail-making B task, the three-dimensional cube copy, and the clock drawing test subtest [32]. The scoring methods for all tests were based on the MoCA scoring system as a combined score of executive function.

The Trail-Making Test (TMT), in which the participant has to connect a set of dots as quickly and accurately as possible, was administered [33]. TMT Part B, which requires participants to connect numbers and letters in alternating sequence, was completed. The participant was given 1 point if completed correctly. TMT measured executive function, in addition to attention, self-regulation, visual speed and processing and mental flexibility, and task shifting.

The participants were told to copy the three-dimensional cube figure from the example, 1 point was given if copied correctly and 0 if incorrect [34].

The Clock Drawing Test (CDT) draws on different elements of a clock, including the clock face, numbers, and arrow-hands [35]. The participants were asked to draw a clock and to place the clock hands to read '10 past 11'. Points are given for numbers placed in the correct position, for the accuracy of the hands denoting the time 11:10, and contour (total score of 3 points if correct). Executive skill demands made of the CDT include planning (goal-directed behaviour), inhibitory control, and cognitive shifting (adopting a new strategy).

2.5. Statistical Analysis

All data were coded and analysed using SAS 9.3 (SAS Institute Inc., Cary, NC, USA). The data are presented as means $\pm$ standard deviations (SD) for continuous variables and frequencies for categorical variables. The inter-group comparison was performed using the chi-square test, and a Student's *t*-test for independent variables to compare OB and NW groups for anthropometric parameters, memory, attention, and executive function. A *p*-value < 0.05 (two-tailed) was considered statistically significant.

3. Results

3.1. Descriptives

Table 1 presents the descriptive statistics of the sample of participants in this study. The sample included 175 obese (*n* = 75) and normal-weight (*n* = 100) women. A statistically significant difference was found for age (*p* < 0.001), with the obese group being older (mean age = 33.76 years). There was a statistically significant difference in health co-morbidity (*p* < 0.001) between the normal-weight and obese group, as shown in Table 1. The obese group and normal-weight group did not differ on emotional adjustment (*p* = 0.07).

Table 1. Descriptive data of the study sample.

Characteristics	Normal-Weight Group (*n* = 100)	Obese Group (*n* = 75)	*p*-Value
Age			
Mean	23.73	33.76	<0.001
SD	4.08	10.21	
Gender			
Female (%)	57.14	42.86	0.000
Health comorbidity (%)			
Yes	7.14	92.86	<0.001
No	61.49	38.57	
Depression			
Mean scores	8.93	10.88	0.07
SD	7.51	9.38	
BMI (kg/m^2)			
Mean BMI	22.27	39.46	0.000
SD	1.91	9.17	

3.2. Association between Cognitive Domain Tasks and Body Mass Index (BMI)

A significant inverse association was found between BMI and performance on the cognitive test of delayed recall (*r* = −0.19, *p* = 0.001), clock drawing test (*r* = −0.15, *p* = 0.01), cube copying test (*r* = −0.16, *p* = 0.005), serial 7's test (*r* = −0.20, *p* = 0.000), and digit span test (*r* = −0.13, *p* = 0.02) (Table 2).

Table 2. Association between body mass index and neurocognitive test performance.

Neurocognitive Test	BMI (n = 175)	
	r	p-Value
Memory-delayed Recall	−0.185 **	0.001
Clock Drawing Test	−0.145 *	0.010
Cube Copy	−0.158 **	0.005
Serial 7s	−0.202 **	0.000
Digit Span Test	−0.133 *	0.018

BMI: Body Mass Index, Significant p-values * $p < 0.05$, ** $p < 0.01$.

3.3. Differences in Neurocognitive Performance for Normal-Weight and Obese Groups

Table 3 reports the results of the performance on cognitive tasks for both the normal-weight and obese group. The Student's t-test results showed statistically significant differences between the obese and normal-weight groups. Participants' performance on cognitive tasks of attention ($t(173)$ = 2.39, p = 0.01), memory, ($t(173)$ = 2.52, p = 0.01), and executive function ($t(173)$ = 2.31, p = 0.02) was significantly lower in the obese group (OB) as compared to the normal-weight group (NW). The lower means scores obtained by the obese individuals indicate their difficulties in each of the tasks on the cognitive domains.

Table 3. Neurocognitive functioning performance by weight group.

Characteristics	Normal-Weight Group		Obese Group		95% CI	
	Mean	SD	Mean	SD	t	p-Value
Executive function	3.67	1.15	3.25	1.20	2.31	0.02
Attention	5.41	0.88	5	1.27	2.38	0.01
Memory	3.74	10.21	3.2	1.49	2.52	0.01

4. Discussion

The present study assessed performance in neurocognitive function with a focus on memory, attention, and executive function in obese and normal-weight individuals.

Similar to Farooq and colleagues [11], who assessed the cognitive performance tasks of attention, memory, and planning executive function of 220 women on the Cambridge Neuropsychological Test, this study found a significant association between the cognitive subdomains and BMI. The study found that body mass index (a proxy measure for adiposity) correlated with several cognitive tests: memory-delayed recall (p = 0.001), clock drawing test (p = 0.010), cube-copying (p = 0.005), serial 7's (p = 0.000), and digit span test (p = 0.018). The tests are known to tap into cognitive domains of attention, memory, and executive function, reflecting frontostriatal functionality. These tests are the most well-researched neuropsychological tests, and they can detect a wide variety of neurodegenerative illnesses, such as mild cognitive impairment and dementia [28]. Similar to the current findings, Fergenbaum et al. [32] found that BMI was associated with poor cognitive performance on the clock drawing test and trail making test. Likewise, poor performance on the digit span test, as a measure of executive function, and the memory-delayed recall test were found to be associated with a high BMI [28,31]. The inverse association between BMI and performance on these cognitive tests could reflect the underlying impact of increased adipose on cognition.

This study found a significant difference in cognitive performance on memory, attention, and executive function tasks, in terms of BMI—a finding supported in several previous studies [11,12,36]. The performance on the memory task was significantly lower for the participants who had a high BMI, corresponding with obese individuals, compared to individuals with a normal weight in this study. Previous research [37] found that individuals with a higher BMI had greater hippocampal atrophy, which is expressed as a reduction in hippocampus volume and is associated with deficits in memory. Similar to

Yang and colleagues [31], this study showed that obese participants performed significantly worse on the modified digit span backwards working memory task. This is an important finding since lower working memory has significant clinical implications. Working memory plays an important role in facilitating adherence to weight management programs. Lower working memory in obesity has been linked to a lower ability to keep goal-relevant information in mind. Previous research [38] found memory deficits in obese individuals was linked to poor appetite control that was associated with lowered orbitofrontal cortical volumes. Furthermore, Yang and colleagues [31] found that a lower performance on working memory tasks in individuals with a higher BMI was associated with increased levels of inflammatory protein (CRP). This supports the role of neuroinflammation in the relationship between obesity and working memory dysfunction [10]. This viewpoint was corroborated by a recent analysis including a decades' worth of epidemiological data, which revealed that adipokine activation in adipose tissue may alter brain function [39]. Adipokines, such as peripheral leptin, interleukin-6 (IL-6), and tumour necrosis factor alpha (TNF-α), are released by adipose tissue and interact directly with certain nuclei, including the hippocampus. Adipokines have been linked to obesity-related inflammation and direct energy metabolism dysfunction. Not only does this assist in the regulation of eating behaviour, but it also supports issues with memory consolidation in obese people, which has been linked to cognitive dysfunction, including learning and memory processing difficulties, and risk of dementia and Alzheimer's disease [6].

In contrast to Gunstad [40] and Farooq and colleagues [11], this study found that obese individuals performed significantly lower on the attention task relative to normal-weight individuals. The finding is similar to Tsai and colleagues who showed that obesity was associated with the reduced modulatory ability of attentional networks [41]. Similar to this study, Cook and colleagues [42] investigated the relationship between obesity and cognitive function in 299 women aged 18–35 years and found a significantly lower performance on attention tasks for obese compared to normal-weight individuals.

As expected, obese individuals performed poorly on executive tasks compared to individuals with a normal weight. This finding is consistent with previous studies [43] that showed higher BMI was associated with executive dysfunction (see review by Yang and colleagues). The current findings are also supported by Dassen and colleagues [44] who showed that obese individuals had a lower performance on executive function measures relative to normal-weight individuals. The researchers found that obese individuals reported a poorer performance on neurocognitive tasks specific to inhibition and self-regulation. Similar to the current findings, a cross-sectional study in Canada assessed neuropsychological function using the Clock Drawing Test and the Trail Making Test as a measure of executive function and found that obese participants were almost fourfold more likely to show poor executive function compared to non-obese individuals [32].

This study finding has important clinical implications. Executive function, which is largely mediated by frontal brain regions, plays an important role in self-regulation, behavioural inhibition, shifting, and goal-directed behaviour [13]. These are important cognitive skills for weight-related behaviours. People with a high BMI (a proxy for obese) were shown to have functional and structural connectivity abnormalities in the frontostriatal system [45]. Importantly, abnormalities in this neuroanatomical system are related to deficits in cognitive flexibility (i.e., a key cognitive control function). For example, a deficit in cognitive flexibility, self-regulation, and inhibitory control has been shown to be associated with higher food intake and less exercise and thus increases the risk for obesity [45]. Likewise, research has shown that persons with obesity exhibit less efficient general and food-specific behavioural inhibition, which is linked to weight control problems [46]. Impairment in shifting ability is an important facet of executive function and has been shown to be associated with pathological eating behaviour in individuals with obesity [47]. Importantly, while obesity and cognitive deficiencies are linked, their influence appears to be restricted to early adulthood, with contradictory findings in older adult populations. Moreover, lowered cognitive performance in midlife obesity has been linked to an increased risk of neurodegenerative pathologies such as Alzheimer's disease [16] and other dementias, emphasizing the need for more research that could help to establish if early interventions and lifestyle modifications targeting at-risk groups can reduce future dementia risk.

This study found no significant difference in the emotional functioning between high-BMI and normal-BMI individuals. Previous studies provide evidence of a link between BMI and depression risk [48]; however, in some studies the link was confined only to individuals with severe obesity. It is important to note, when trying to understand this inconsistency in research, that in many African cultures overweight or large body size has been associated with richness, health, strength, and fertility [49]. Perhaps the cultural perception of body image and body size perception is an important mediator of depression risk.

Importantly, in this study participants with a high BMI were significantly older. There was an average age difference of 10 years between participants with a high BMI, corresponding with obesity (age range 20–54 years), and those with a normal BMI (age range 20–41 years). Additionally, the majority (>90%) of the obese individuals reported cardiometabolic disease comorbidities such as diabetes and hypertension. This finding backs up earlier studies that showed a link between a higher BMI and an increased risk of hypertension and diabetes, with the risk peaking between the ages of 18 and 53 [50]. Previous research found that age and metabolic disorder, both independently and in combination, may increase the risk for cognitive decline in obese individuals, especially in midlife [51].

This study builds on the existing evidence that obesity impacts performance on cognitive tasks related to memory, attention, and executive function; however, the exact underlying mechanism by which obesity affects cognition is far from clear.

Important to note, data on the subjective cognitive function or self-reported daily functioning capacity of the participants was not collected in this study. It is common practice for women in this community to be the primary caretakers of the family while still juggling fulltime employment. These roles require them to draw on higher order cognitive functioning skills on a daily basis. It would therefore be important in future studies to reconcile these roles and responsibilities with performances on standard cognitive tests to get a better understanding of cognitive testing results. This is an important limitation of the study that must be kept in mind when interpreting the findings of this study. There is considerably more evidence from cognitive neuroscience work that cultural experiences and behavioural practices affect neural structure and function [52]. It is critical that the meaning of culture in the context of neuropsychological testing be properly stated in neurocognitive science research. Focusing on the fundamental neuroscience underlying how different aspects of culture influence cognitive test performance and how it relates to brain function will only improve study findings. Another limitation of this study is the cross-sectional design that precludes drawing causative inferences. Third, the use of the BMI metric as the only parameter of adiposity was also a limitation of this study. Fourth, the study used a small purposively selected community-based sample which therefore may not be representative; as such, the findings must be interpreted with caution. Finally, while the brief neurocognitive tests utilized in the study are often used in clinical practice in South Africa, they are not normalized for this population. The use of a detailed neuropsychological battery with a cross-cultural focus could ameliorate this limitation in future studies. The absence of self-reported functioning signals a cautious interpretation of the findings. Future comparative study designs could include the participants' perception of their personal functional state to enhance the findings.

5. Conclusions

Overall, in this study individuals with a higher BMI, corresponding with obesity, performed poorly on tasks of memory, attention, and executive function relative to normal-weight individuals. Poor performance on these tasks reflects overall deficits in these neurocognitive domains. The cognitive performance of obese individuals could reflect the functional and structural brain changes that form part of the pathophysiological effects of obesity on the brain. Because cognitive capabilities are considered predictors of eating and body-weight behaviour change, the findings of this study have substantial treatment implications. Additionally, because obesity is a modifiable risk factor for cognitive decline, dementia, and Alzheimer's disease, early identification and management are more likely to minimize the chance of getting these diseases later in life. Adults with a high BMI, corresponding with obesity, may benefit from interventions designed to lower the risk of cognitive loss. The study highlighted important limitations that could be addressed in future studies. Further knowledge on the underlying mechanism via which obesity impacts the brain is needed. The current study findings could be enhanced in future longitudinal studies that combine objective and subjective cognition, neuroimaging, and biological markers with a relatively large cohort reflecting the diverse population of the context.

Funding: This research received no external funding.

Institutional Review Board Statement: The study was conducted in accordance with the Declaration of Helsinki and was approved by the Institutional Ethics Committee of Sefako Makgatho Health Sciences University.

Informed Consent Statement: Informed consent was obtained from all subjects involved in the study.

Data Availability Statement: The data presented in this study are available on reasonable request from the corresponding author. The data are not publicly available due to ethical reasons related to subjects' confidentiality.

Acknowledgments: The subjects are acknowledged for voluntarily agreeing to take part in the study. A special thank you to the student psychologist who conducted the assessments. Thank you Thandokuhle Nxiweni for assisting with the literature search.

Conflicts of Interest: The author declares no conflict of interest.

References

1. Dave, J. Morbid obesity in South Africa: Considerations and solutions. *S. Afr. J. Surg.* **2020**, *58*, 110–112. [CrossRef]
2. Hill, J.O.; Wyatt, H.R.; Peters, J.C. Energy Balance and Obesity. *Circulation* **2012**, *126*, 126–132. [CrossRef] [PubMed]
3. Flores-Dorantes, M.T.; Díaz-López, Y.E.; Gutiérrez-Aguilar, R. Environment and Gene Association with Obesity and Their Impact on Neurodegenerative and Neurodevelopmental Diseases. *Front. Neurosci.* **2020**, *14*, 863. [CrossRef] [PubMed]
4. Abdelaal, M.; le Roux, C.W.; Docherty, N.G. Morbidity and mortality associated with obesity. *Ann. Transl. Med.* **2017**, *5*, 161–172. [CrossRef] [PubMed]
5. Ely, A.V.; Williams, K.D.; Condiracci, C. "Obesity Paradox" Evident in the Relationship between Cognitive Function and Hospitalizations in Heart Failure Patients: A Pilot Study. *J. Card. Fail.* **2019**, *25*, S104. [CrossRef]
6. Tang, X.; Zhao, W.; Lu, M.; Zhang, X.; Zhang, P.; Xin, Z.; Sun, R.; Tian, W.; Cardoso, M.A.; Yang, J.; et al. Relationship between Central Obesity and the incidence of Cognitive Impairment and Dementia from Cohort Studies Involving 5,060,687 Participants. *Neurosci. Biobehav. Rev.* **2021**, *130*, 301–313. [CrossRef]
7. Buie, J.J.; Watson, L.S.; Smith, C.J.; Sims-Robinson, C. Obesity-related cognitive impairment: The role of endothelial dysfunction. *Neurobiol. Dis.* **2019**, *132*, 104580. [CrossRef]
8. Laurent, J.S.; Watts, R.; Adise, S.; Allgaier, N.; Chaarani, B.; Garavan, H.; Potter, A.; Mackey, S. Associations Among Body Mass Index, Cortical Thickness, and Executive Function in Children. *JAMA Pediatr.* **2020**, *174*, 170–177. [CrossRef]
9. Fernández-Sánchez, A.; Madrigal-Santillán, E.; Bautista, M.; Esquivel-Soto, J.; Morales-González, Á.; Esquivel-Chirino, C.; Durante-Montiel, I.; Sánchez-Rivera, G.; Valadez-Vega, C.; Morales-González, J.A. Inflammation, Oxidative Stress, and Obesity. *Int. J. Mol. Sci.* **2011**, *12*, 3117–3132. [CrossRef]
10. Arnoriaga-Rodríguez, M.; Mayneris-Perxachs, J.; Burokas, A.; Contreras-Rodríguez, O.; Blasco, G.; Coll, C.; Biarnés, C.; Miranda-Olivos, R.; Latorre, J.; Moreno-Navarrete, J.-M.; et al. Obesity Impairs Short-Term and Working Memory through Gut Microbial Metabolism of Aromatic Amino Acids. *Cell Metab.* **2020**, *32*, 548–560.e7. [CrossRef]
11. Farooq, A.; Gibson, A.-M.J.; Reilly, J.; Gaoua, N. The Association between Obesity and Cognitive Function in Otherwise Healthy Premenopausal Arab Women. *J. Obes.* **2018**, *2018*, 1741962. [CrossRef] [PubMed]
12. Cheke, L.G.; Simons, J.S.; Clayton, N.S. Higher body mass index is associated with episodic memory deficits in young adults. *Q. J. Exp. Psychol.* **2006**, *69*, 2305–2316. [CrossRef] [PubMed]
13. Favieri, F.; Forte, G.; Casagrande, M. The Executive Functions in Overweight and Obesity: A Systematic Review of Neuropsychological Cross-Sectional and Longitudinal Studies. *Front. Psychol.* **2019**, *10*, 2126–2152. [CrossRef] [PubMed]
14. Willeumier, K.C.; Taylor, D.V.; Amen, D.G. Elevated BMI Is Associated With Decreased Blood Flow in the Prefrontal Cortex Using SPECT Imaging in Healthy Adults. *Obesity* **2011**, *19*, 1095–1097. [CrossRef]

15. Sargénius, H.L.; Lydersen, S.; Hestad, K. Neuropsychological function in individuals with morbid obesity: A cross-sectional study. *BMC Obes.* **2017**, *4*, 6. [CrossRef]
16. Nguyen, J.C.D.; Killcross, A.S.; Jenkins, T.A. Obesity and cognitive decline: Role of inflammation and vascular changes. *Front. Neurosci.* **2014**, *19*, 375. [CrossRef]
17. Nglazi, M.D.; Ataguba, J.E.O. Overweight and obesity in non-pregnant women of childbearing age in South Africa: Subgroup regression analyses of survey data from 1998 to 2017. *BMC Public Health* **2022**, *22*, 395. [CrossRef]
18. Houle, B.; Gaziano, T.; Farrell, M.; Gómez-Olivé, F.X.; Kobayashi, L.C.; Crowther, N.J.; Wade, A.N.; Montana, L.; Wagner, R.G.; Berkman, L.; et al. Cognitive function and cardiometabolic disease risk factors in rural South Africa: Baseline evidence from the HAALSI study. *BMC Public Health* **2019**, *19*, 1579. [CrossRef]
19. Amugsi, D.A.; Dimbuene, Z.T.; Mberu, B.; Muthuri, S.; Ezeh, A.C. Prevalence and time trends in overweight and obesity among urban women: An analysis of demographic and health surveys data from 24 African countries, 1991–2014. *BMJ Open* **2017**, *7*, e017344. [CrossRef]
20. Belahsen, R.; Mziwira, M.; Fertat, F. Anthropometry of women of childbearing age in Morocco: Body composition and prevalence of overweight and obesity. *Public Health Nutr.* **2004**, *7*, 523–530. [CrossRef]
21. Asare, G.A.; Santa, S.; Ngala, R.A.; Asiedu, B.; Afriyie, D.; Amoah, A.G. Effect of hormonal contraceptives on lipid profile and the risk indices for disease in a Ghanaian community. *Int. J. Womens Health* **2014**, *6*, 597–603. [CrossRef] [PubMed]
22. Zerihun, M.F.; Malik, T.; Ferede, Y.M.; Bekele, T.; Yeshaw, Y. Changes in body weight and blood pressure among women using Depo-Provera injection in Northwest Ethiopia. *BMC Res. Notes* **2019**, *12*, 512. [CrossRef] [PubMed]
23. Le, Y.C.L.; Rahman, M.; Berenson, A.B. Early Weight Gain Predicting Later Weight Gain Among Depot Medroxyprogesterone Acetate Users. *Obstet. Gynecol.* **2009**, *114*, 279–284. [CrossRef] [PubMed]
24. Robinson, J.A.; Burke, A.E. Obesity and hormonal contraceptive efficacy. *Womens Health Lond. Engl.* **2013**, *9*, 453–466. [CrossRef] [PubMed]
25. Anjum, I.; Fayyaz, M.; Wajid, A.; Sohail, W.; Ali, A. Does Obesity Increase the Risk of Dementia: A Literature Review. *Cureus* **2018**, *10*, e2660. [CrossRef]
26. Hsu, C.L.; Voss, M.W.; Best, J.R.; Handy, T.C.; Madden, K.; Bolandzadeh, N.; Liu-Ambrose, T. Elevated body mass index and maintenance of cognitive function in late life: Exploring underlying neural mechanisms. *Front. Aging Neurosci.* **2015**, *7*, 155. [CrossRef]
27. Fellows, R.P.; Schmitter-Edgecombe, M. Independent and Differential Effects of Obesity and Hypertension on Cognitive and Functional Abilities. *Arch. Clin. Neuropsychol.* **2018**, *33*, 24–35. [CrossRef]
28. Sui, S.X.; Pasco, J.A. Obesity and Brain Function: The Brain–Body Crosstalk. *Medicina* **2020**, *56*, 499–508. [CrossRef]
29. Restivo, M.R.; McKinnon, M.C.; Frey, B.N.; Hall, G.B.; Taylor, V.H. Effect of obesity on cognition in adults with and without a mood disorder: Study design and methods. *BMJ Open* **2016**, *6*, e009347. [CrossRef]
30. Hayden, M.; Brown, W.; Brennan, L.; O'Brien, P. Validity of the Beck Depression Inventory as a Screening Tool for a Clinical Mood Disorder in Bariatric Surgery Candidates. *Obes. Surg.* **2012**, *22*, 1666–1675. [CrossRef]
31. Yang, Y.; Shields, G.S.; Wu, Q.; Liu, Y.; Chen, H.; Guo, C. The association between obesity and lower working memory is mediated by inflammation: Findings from a nationally representative dataset of U.S. adults. *Brain Behav. Immun.* **2020**, *84*, 173–179. [CrossRef] [PubMed]
32. Fergenbaum, J.H.; Bruce, S.; Lou, W.; Hanley, A.J.G.; Greenwood, C.; Young, T.K. Obesity and Lowered Cognitive Performance in a Canadian First Nations Population. *Obesity* **2009**, *17*, 1957–1963. [CrossRef] [PubMed]
33. Wagner, S.; Helmreich, I.; Dahmen, N.; Lieb, K.; Tadic, A. Reliability of Three Alternate Forms of the Trail Making Tests A and B. *Arch. Clin. Neuropsychol. Off. J. Natl. Acad. Neuropsychol.* **2011**, *26*, 314–321. [CrossRef] [PubMed]
34. Charernboon, T. Diagnostic Accuracy of the Overlapping Infinity Loops, Wire Cube, and Clock Drawing Tests for Cognitive Impairment in Mild Cognitive Impairment and Dementia. *Int. J. Alzheimers Dis.* **2017**, *2017*, 5289239. [CrossRef] [PubMed]
35. Mendes-Santos, L.C.; Mograbi, D.; Spenciere, B.; Charchat-Fichman, H. Specific algorithm method of scoring the Clock Drawing Test applied in cognitively normal elderly. *Dement. Neuropsychol.* **2015**, *9*, 128–135. [CrossRef]
36. West, R.K.; Ravona-Springer, R.; Sharvit-Ginon, I.; Ganmore, I.; Manzali, S.; Tirosh, A.; Golan, S.; Boccara, E.; Heymann, A.; Beeri, M.S. Long-term trajectories and current BMI are associated with poorer cognitive functioning in middle-aged adults at high Alzheimer's disease risk. *Alzheimers Dement. Diagn. Assess Dis. Monit.* **2021**, *13*, e12247. [CrossRef]
37. Han, Y.P.; Tang, X.; Han, M.; Yang, J.; Cardoso, M.A.; Zhou, J.; Simo, R. Relationship between obesity and structural brain abnormality: Accumulated evidence from observational studies. *Ageing Res. Rev.* **2021**, *71*, 101445. [CrossRef]
38. Maayan, L.; Hoogendoorn, C.; Sweat, V.; Convit, A. Disinhibited eating in obese adolescents is associated with orbitofrontal volume reductions and executive dysfunction. *Obes. Silver Spring Md.* **2011**, *19*, 1382–1387. [CrossRef]
39. Arnoldussen, I.A.C.; Gustafson, D.R.; Leijsen, E.M.C.; de Leeuw, F.E.; Kiliaan, A.J. Adiposity is related to cerebrovascular and brain volumetry outcomes in the RUN DMC study. *Neurology* **2019**, *93*, e864–e878. [CrossRef]
40. Gunstad, J.; Spitznagel, M.B.; Paul, R.H.; Cohen, R.A.; Kohn, M.; Luyster, F.S.; Clark, R.; Williams, L.M.; Gordon, E. Body mass index and neuropsychological function in healthy children and adolescents. *Appetite* **2008**, *50*, 246–251. [CrossRef]
41. Tsai, C.L.; Chen, F.C.; Pan, C.Y.; Tseng, Y.T. The Neurocognitive Performance of Visuospatial Attention in Children with Obesity. *Front. Psychol.* **2016**, *7*, 1033. [CrossRef] [PubMed]

42. Cook, R.L.; O'Dwyer, N.J.; Donges, C.E.; Parker, H.M.; Cheng, H.L.; Steinbeck, K.S.; Cox, E.P.; Franklin, J.L.; Garg, M.L.; Rooney, K.B.; et al. Relationship between Obesity and Cognitive Function in Young Women: The Food, Mood and Mind Study. *J. Obes.* **2017**, *2017*, e5923862. [CrossRef] [PubMed]
43. Yang, Y.; Shields, G.S.; Guo, C.; Liu, Y. Executive function performance in obesity and overweight individuals: A meta-analysis and review. *Neurosci. Biobehav. Rev.* **2018**, *84*, 225–244. [CrossRef] [PubMed]
44. Dassen, F.C.M.; Houben, K.; Allom, V.; Jansen, A. Self-regulation and obesity: The role of executive function and delay discounting in the prediction of weight loss. *J. Behav. Med.* **2018**, *41*, 806–818. [CrossRef]
45. Gómez-Acosta, A.; Londoño-Pérez, C. Emotion regulation and healthy behaviors of the body energy balance in adults: A review of evidence. *Acta Colomb. Psicol.* **2020**, *23*, 349–365. [CrossRef]
46. Houben, K.; Nederkoorn, C.; Jansen, A. Eating on impulse: The relation between overweight and food-specific inhibitory control: Food-Specific Inhibition and Overweight. *Obesity* **2014**, *22*, E6–E8. [CrossRef]
47. Segura-Serralta, M.; Ciscar, S.; Blasco, L.; Oltra-Cucarella, J.; Roncero, M.; Espert, R.; Elvira, V.; Pinedo-Esteban, R.; Perpiñá, C. Contribution of executive functions to eating behaviours in obesity and eating disorders. *Behav. Cogn. Psychother.* **2020**, *48*, 725–733. [CrossRef]
48. Kranjac, A.W.; Nie, J.; Trevisan, M.; Freudenheim, J.L. Depression and Body Mass Index, Differences by Education: Evidence from a Population-based Study of Adult Women in the U.S. Buffalo-Niagara Region. *Obes. Res. Clin. Pract.* **2017**, *11*, 63–71. [CrossRef]
49. Naigaga, D.A.; Jahanlu, D.; Claudius, H.M.; Gjerlaug, A.K.; Barikmo, I.; Henjum, S. Body size perceptions and preferences favor overweight in adult Saharawi refugees. *Nutr. J.* **2018**, *17*, 17. [CrossRef]
50. Keage, H.A.D.; Feuerriegel, D.; Greaves, D.; Tregoweth, E.; Coussens, S.; Smith, A.E. Increasing Objective Cardiometabolic Burden Associated with Attenuations in the P3b Event-Related Potential Component in Older Adults. *Front. Neurol.* **2020**, *11*, 643. [CrossRef]
51. Angoff, R.; Himali, J.J.; Maillard, P.; Aparicio, H.J.; Vasan, R.S.; Seshadri, S.; Beiser, A.S.; Tsao, C.W. Relations of Metabolic Health and Obesity to Brain Aging in Young to Middle-Aged Adults. *J. Am. Heart Assoc.* **2022**, *11*, e022107. [CrossRef] [PubMed]
52. Park, D.C.; Huang, C.M. Culture Wires the Brain: A Cognitive Neuroscience Perspective. *Perspect. Psychol. Sci. J. Assoc. Psychol. Sci.* **2010**, *5*, 391–400. [CrossRef] [PubMed]

Article

Parasympathetic and Sympathetic Monitoring Identifies Earliest Signs of Autonomic Neuropathy

Nicholas L. DePace [1,2], Luis Santos [3], Ramona Munoz [1], Ghufran Ahmad [1], Ashish Verma [1], Cesar Acosta [4], Karolina Kaczmarski [1], Nicholas DePace, Jr. [1], Michael E. Goldis [5] and Joe Colombo [1,2,6,*]

[1] Franklin Cardiovascular, Autonomic Dysfunction and POTS Center, Sicklerville, NJ 08081, USA; nicholasdepace@aol.com (N.L.D.); rmunoz@franklincardio.com (R.M.); ghufran.kmc@gmail.com (G.A.); ashish@ashishverma.net (A.V.); kikaczmarski@gmail.com (K.K.); nicholas.depace@gmail.com (N.D.J.)
[2] NeuroCardiology Research Center, Sicklerville, NJ 08081, USA
[3] New Jersey Heart, Sicklerville, NJ 08081, USA; drlou214@icloud.com
[4] Wyckoff Heights Medical Hospital, Brooklyn, NY 11237, USA; drcesjr@gmail.com
[5] Primary Care and Geriatrics, Stratford, NJ 08084, USA; goldisgeriatrics@yahoo.com
[6] Physio PS, Inc., Atlanta, GA 30339, USA
* Correspondence: jcolombo@physiops.com

Citation: DePace, N.L.; Santos, L.; Munoz, R.; Ahmad, G.; Verma, A.; Acosta, C.; Kaczmarski, K.; DePace, N., Jr.; Goldis, M.E.; Colombo, J. Parasympathetic and Sympathetic Monitoring Identifies Earliest Signs of Autonomic Neuropathy. *NeuroSci* **2022**, *3*, 408–418. https://doi.org/10.3390/neurosci3030030

Academic Editor: Patrizia Polverino de Laureto

Received: 20 June 2022
Accepted: 11 July 2022
Published: 13 July 2022

Publisher's Note: MDPI stays neutral with regard to jurisdictional claims in published maps and institutional affiliations.

Abstract: The progression of autonomic dysfunction from peripheral autonomic neuropathy (PAN) to cardiovascular autonomic neuropathy, including diabetic autonomic neuropathy and advanced autonomic dysfunction, increases morbidity and mortality risks. PAN is the earliest stage of autonomic neuropathy. It typically involves small fiber disorder and often is an early component. Small fiber disorder (SFD) is an inflammation of the C-nerve fibers. Currently, the most universally utilized diagnostic test for SFD as an indicator of PAN is galvanic skin response (GSR), as it is less invasive than skin biopsy. It is important to correlate a patient's symptoms with several autonomic diagnostic tests so as not to treat patients with normal findings unnecessarily. At a large suburban northeastern United States (Sicklerville, NJ) autonomic clinic, 340 consecutive patients were tested with parasympathetic and sympathetic (P&S) monitoring (P&S Monitor 4.0; Physio PS, Inc., Atlanta, GA, USA) with cardiorespiratory analyses, and TMFlow (Omron Corp., Hoffman Estates, Chicago, IL, USA) with LD Technology sudomotor test (SweatC™). This is a prospective, nonrandomized, observational, population study. All patients were less than 60 y/o and were consecutively tested, analyzed and followed from February 2018 through May 2020. P&S Monitoring is based on cardiorespiratory analyses and SweatC™ sudomotor testing is based on GSR. Overall, regardless of the stage of autonomic neuropathy, SweatC™ and P&S Monitoring are in concordance for 306/340 (90.0%) of patients from this cohort. The result is an 89.4% negative predictive value of any P&S disorder if the sudomotor GSR test is negative and a positive predictive value of 90.4% if the sudomotor testing is positive. In detecting early stages of autonomic neuropathy, P&S Monitoring was equivalent to sudomotor testing with high sensitivity and specificity and high negative and positive predictive values. Therefore, either testing modality may be used to risk stratify patients with suspected autonomic dysfunction, including the earliest stages of PAN and SFD. Moreover, when these testing modalities were normal, their high negative predictive values aid in excluding an underlying autonomic nervous system dysfunction.

Keywords: peripheral autonomic neuropathy; advanced autonomic dysfunction; diabetic autonomic neuropathy; cardiovascular autonomic neuropathy; small fiber disorder; sudomotor testing

1. Introduction

The progression of autonomic dysfunction from peripheral autonomic neuropathy (PAN) to cardiovascular autonomic neuropathy, including diabetic autonomic neuropathy and advanced autonomic dysfunction, increases morbidity and mortality risks [1,2]. Due to disagreements as to whether autonomic neuropathy is a function of the aging process and its progression in patients under the age of 65, this study includes only patients under the age of 60 to omit the geriatric population. The increase in morbidity and mortality risks associated with the geriatric population and underlying autonomic dysfunction increases the risks of cardiovascular disease and renal disease as well as various multi-organ system disorders that contribute to numerous symptoms [3–15], thereby increasing medication-load, hospitalizations and healthcare costs. Earlier detection of autonomic dysfunction enables earlier treatment and a higher likelihood of slowing or staying the progression of autonomic dysfunction [16]. Autonomic neuropathy often includes orthostatic dysfunction [1,2] that affects cardiac and cerebral perfusion, leading to clinical symptoms including lightheadedness, dizziness, brain fog, cognitive and memory difficulties, sleep dysfunction, tension and migraine headache disorders and cranial sensory dysfunction. These symptoms alert clinicians to the need to test for autonomic dysfunction.

Autonomic dysfunction increases with age [16] and is accelerated by chronic disease and trauma, regardless of whether the trauma is psychologic or physiologic [16]. The accepted stages of autonomic neuropathy are—in order of severity: PAN, diabetic autonomic neuropathy and cardiovascular autonomic neuropathy [1,2]. In a standard autonomic function test, deep breathing, Valsalva, and head-up postural change (tilt-test or standing) are the challenges used to determine autonomic function as compared with the resting baseline. The earliest stage of autonomic dysfunction (PAN) is indicated by either or both deep breathing (see Figure 1) or Valsalva (see Figure 2) abnormalities. Diabetic autonomic neuropathy is indicated when either or both of the resting autonomic (parasympathetic or sympathetic) responses fall below normal, but the parasympathetic response is still >0.1 bpm^2 (see Figure 3) [16]. Diabetic autonomic neuropathy has been labeled advanced autonomic dysfunction for the same stage of autonomic neuropathy in patients not diagnosed with diabetes (see Figure 3) [16]. Cardiovascular autonomic neuropathy is indicated when the resting parasympathetic measure is extremely low (<0.1 bpm^2, see Figure 3) [16]. If autonomic dysfunction is detected early and treated [17], its progression may be slowed or stayed, regardless of the stage of autonomic neuropathy.

As PAN is the early stage of autonomic neuropathy, it typically involves small fiber disorder (SFD), often as an early component. SFD is an inflammation or deficiency of the C-nerve fibers which carry sympathetic and pain signals to and from the periphery. The inflammatory state of SFD is typically the early stage and the deficiency state of SFD is the later stage. The sympathetic nerve fibers affect peripheral vasoconstriction and sweat gland function, thereby affecting temperature control and wound healing. Currently, the preferred test for SFD as an indicator of PAN is galvanic skin response (GSR), as it is much more readily available, and far less invasive than a skin biopsy. GSR measures are used to indicate small fiber function [19–22]. GSR succeeded earlier tests, including Q-SART, Q-Sweat and thermoregulatory sweat testing as it is less time- and technician-intensive. There are multiple galvanic skin conduction testing modalities to assess sweat gland function [23–29]. SweatC™ is a GSR device and does not use, nor is it, an electrochemical skin response device. It captures quantitative sudomotor responses to assess the integrity of the post-ganglionic sudomotor nerves along the axon reflex. SweatC™ was used exclusively in this study as a type of quantitative sudomotor axon reflex test. While GSR has become a standard, there are still concerns that non-invasive sweat gland function may fail to demonstrate a conclusive measure of small fiber function [30,31]. Two technologies, Quantitative Sudomotor Axon Reflex Test (QSART) and Sympathetic Skin Response (SSR), both have overwhelming amounts of data supporting their use as medically necessary for the evaluation of autonomic dysfunction [32]. This study considers a complimentary and alternate approach to detecting PAN and thereby SFD or C-nerve Fiber function and compares it to sudomotor function as the current standard.

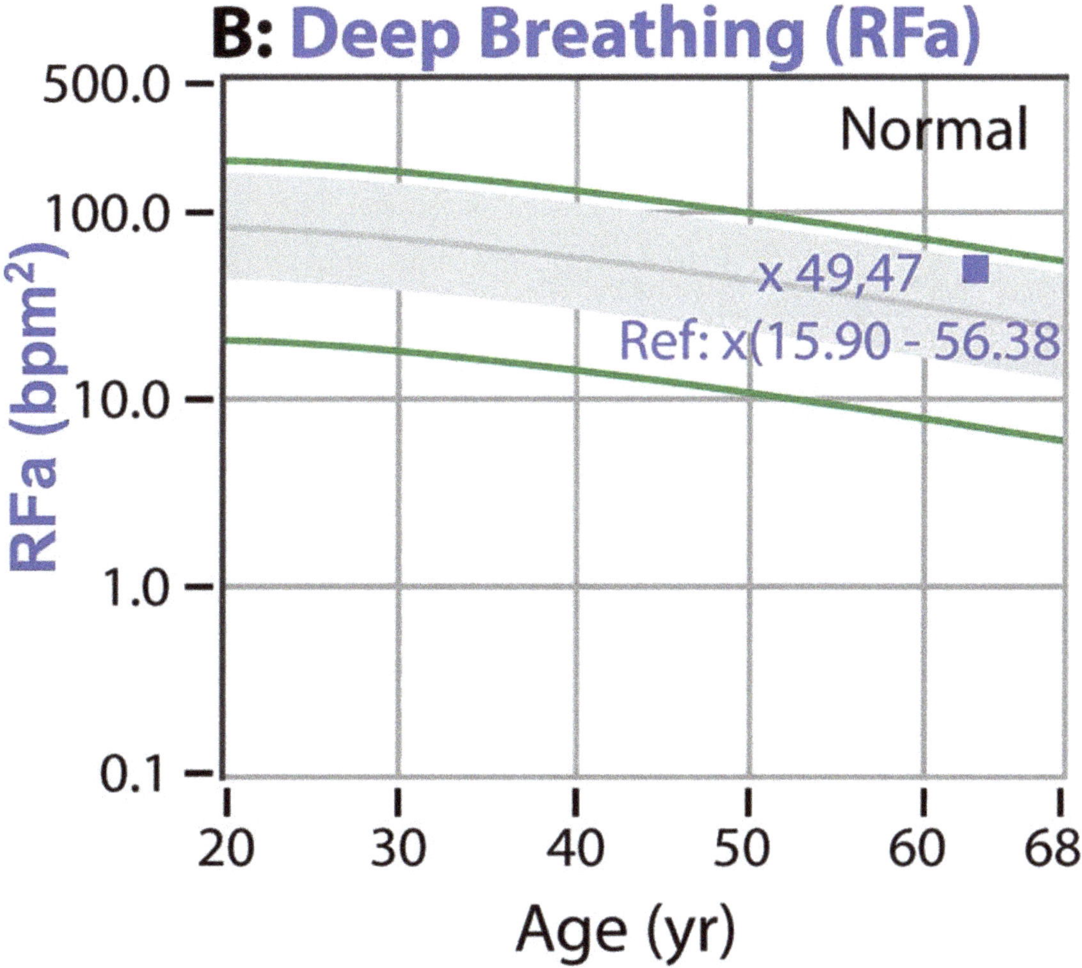

Figure 1. "Normal Response to Deep Breathing." The deep breathing challenge is a well-known parasympathetic stimulus, and arguably one of the strongest of parasympathetic stimuli. The figure displays a deep breathing response plot demonstrating a normal parasympathetic response for a 63 y/o patient. The gray area indicates the normal region [18]. These data are both age-and baseline-adjusted [16]. The 'x' preceding a number indicates that that number is baseline adjusted. The area between the green lines and the gray indicates borderline normal responses. Borderline low to low deep breathing responses are some of the earliest signs of PAN and risk of sudomotor abnormality or SFD.

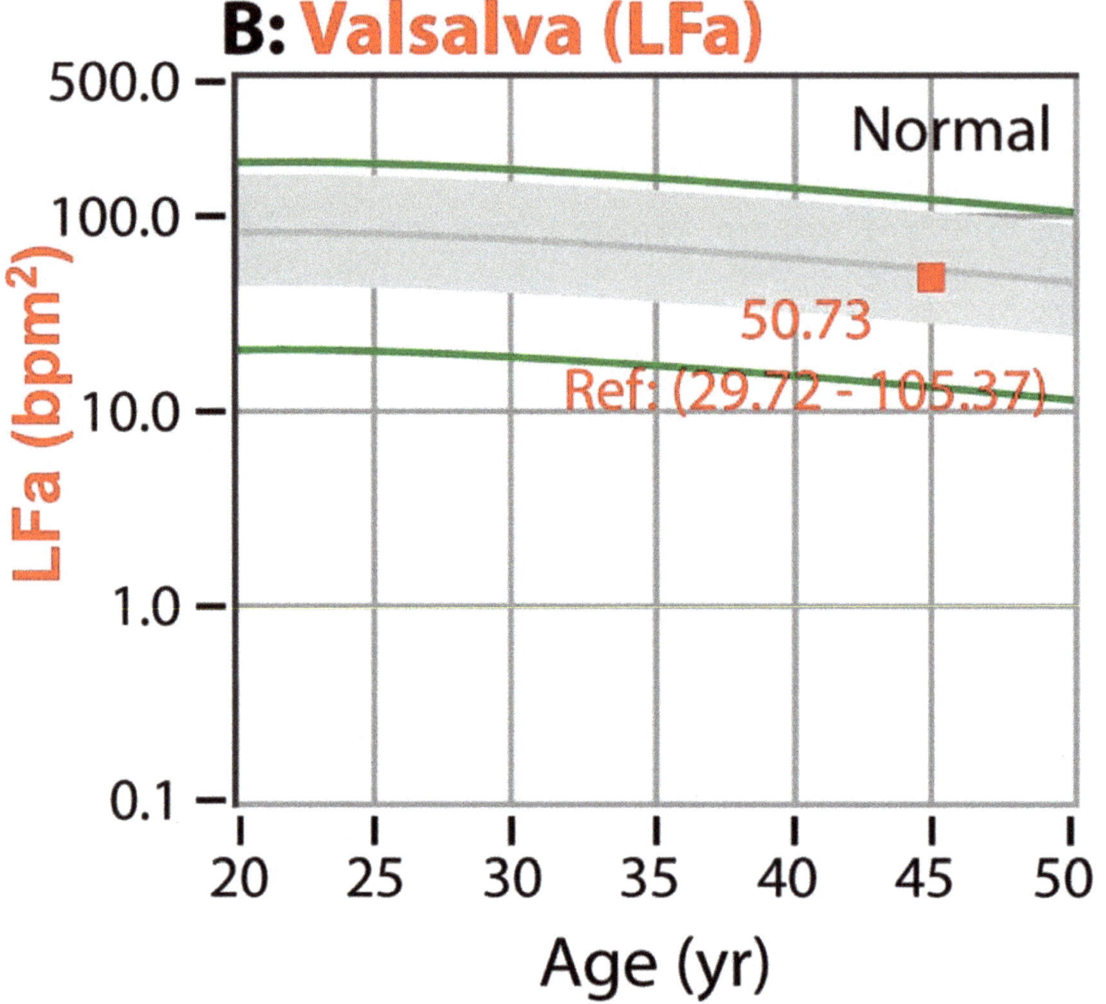

Figure 2. "Normal Response to Valsalva." The Valsalva challenge is a series of short (≤15 s) Valsalva maneuvers. Short Valsalva maneuvers are well-known and potent sympathetic stimuli. The figure displays a Valsalva response plot demonstrating a normal sympathetic response for a 45 y/o patient. The gray area indicates the normal region [18]. These data are both age- and baseline-adjusted [16]. The area between the green lines and the gray indicates borderline normal responses. Borderline low to low Valsalva responses are some of the earliest signs of PAN and risk of sudomotor or SFD.

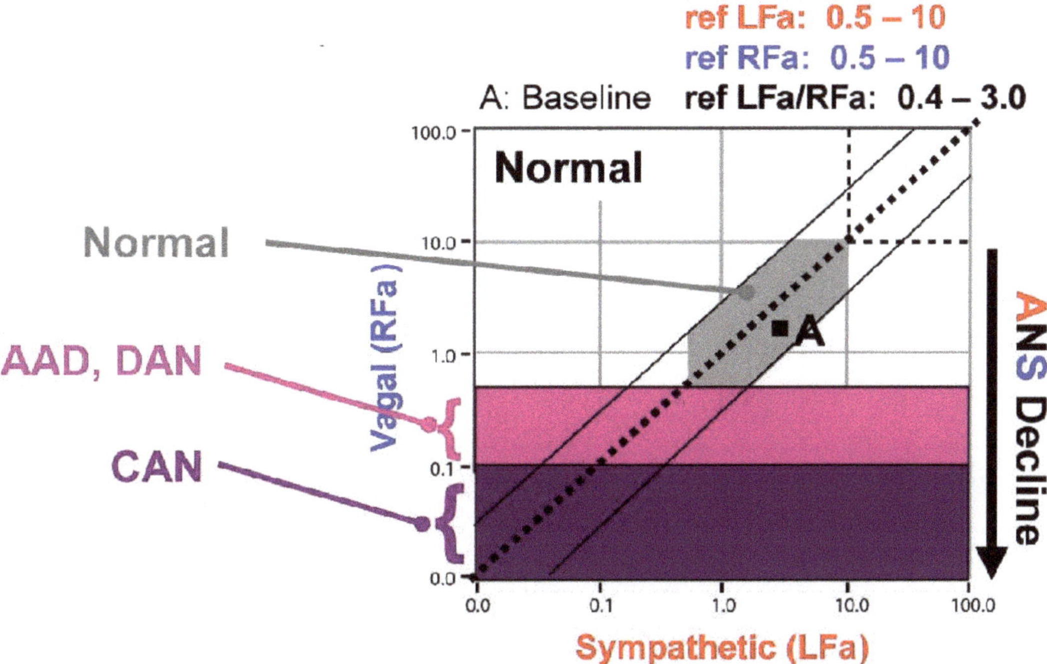

Figure 3. "Normal at Rest" An example resting (baseline) parasympathetic and sympathetic (P&S) response plot. The gray area depicts the normal response region. The purple highlighted areas depict the definitions of advanced autonomic dysfunction (AAD, light purple), diabetic autonomic neuropathy (DAN, also light purple) and cardiovascular autonomic neuropathy (CAN, dark purple). AAD and DAN indicate morbidity risk and CAN indicates mortality risk. Risk is stratified by sympathovagal balance. the space between the two outer diagonal lines defines normal sympathovagal balance ("LFa/RFa"). Regardless of resting autonomic state, normal sympathovagal balance normalizes morbidity and mortality risks. Above and to the left of the upper diagonal line indicates low sympathovagal balance which is a resting parasympathetic excess. Below and to the right of the lower diagonal line indicates high sympathovagal balance which is a resting sympathetic excess.

2. Methods

From a suburban cardiovascular and autonomic dysfunction practice in the northeastern United States (Sicklerville, NJ, USA), with patients drawn from around the world, 340 patients presenting with more than four autonomic symptoms [33] were consecutively tested and followed between February 2018 to September 2020. This cohort included 225 Females (66.0%), with an average age of 36.5 years (ranging from 14 to 59 years), and an average BMI of 26.7 #/in^2. All patients were tested with P&S Monitoring (P&S Monitor 4.0; Physio PS, Inc., Atlanta, GA, USA), which includes cardiorespiratory analysis, and TMFlow (Omron Corp., Hoffman Estates, Chicago, IL, USA), which includes the LD Technology Sudomotor SweatC™ test. This is a prospective, nonrandomized, observational, population study. PAN and SFD results from both test modalities are compared. All patients provided consent for their data to be included in this large population study and patient data were maintained according to HIPAA guidelines.

P&S Monitoring collects EKG, respiratory activity from chest electrodes and BP during four challenges: (1) rest (baseline, 5-min), (2) deep breathing (0.1 Hz, a parasympathetic challenge, 1-min), (3) short Valsalva maneuvers (<15 s, as a sympathetic challenge, 1:35-min), and (4) head-up postural change (stand, which is equivalent to tilt, 5-min [34]). With spectral analyses these data are analyzed, and independent and simultaneous P&S activity is measured throughout the clinical study [16]. Weakness in response to either or both deep breathing and Valsalva (collectively known as the breathing challenges) are the first signs of PAN, including SFD, perhaps even before overt symptoms of SFD. Normal and abnormal P&S response plots are depicted in Figures 1–4, respectively [16].

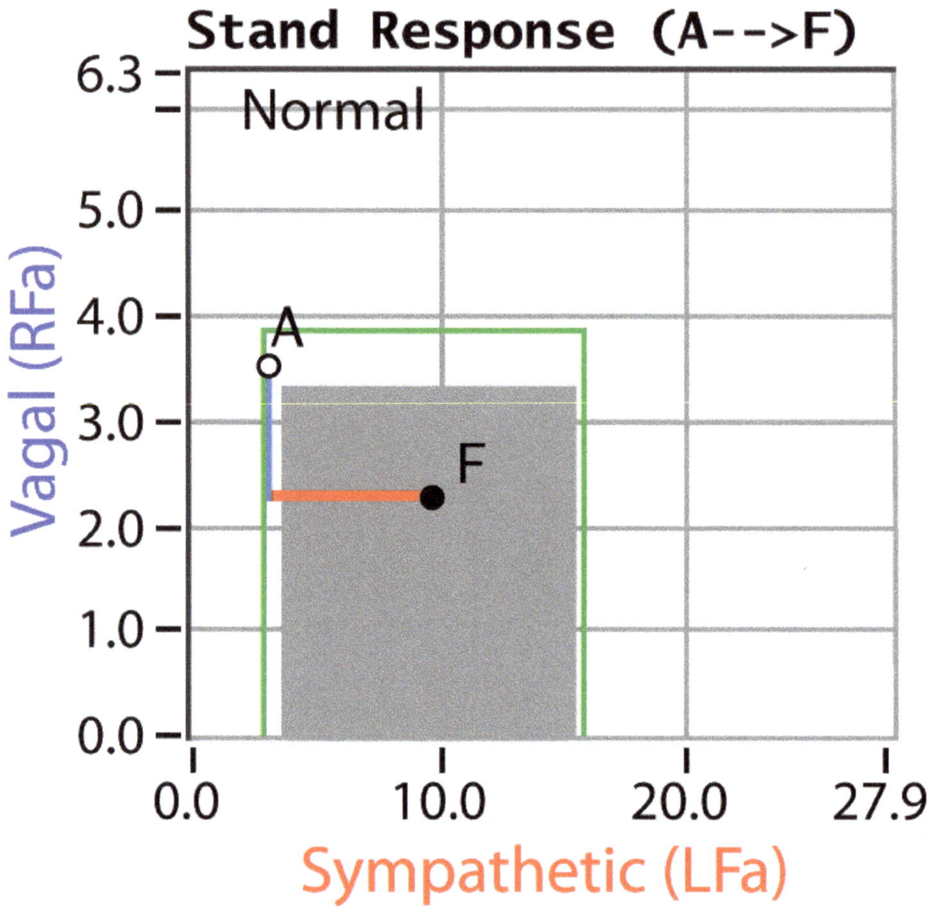

Figure 4. "Normal Response to Stand or Head-Up Postural Change." The stand challenge response provides information regarding causes of lightheadedness (e.g., orthostatic dysfunction as associated with and abnormally low sympathetic response, or syncope as associated with an abnormally high parasympathetic response), as well as an indication of the ability of the two autonomic branches to coordinate responses, not only to postural change but in the control and coordination of organs and organ systems [16]. The figure displays a stand response plot demonstrating a normal Parasympathetic (a decreased response from rest, 'A', to stand, 'F') with a normal Sympathetic response (an increased response from rest, 'A', to stand, 'F'). The gray area indicates the normal region. These data are baseline-adjusted. The area between the green lines and the gray indicates borderline normal responses. The symptoms associated with abnormal stand responses are often the first symptoms recognized as the results of autonomic dysfunction [16].

LD Technology SweatC™ test is not an electrochemical skin response. It does not use stainless steel electrodes for the hands and feet, nor does it use incremental voltages. Instead, it uses constant voltage and cloth (textile) electrodes placed on the feet, thus no electrochemical reactions may be obtained, and therefore it cannot induce any reverse ionophoresis. Sudomotor testing is based on electrochemical skin conductance analysis of the function of sweat glands from the bottom of the feet measured for 2 min [31]. Low sudomotor sweat peaks indicate decreased density of active cholinergic nerve fibers. Elevated sudomotor sweat peaks indicate inflammation.

The autonomic testing environment was a quiet, out of the way examination space, maintained at 70 °F and patients were permitted up to an hour to rest and acclimate. Technicians took steps to help patients remain calm and feel secure throughout the tests, including diming the lights for those who are light-sensitive. Pearson correlation and Student's *t*-test statistics are based on SPSS v. 20.

3. Results

The SweatC™ sudomotor test and the P&S Monitor indication of PAN both are indications for SFD. Table 1 provides a breakdown of the results from the two tests. From Sudomotor testing, 144 patients demonstrated abnormal results, indicating SFD from either inflammation or from depletion. Of the 144 patients demonstrating abnormal SweatC™, 82 demonstrated inflammation (high SweatC™ responses), and 67 demonstrated depletion (low SweatC™ responses); five patients demonstrated high and low SweatC™ responses, one foot each. The remaining 196 patients demonstrated normal SweatC™ responses. From P&S Monitoring, 122 patients demonstrated PAN (low breathing challenge responses), and 86 patients demonstrated advanced autonomic dysfunction, including PAN, totaling 208 abnormal indications from P&S Monitoring. The remaining 132 patients did not demonstrate autonomic neuropathy. Pearson's Correlation and Student's *t*-test indicate a statistically significant similarity between the results of the two tests (see Table 1).

Table 1. Sudomotor and P&S responses.

Total Pop. = 340	Inflammation/PAN	Depletion/AN	Total Abn	Total Nml
Sudomotor	82	67	144 *	196
P&S	122	86	208	132
r	0.960	0.894	0.802	0.802
p	0.051	<0.001	<0.001	<0.001

* Five (5) patients demonstrated both inflammation and depletion.

Table 2 compares the two technologies, considering only the specific indication for PAN (abnormal breathing challenge responses). Both testing modalities indicate the same 130 (38.3%) patients as SFD positive, and 118 (34.7%) patients negative for SFD. There are only 14 (4.1%) patients for which SweatC™ did not indicate SFD and P&S Monitoring did indicate SFD (classified as false negative). There are 78 (34.7%) patients for which SweatC™ did indicate SFD and P&S Monitoring did not indicate SFD (classified as false positive). In summary, when considering only the specific indication for PAN, both technologies have a high (89.4%) negative predictive value and a low positive predictive value (62.5%).

Table 2. Comparison of SFD indications from SweatC™ sudomotor testing (sudomotor) and P&S monitoring (P&S), for the entire cohort, based on the specific definition of PAN. See text for details.

n = 340	P&S Positive	P&S Negative
Sudomotor Positive	130 (38.3%)	78 (22.9%)
Sudomotor Negative	14 (4.1%)	118 (34.7%)

Upon further investigation, 58 (17.1% of the cohort) of the 78 patients for which SweatC™ did indicate SFD and P&S Monitoring did not (false positives), P&S Monitoring also documented advanced autonomic dysfunction, diabetic autonomic neuropathy or cardiovascular autonomic neuropathy. These were the only patients within the cohort to demonstrate these advanced stages of autonomic neuropathy. These advanced stages of autonomic neuropathy inherently include PAN as an earlier phase. However, the P&S Monitor computation of the deep breathing challenge and Valsalva challenge responses are baseline adjusted [16] for clinical accuracy. This adjustment may artificially normalize the challenge responses, relatively speaking. Removing the baseline adjustment to demonstrate that PAN is an inherent part of the more advanced stages of autonomic neuropathy, reveals that all 58 of those patients do indeed demonstrate PAN. Table 3 reanalyzes the results of the cohort with these 58 patients reclassified as SFD. As a result, the percent positive increases to 55.3%, and the percent false positive drops to 5.9% (the false negative and negative percentages do not change; therefore, the negative predictive value remains 89.4%) but the positive predictive value is now 90.4%.

Table 3. Comparison of SFD indications from SweatC™ sudomotor testing (sudomotor) and P&S Monitoring (P&S) considering that the more advanced autonomic neuropathies also involved PAN. In other words, reclassifying patients demonstrating advanced autonomic dysfunction, diabetic autonomic neuropathy or cardiovascular autonomic neuropathy as also demonstrating peripheral autonomic neuropathy and therefore small fiber disorder (P&S Positive). See text for details.

$n = 340$	P&S Positive	P&S Negative
Sudomotor Positive	188 (55.3%)	20 (5.9%)
Sudomotor Negative	14 (4.1%)	118 (34.7%)

With this correction there is a high concordance rate (306/340 or 90.0%) and association between SweatC™ sudomotor positive indications and P&S Monitoring for all autonomic dysfunctions. In other words, abnormal physiology of small fibers is assessed by all types of P&S malfunction 90% of the time. Of the remaining 34 patients, 20 are classified as false positive, where SweatC™ indicates SFD and P&S Monitoring does not, and there continue to be 14 patients classified as false negative. The results from the re-analysis are: (1) a specificity of 85.5% (118/138 patients), (2) a sensitivity of 93.1% (188/202 patients), (3) a positive predictive value of 90.4% (188/208 patients) and (4) a negative predictive value of 89.4% (118/132 patients).

4. Discussion

In this study it is important to note that P&S Monitoring does not differentiate small fiber inflammation from deficiency. Both forms of SFD are included in the analyses since both are involved in SFD. In addition, for the purposes of this study, GSR is considered a very accurate test of SFD and is used as the gold standard even though there is some controversy. Sudomotor testing, as from SweatC™, measures the GSR at the positive electrode and the sweat peaks estimate of cholinergic nerve fiber density. Sweat peaks are calculated from the peak amplitude of the GSR at the positive electrode of patients resulting from sweating. Although sweat glands are controlled by the sympathetics, sweating may be influenced by other factors, including daily hydration (or dehydration). While many autonomic patients drink plenty of water a day they remain largely dehydrated because the water does not stay in their vasculature. Furthermore, many drinks intended to hydrate (i.e., sports drinks), actually dehydrate due to sugar, sugar substitutes, caffeine and alcohol (especially if one is not running around burning the sugar). In addition, many patients diagnosed with hypertension are also prescribed diuretics, when the hypertension may be compensatory for orthostatic dysfunction. Significant dehydration may lead to anhidrosis and false positives, as defined above.

P&S Monitoring differs from all other autonomic monitors in that it is uniquely capable of measuring the two individual autonomic branches independently and simultaneously without assumption and approximation [35–38]. P&S monitoring permits follow-up testing and includes indications for PAN (including small C-fiber disorder) as well as P&S dysfunctions (including autonomic neuropathies) not detected by typical autonomic monitors. Such P&S dysfunctions include: (1) sympathetic withdrawal (an alpha-adrenergic insufficiency upon assuming a head-up posture, associated with orthostatic dysfunction; see Figure 4) [39] and (2) parasympathetic excess (an excessive cholinergic response to a stress, as modeled by Valsalva challenge or upon assuming a head-up posture, associated with parasympathetic or vagal over-reactions) [40]. Both of these autonomic dysfunctions may result in poor cerebral perfusion and symptoms of lightheadedness, some of the first symptoms reported that are associated with autonomic dysfunction.

The percent of patients classified as false negative (P&S monitoring indicating SFD and sudomotor testing not indicating SFD), while very low, may highlight the fact that P&S monitoring is typically a leading indicator of autonomic dysfunction and neuropathy, involving abnormal sudomotor results. Abnormal sudomotor testing must wait until SFD is present, and further, wait until it is significant enough before detecting significantly abnormal changes in skin conductance. P&S monitoring may detect these changes in the autonomic nervous system earlier, before resulting symptoms are significant. This may help to treat earlier and prevent symptoms and possibly slow the progression of autonomic dysfunction.

The percent of patients classified as false positive (P&S monitoring not indicating SFD and positive sudomotor testing indicating SFD) while also low, may highlight the fact that there are multiple reasons for sweating disorders, not all related to small fibers, such as connective tissue disorders, and medications, including some anti-depressants, antipsychotics, antihypertensives and opioids, all of which are common in today's culture.

Hypothetically, inflammation is an earlier stage of SFD and deficiency represents a more advanced stage of SFD. Typically, inflammation precedes deficiency in cells. The same is presumed here. Other nerve fibers may also be involved, including A-beta and A-delta nerve fibers. A-beta fibers carry touch information and feedback to the autonomic nervous system to signal incoming sensory information. A-delta fibers carry pain and temperature information and are known to be affected by blood flow [41]. The larger, more myelinated A-fibers are faster to respond than C-fibers and typically signal the acute ("sharp, specific") pain information and the C-fibers carry the chronic ("dull, diffuse") pain information.

Further studies are required to determine (1) if inflammation is associated with early stages of SFD or autonomic neuropathy (i.e., PAN, potentially more treatable if detected early); and (2) if deficiency is associated with later stages of SFD or autonomic neuropathy (i.e., advanced autonomic neuropathy or cardiovascular autonomic neuropathy, and may not be readily treatable). We believe that early detection may provide an advantage for reversal of SFD as well as autonomic neuropathy. Although further prospective studies are indicated, either test may be used alternatively by itself, which would be a cost saving measure to assess for autonomic dysfunction and evaluate for the presence of underlying risk factors (e.g., diabetes mellitus). Further prospective studies are needed.

5. Conclusions

In detecting SFD as an early stage of Autonomic Neuropathy, including Diabetic Autonomic Neuropathy, P&S Monitoring is comparable to sudomotor testing with high sensitivity, specificity and high positive and negative predictive values. Therefore, either testing modality may be used to risk stratify patients with suspected autonomic dysfunction, including the earliest stage, PAN, involving SFD. Likewise, these testing modalities when normal, with their high negative predictive values, may help to exclude an existing autonomic dysfunction. Further prospective studies are needed to assess if any one study is sufficient to objectively diagnose patients with symptoms of autonomic dysfunction.

Author Contributions: All authors contributed significantly to the data collection, data analysis and development of this manuscript as confirmed by J.C. and N.L.D., Conceptualization and methodology, J.C. and N.L.D.; software, J.C.; validation, formal analysis, and investigation, J.C., N.L.D. and L.S.; resources, and data curation and collection, R.M., G.A., A.V., C.A., K.K., N.D.J. and M.E.G.; writing—original draft preparation, J.C.; writing—review and editing, N.L.D. and L.S.; visualization, supervision, project administration, J.C. and N.L.D.; funding acquisition, N/A. All authors have read and agreed to the published version of the manuscript.

Funding: This research received no external funding.

Institutional Review Board Statement: All patients provided consent for their data to be included in this large population study and patient data were maintained according to HIPAA guidelines.

Informed Consent Statement: Patient consent was waived due to all patients attending these clinics are informed that their data may be used for large population clinical studies, unless the patient objects. None of the 340 patients included in this study objected.

Data Availability Statement: All data are HIPAA protected and may be made available upon request.

Conflicts of Interest: Colombo is founder, and part owner of Physio PS, Inc., the provider of P&S Monitoring. There is no conflict with NeuroCardiology Research Corp., as it does not own anything involved in this study.

References

1. Vinik, A.I.; Maser, R.E.; Nakave, A.A. Diabetic cardiovascular autonomic nerve dysfunction. *US Endocr. Dis.* **2007**, *2*, 2–9. [CrossRef]
2. Vinik, A.; Ziegler, D. Diabetic cardiovascular autonomic neuropathy. *Circulation* **2007**, *115*, 387–397. [CrossRef] [PubMed]
3. DePace, N.L.; Mears, J.P.; Yayac, M.; Colombo, J. Cardiac autonomic testing and diagnosing heart disease. "A clinical perspective". *Heart Int.* **2014**, *9*, 37–44. [CrossRef]
4. DePace, N.L.; Mears, J.P.; Yayac, M.; Colombo, J. Cardiac autonomic testing and treating heart disease. "A clinical perspective". *Heart Int.* **2014**, *9*, 45–52. [CrossRef]
5. Bullinga, J.R.; Alharethi, R.; Schram, M.S.; Bristow, M.R.; Gilbert, E.M. Changes in heart rate variability are correlated to hemodynamic improvement with chronic CARVEDILOL therapy in heart failure. *J. Card Fail.* **2005**, *11*, 693–699. [CrossRef] [PubMed]
6. Fatoni, C.; Raffa, S.; Regoli, F.; Giraldi, F.; La Rovere, M.T.; Prentice, J.; Pastori, F.; Fratini, S.; Salerno-Uriarte, J.A.; Klein, H.U.; et al. Cardiac resynchronization therapy improves heart rate profile and heart rate variability of patients with moderate to severe heart failure. *J. Am. Coll Cardiol.* **2005**, *46*, 1875–1882. [CrossRef] [PubMed]
7. Fathizadeh, P.; Shoemaker, W.C.; Woo, C.C.J.; Colombo, J. Autonomic activity in trauma patients based on variability of heart rate and respiratory rate. *Crit Care Med.* **2004**, *32*, 1300–1305. [CrossRef]
8. Chen, J.Y.; Fung, J.W.; Yu, C.M. The mechanisms of atrial fibrillation. *J. Cardiovasc. Electrophysiol.* **2006**, *17* (Suppl. 3), S2–S7. [CrossRef]
9. Copie, X.; Lamaison, D.; Salvador, M.; Sadoul, N.; DaCosta, A.; Faucher, L.; Legal, F.; Le Heuzey, J.Y.; VALID Investigators. Heart rate variability before ventricular arrhythmias in patients with coronary artery disease and an implantable cardioverter defibrillator. *Ann. Noninvasive Electrocardiol.* **2003**, *8*, 179–184. [CrossRef]
10. Alter, P.; Grimm, W.; Vollrath, A.; Czerny, F.; Maisch, B. Heart rate variability in patients with cardiac hypertrophy–relation to left ventricular mass and etiology. *Am. Heart J.* **2006**, *151*, 829–836. [CrossRef]
11. Debono, M.; Cachia, E. The impact of cardiovascular autonomic neuropathy in diabetes: Is it associated with left ventricular dysfunction? *Auton. Neurosci.* **2007**, *132*, 1–7. [CrossRef] [PubMed]
12. Just, H. Peripheral adaptations in congestive heart failure: A review. *Am. J. Med.* **1991**, *90*, 23S–26S. [CrossRef]
13. Nakamura, K.; Matsumura, K.; Kobayashi, S.; Kaneko, T. Sympathetic premotor neurons mediating thermoregulatory functions. *Neurosci. Res.* **2005**, *51*, 1–8. [CrossRef] [PubMed]
14. Manfrini, O.; Morgagni, G.; Pizzi, C.; Fontana, F.; Bugiardini, R. Changes in autonomic nervous system activity: Spontaneous versus balloon-induced myocardial ischaemia. *Eur. Heart J.* **2004**, *25*, 1502–1508. [CrossRef] [PubMed]
15. Clarke, B.; Ewing, D.; Campbell, I. Diabetic autonomic neuropathy. *Diabetologia* **1979**, *17*, 195–212. [CrossRef]
16. Colombo, J.; Arora, R.R.; DePace, N.L.; Vinik, A.I. *Clinical Autonomic Dysfunction: Measurement, Indications, Therapies, and Outcomes*; Springer Science + Business Media: New York, NY, USA, 2014.
17. Vinik, A.I.; Murray, G.L. Autonomic neuropathy is treatable. *US Endocrinol.* **2008**, *2*, 82–84. [CrossRef]

18. Akinola, A.; Bleasdale-Barr, K.; Everall, L. Mathias CJInvestigation of Autonomic Disorders: Appendix, I. In *Autonomic Failure: A Textbook of Clinical Disorders of the Autonomic Nervous System*; Mathias, C.J., Bannister, R., Eds.; Oxford Medical Publications: London, UK, 1999.

19. Vinik, A.I.; Maser, R.E.; Mitchell, B.D.; Freeman, R. Diabetic autonomic neuropathy. *Diabetes Care* **2003**, *26*, 1553–1579. [CrossRef]

20. Novak, P. Electrochemical skin conductance: A systematic review. *Clin. Auton. Res.* **2019**, *29*, 17–29. [CrossRef] [PubMed]

21. Chae, C.S.; Park, G.Y.; Choi, Y.M.; Jung, S.; Kim, S.; Sohn, D.; Im, S. Rapid, Objective and Non-invasive Diagnosis of Sudomotor Dysfunction in Patients with Lower Extremity Dysesthesia: A Cross-Sectional Study. *Ann. Rehabil. Med.* **2017**, *41*, 1028–1038. [CrossRef]

22. Smith, A.G.; Lessard, M.; Reyna, S.; Doudova, M.; Singleton, J.R. The diagnostic utility of Sudoscan for distal symmetric peripheral neuropathy. *J Diabetes Complicat.* **2014**, *28*, 511–516. [CrossRef]

23. Low, P.A.; Caskey, P.E.; Tuck, R.R.; Fealey, R.D.; Dyck, P.J. Quantitative sudomotor axon reflex test in normal and neuropathic subjects. *Ann. Neurol.* **1983**, *14*, 573–580. [CrossRef] [PubMed]

24. Illigens, B.M.; Gibbons, C.H. Sweat testing to evaluate autonomic function. *Clin. Auton. Res.* **2009**, *19*, 79–87. [CrossRef] [PubMed]

25. Vinik, A.I.; Nevoret, M.L.; Casellini, C. The New Age of Sudomotor Function Testing: A Sensitive and Specific Biomarker for Diagnosis, Estimation of Severity, Monitoring Progression, and Regression in Response to Intervention. *Front. Endocrinol.* **2015**, *6*, 94. [CrossRef]

26. Vinik, A.I.; Nevoret, M.; Casellini, C.; Parson, H. Neurovascular function and sudorimetry in health and disease. *Curr. Diab. Rep.* **2013**, *13*, 517–532. [CrossRef] [PubMed]

27. Vinik, A.I.; Casellini, C.; Névoret, M.L. Alternative Quantitative Tools in the Assessment of Diabetic Peripheral and Autonomic Neuropathy. *Int. Rev. Neurobiol.* **2016**, *127*, 235–285. [CrossRef] [PubMed]

28. Vinik, A.I.; Smith, A.G.; Singleton, J.R.; Callaghan, B.; Freedman, B.I.; Tuomilehto, J.; Bordier, L.; Bauduceau, B.; Roche, F. Normative Values for Electrochemical Skin Conductances and Impact of Ethnicity on Quantitative Assessment of Sudomotor Function. *Diabetes Technol. Ther.* **2016**, *18*, 391–398. [CrossRef] [PubMed]

29. Casellini, C.M.; Parson, H.K.; Richardson, M.S.; Nevoret, M.L.; Vinik, A.I. Sudoscan, a noninvasive tool for detecting diabetic small fiber neuropathy and autonomic dysfunction. *Diabetes Technol. Ther.* **2013**, *15*, 948–953. [CrossRef]

30. Rajan, S.; Campagnolo, M.; Callaghan, B.; Gibbons, C.H. Sudomotor function testing by electrochemical skin conductance: Does it really measure sudomotor function? *Clin. Auton. Res.* **2019**, *29*, 31–39. [CrossRef] [PubMed]

31. Vinik, A.I.; Casellini, C.M.; Parson, H.K. Electrochemical skin conductance to measure sudomotor function: The importance of not misinterpreting the evidence. *Clin. Auton. Res.* **2019**, *29*, 13–15. [CrossRef] [PubMed]

32. Gibbons, C.H.; Cheshire, W.P.; Fife, T.D. *American Academy of Neurology Autonomic Testing Model Coverage Policy*; American Academy of Neurology: Minneapolis, MN, USA, 2014.

33. Colombo, J.; Weintraub, M.I.; Munoz, R.; Verma, A.; Ahmed, G.; Kaczmarski, K.; Santos, L.; DePace, N.L. Long-COVID and the Autonomic Nervous System: The journey from Dysautonomia to Therapeutic Neuro-Modulation, Analysis of 152 Patient Retrospectives. *NeuroSci* **2022**, *3*, 21. [CrossRef]

34. Bloomfield, D.M.; Kaufman, E.S.; Bigger, J.T., Jr.; Fleiss, J.; Rolnitzky, L.; Steinman, R. Passive head-up tilt and actively standing up produce similar overall changes in autonomic balance. *Am Heart J.* **1997**, *134 2 Pt 1*, 316–320. [CrossRef]

35. Akselrod, S.; Gordon, S.; Ubel, F.A.; Shannon, D.C.; Berger, A.C.; Cohen, R.J. Power spectrum analysis of heart rate fluctuations: A quantitative probe of beat-to- beat cardiovascular control. *Science* **1981**, *213*, 213–220. [CrossRef] [PubMed]

36. Akselrod, S.; Gordon, D.; Madwed, J.B.; Snidman, N.C.; Shannon, D.C.; Cohen, R.J. Hemodynamic regulation: Investigation by spectra analysis. *Am. J. Physiol.* **1985**, *249*, H867–H875. [CrossRef] [PubMed]

37. Akselrod, S.; Eliash, S.; Oz, O.; Cohen, S. Hemodynamic regulation in SHR: Investigation by spectral analysis. *Am. J. Physiol.* **1987**, *253*, H176–H183. [CrossRef] [PubMed]

38. Akselrod, S. Spectral analysis of fluctuations in cardiovascular parameters: A quantitative tool for the investigation of autonomic control. *Trends Pharmacol. Sci* **1988**, *9*, 6–9. [CrossRef]

39. Arora, R.R.; Bulgarelli, R.J.; Ghosh-Dastidar, S.; Colombo, J. Autonomic mechanisms and therapeutic implications of postural diabetic cardiovascular abnormalities. *J. Diabetes Sci. Technol.* **2008**, *2*, 568–571. [CrossRef]

40. Tobias, H.; Vinitsky, A.; Bulgarelli, R.J.; Ghosh-Dastidar, S.; Colombo, J. Autonomic nervous system monitoring of patients with excess parasympathetic responses to sympathetic challenges—Clinical observations. *US Neurol.* **2010**, *5*, 62–66. [CrossRef]

41. Torebjörk, H.E.; Hallin, R.G. Perceptual changes accompanying controlled preferential blocking of A and C fibre responses in intact human skin nerves. *Exp. Brain Res.* **1973**, *16*, 321–332. [CrossRef]

Article

Encoding of Race Categories by Single Neurons in the Human Brain

André B. Valdez [1], Megan H. Papesh [2], David M. Treiman [3], Stephen D. Goldinger [4] and Peter N. Steinmetz [1,*]

[1] Neurtex Brain Research Institute, 8300 Douglas, Suite 800, Dallas, TX 75225, USA
[2] Department of Psychology, New Mexico State University, Las Cruces, NM 88003, USA
[3] Department of Neurology, Barrow Neurological Institute, Phoenix, AZ 85013, USA
[4] Department of Psychology, Arizona State University, Tempe, AZ 85287, USA
* Correspondence: peternsteinmetz@steinmetz.org; Tel.: +1-(972)-752-1910

Abstract: Previous research has suggested that race-specific features are automatically processed during face perception, often with out-group faces treated categorically. Functional imaging has illuminated the hemodynamic correlates of this process, with fewer studies examining single-neuron responses. In the present experiment, epilepsy patients undergoing microwire recordings in preparation for surgical treatment were shown realistic computer-generated human faces, which they classified according to the emotional expression shown. Racial categories of the stimulus faces varied independently of the emotion shown, being irrelevant to the patients' primary task. Nevertheless, we observed race-driven changes in neural firing rates in the amygdala, anterior cingulate cortex, and hippocampus. These responses were broadly distributed, with the firing rates of 28% of recorded neurons in the amygdala and 45% in the anterior cingulate cortex predicting one or more racial categories. Nearly equal proportions of neurons responded to White and Black faces (24% vs. 22% in the amygdala and 26% vs. 28% in the anterior cingulate cortex). A smaller fraction (12%) of race-responsive neurons in the hippocampus predicted only White faces. Our results imply a distributed representation of race in brain areas involved in affective judgments, decision making, and memory. They also support the hypothesis that race-specific cues are perceptually coded even when those cues are task-irrelevant.

Keywords: human single neuron; hippocampus; amygdala; human race perception

Citation: Valdez, A.B.; Papesh, M.H.; Treiman, D.M.; Goldinger, S.D.; Steinmetz, P.N. Encoding of Race Categories by Single Neurons in the Human Brain. *NeuroSci* **2022**, 3, 419–439. https://doi.org/10.3390/neurosci3030031

Academic Editor: Szczepan Paszkiel

Received: 31 May 2022
Accepted: 19 July 2022
Published: 5 August 2022

Publisher's Note: MDPI stays neutral with regard to jurisdictional claims in published maps and institutional affiliations.

1. Introduction

Group membership can be based on many attributes, including race and affects brain activity involved in different types of perception (e.g., related to words, actions, and faces [1])(There is a current debate over whether "race" or "ethnicity" should be used to refer to the phenotypic characteristics that are varied in this experiment. We use the term "race" to refer to such differences, remaining consistent with much of the previous neuroimaging literature.). Multiple studies in neuroscience have shown that people process cues about in-group vs out-group members differently [2–5], which may correlate with biased perceptions, attitudes, and behaviors.

Studies using functional magnetic resonance imaging (fMRI) have shown that brain areas involved in face processing produce less hemodynamic activation to out-group faces [6], while the amygdala (which is involved in the perception of face gaze, affect, and familiarity [7–13]) produces early [14] and sustained [15] hemodynamic changes in response to out-group faces. Other brain areas, like the prefrontal cortex, play a role in accessing social knowledge [16] and may regulate race evaluation [14]. For example, there is increased frontal activity for remembered in-group faces early during memory retrieval [17]. Functional imaging has further shown that activity in the anterior cingulate cortex (ACC), which is involved in decision making and affect, tracks out-group versus in-group trust [18], reflects the observation of in-group non-verbal actions [19], and varies when a person observes painful stimulation applied to in-group versus out-group people [20].

Imaging studies have thus provided evidence, via changes in regional blood flow, regarding the neural correlates of racial processing in face perception. Collectively, these studies tend to show different hemodynamic responses (whether excitatory or inhibitory) to racial in-group versus out-group people across tasks (e.g., face recognition and classification, observation of social interactions, and observation of applied stimulation to faces). By contrast, the present investigation tested whether single-neuron firing rates are sensitive to variations in race features during a face classification procedure. Several previous single-neuron recording studies have focused on the processing of emotion in facial expressions [21–24] and gender and affect [25]. However, due to the scarcity of unit-level studies of racial and other social cues [21,22] and inconsistencies in fMRI findings e.g., out-group-elicited bilateral amygdala activity in some cases [15,26] and right-lateralized activity in others [14], we sought to gather more direct evidence of neural processing in race perception.

The present study was designed as an initial exploratory study of how the activity of single neurons in the human brain change their firing rate depending on unattended perceptual cues of race. In this study, subjects viewed realistic synthetic faces which were varied in both race and the emotion depicted. Subjects classified the faces as having a positive or negative expression and variations in racial features were irrelevant to the patients' task and were orthogonal to the depicted expressions. The volunteers were patients with epilepsy, undergoing clinical microwire recording to locate and surgically assess seizure onset zones. In these patients, we recorded neurons in the amygdala, ACC, hippocampus, and ventromedial prefrontal cortex (vmPFC). How neural firing rates in this experiment reflect the gender of the participants has previously been reported [27]. This paper reports how the firing rates depend on the race of the presented face.

Given the hypothesis that race is perceptually processed as an automatic, categorical response [28], we predicted that we would observe race-selective neural firing, despite attention being oriented toward an orthogonal stimulus dimension. In addition, we examined whether neurons in these brain areas would show in-group or out-group selectivity, as observed in fMRI, which could suggest a potential correlation between unit-level coding and responses in regional blood flow [25,29,30].

To anticipate our results, we indeed observed race-signaling neurons in the hippocampus, ACC, and amygdala, but not the vmPFC. Thus, despite race variations being task-irrelevant, notable proportions of recorded neurons responded in a race-selective manner. Regression models showed that neural firing rates in the amygdala and ACC predicted race categories in a distributed fashion, i.e., with apparently broad tuning for several race categories and low lifetime sparsity, or a large proportion of stimuli eliciting a significant response across the brain areas [31]. In addition, nearly equal proportions of neurons in the amygdala and ACC responded to white and black faces, in contrast to the outgroup-specific responses in these brain areas reported in the fMRI literature. Our results suggest that neurons were broadly sensitive to perceptual features indicating race, rather than being highly selective for either in-groups or out-groups.

2. Method

Participants and Exclusions. We recorded single-neuron activity from 14 patients at the Barrow Neurological Institute. The original sample included 10 Caucasian White, two Caucasian Hispanic, and two African-American Black patients (see further details in Appendix A). All patients had drug-resistant epilepsy and were being evaluated via microwire recording for possible resection of epileptogenic foci. Each patient consented both verbally and in writing to participate per a protocol approved by the Institutional Review Board of Saint Joseph's Hospital and Medical Center in October 2005, August 2007, May 2009, November 2010, October 2012, and February 2013. The patients were invited to complete multiple experimental sessions. Since microwires shift slightly as patients spend time in the hospital, repeated sessions were treated as independent samples (i.e., recordings from different neurons). The number of patients was determined by clinical scheduling and recordings were continued until more than 400 single unit recordings had been obtained.

During the 7–10 days that patients were in the Epilepsy Monitoring Unit and performing these experiments, they were gradually tapered off their anti-epileptic medications in order to increase the probability of them having seizures. It is thus possible that these medications may have affected their cognitive performance on the experimental tasks.

In cognitive experiments on race perception, researchers typically strive to recruit near-equal participant groups matching the racial backgrounds depicted in the stimulus faces (e.g., [32]). Given our patient population, such balanced sampling was not practical: the Barrow Neurological Institute (BNI) only evaluates 8–10 microwire patients per year and fewer than 6% of Arizona residents are African-American, making it nearly impossible to collect a robust African-American sample. To guide our approach, we conducted initial analyses, comparing White and Black participants with respect to fractions of neurons per brain area that differentially responded, based on stimulus race. This analysis involved first performing one-way ANOVAs for an effect of stimulus race on the response of individual neurons. Next, the fractions of neurons in each brain area with a significant response were compared between subject races using Fisher's Exact Test for fractions [33]. This test showed no significant interaction between subject and stimulus races, perhaps reflecting the small African-American sample. For maximum clarity, data analyses reported in the main text exclude the two African-American patients, allowing a clear division of "in-group" and "out-group" stimulus faces for the remaining participants. Results including all participants appear in Appendix B and were quite similar to the results presented in the main text. Finally, we excluded data from one patient, and two sessions from other patients, because of technical errors. Following exclusions, the sample size for all analyses was 11 patients, although our analyses treat recorded neurons as the fundamental unit of analysis. In an earlier report, Newhoff et al. [27] published results based on these data (focusing on facial affect), without excluding any patients or sessions. Our current focus on race perception required more stringent inclusion criteria.

Microwire Bundles and Recording System. Extracellular action potentials corresponding to single-neuron activity were recorded from microwire tips implanted bilaterally along with clinical depth electrodes used to record clinical field potentials [34,35]. The implantation sites were chosen according to clinical criteria, which limits the potential recording sites. For these patients, and per standard clinical practice at the BNI, the sites included the bilateral hippocampus, prefrontal cortex, anterior cingulate cortex, and amygdala. In the hippocampus, the wires were targeted to the midbody of the hippocampus, just behind the head of the hippocampus, opposite the apex of the cerebral peduncle. In the prefrontal cortex, the wires were targeted to the ventromedial prefrontal cortex, below the anterior cingulate gyrus. In the anterior cingulate, the wires were targeted to parts above and behind the genu of the corpus callosum. In the amygdala, the wires were targeted to be in the center of that structure.

Each recording site received a bundle of nine 38 µm-diameter platinum-iridium microwires (California Fine Wire, Grover Beach, CA, USA), implanted stereotactically (Medtronic StealthStation, Minneapolis, MN, USA) using a 1.5-T structural MRI. Each microwire typically had an impedance of 300 kΩ at 1000 Hz. Electrodes were placed through a skull bolt with a custom frame to align the depth electrode along the chosen trajectory. The error in tip placement using this technique is estimated to be ±3 mm based on manual inspection of the pre-operative MRI and post-operative CT and prior work [36,37]. While this technique, lacking co-registration of a post-operative CT with the preoperative targets, cannot guarantee placement within the targeted brain structures, the majority of microwire bundles are estimated to be in the targeted structures. We, therefore, refer to the position of the microwires as within their targeted structures throughout the results presented here; we address this limitation in the General Discussion.

The system used to record extracellular potentials from microwires has been previously described [38]. In brief, each bundle of nine microwires was attached to a custom-designed head-stage amplifier which provided preconditioning with a high pass filter (0.5 Hz), followed by a 10 kHz antialiasing filter and a computer-controlled 1–16× adjustable gain amplifier of the difference between eight of the microwires and a 9th microwire used as reference. The head-stage amplifiers were connected to custom-designed signal conditioning electronics and Data Translation A/D convertors (DT9834) digitized the signal at a sampling rate of 29,412 Hz. The digitized signals were recorded on a laptop computer (IBM Thinkpad, Armonk, NY, USA) for subsequent analysis using a custom C++ program.

Stimuli and Validation. Participants were asked to quickly classify the emotional expressions of synthetic male faces, created using FaceGen Modeler [39], a computer program that generates realistic synthetic faces from imported photographs. We selected original photographs from PAL and Colour FERET databases [40,41]. From a set of 20 White male faces, we generated computerized models equated for size and brightness, all with identical white backgrounds. Once a face model was created, it could be manipulated for emotional expression (using up to 40 parameters that modify the face asymmetrically) and morphed to create racial continua from White to Black. In this manner, we initially created 20 sets of 12 faces; an example set is shown in Figure 1. For each face, we first created subjectively positive (happy), negative (angry), and neutral versions, using different parameter combinations per set to allow for natural variation. After facial expressions were set, those relevant parameter values were frozen, and faces were morphed in four steps from "clearly White" to "clearly Black" (with "ambiguous" values in-between). For this initial study, the morphing process combined changes in skin tone and facial morphology, consistent with different ethnic backgrounds. Thus, this study cannot separate the effects of skin tone and facial morphology on the neural responses. For each affective expression, the same parameter values were selected to represent skin tone and morphological features. In total, participants viewed 120 faces per session (60 ambiguous, 30 Black, 30 White). We created two sets of 120 faces which could be used in separate sessions.

In the interest of clarity, we apply different terminology when referring to human participants versus computerized faces and response labels. When referring to experimental volunteers, we use the terms Caucasian and African-American. When referring to computerized faces or response options, we use the labels White and Black.

Since we created faces using subjective judgment, we validated the affect and race dimensions in two pilot experiments with Arizona State University students as volunteers. All self-reported having a normal or corrected-to-normal vision. In the first validation test, participants (n = 25) were shown all faces in random order, quickly classifying each as "positive" or "negative" in facial expression, with no option to respond "neutral." Variations in the apparent race were included but were orthogonal to the task. "Positive" faces were classified as positive in 93.2% of trials (mean correct RT = 515 ms). "Negative" faces were classified as negative in 98.1% of trials (mean correct RT = 499 ms). "Neutral" faces were classified as positive in 41.5% of trials (mean RT = 688 ms) and as negative in 58.5% of trials (mean RT = 601 ms). The results verified that positive and negative expressions were classified quickly and accurately, and that neutral faces were more challenging. There was also a slight bias toward interpreting neutral expressions as being negative.

To validate the race dimension, we asked 33 undergraduate volunteers to quickly classify each stimulus face as "White" or "Black," with no option to respond "ambiguous." Participants were instructed to ignore variations in affective expressions. Classification agreement was high, and RTs were short when the non-ambiguous faces were classified: White faces were correctly classified in 98.9% of trials (RT = 544 ms) and Black faces were correctly classified in 97.8% of trials (RT = 518 ms). Among the intermediate items, ambiguous White faces were predominantly classified as White (73.4% of trials; RT = 649 ms) but were occasionally classified as Black (26.3% of trials; RT = 712 ms). The ambiguous Black faces were predominantly classified as Black (78.3% of trials; RT = 610 ms) but were occasionally classified as White (21.7% of trials; RT = 709 ms). Thus, classification decisions generally followed the intended categories, with slower RTs to ambiguous faces.

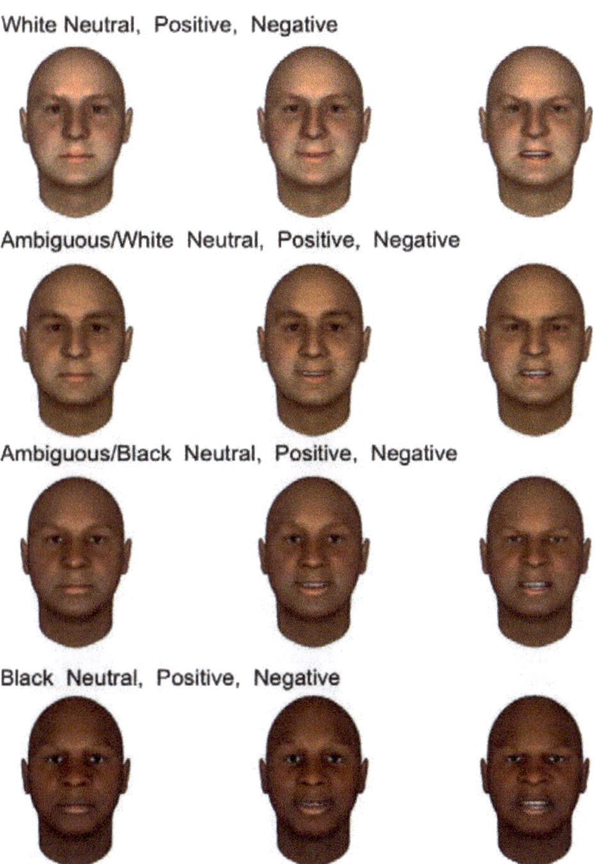

Figure 1. An example set of 12 FaceGen faces created from an original photograph. From left to right, columns show faces with neutral, positive, and negative expressions. With affect parameters frozen, faces were morphed along the race dimension, which involved simultaneous changes in both skin tone and facial morphology. From top to bottom, faces ranged from White to Black, with intermediate ambiguous versions.

Experimental Procedure. In testing sessions with the hospital patients, face images were presented using a laptop computer (Apple Computer, Cupertino, CA, USA) with a 15-inch screen. Each patient was seated in a hospital bed, with the laptop placed on an overbed table. Face images occupied a square of 0.1 m on each side. At a distance of ~0.5 m, this area subtends ~11° of visual arc. Responses were collected from a trackball with large buttons (ExpertMouse, Kensington, Redwood Shores, CA, USA) to increase participant comfort and provide electrical isolation from the laptop switching power supply. As shown in Figure 2, each face was shown for one second, including the relevant analysis period (200–1000 ms after stimulus onset). Once the face disappeared, participants had two seconds to classify its affect. Stimulus presentation, response collection, and synchronization with the neural recordings were performed using a JAVA (Sun Microsystems, Santa Clara, CA, USA) program developed by our laboratory. Each face was presented six times, for a total of 720 trials.

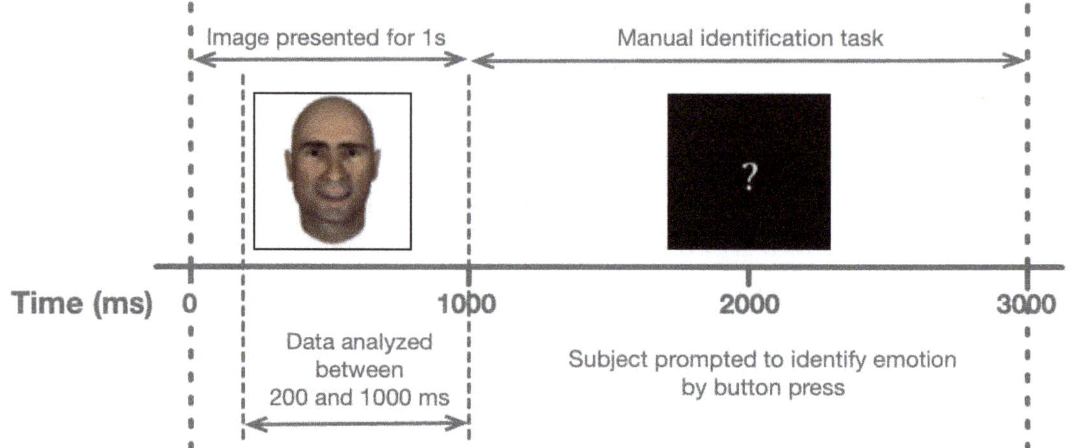

Figure 2. Schematic trial structure. The horizontal axis represents the passage of time in milliseconds. Dashed lines indicate key stages per trial and labeled arrow bars distinguish between the recording and behavioral segments. From Newhoff et al., 2015, Figure 2.

Spike Sorting and Neuronal Responses. In total, we recorded from 1024 channels in 25 experimental sessions. None of the sessions were recorded while patients experienced seizures. Analyzed recording channels included those from brain areas subsequently identified as having clinical seizure foci, as well as those without. We used established spike-sorting methods [38] to isolate single-neuron activity in each recording. In brief, possible action potentials (events) were detected by filtering with a bandpass filter, 300–3000 Hz, followed by a two-sided threshold detector (threshold = 2.8 times each channel's standard deviation) to identify event times. Event waveform shapes surrounding each event were extracted with the event time aligned at the 9th of 32 samples. All events captured from a particular channel were separated into groups of similar waveform shapes (clusters) using the open-source clustering program, KlustaKwik (http://klustakwik.sourceforge.net, accessed on 30 August 2012), which applies a modified version of the Govaert-Celeux expectation maximization algorithm [42,43]. The first principal component of all event shapes recorded from a channel was the waveform feature used for sorting. After sorting, each cluster was classified as being noise, multi-unit activity (MUA), or single-unit activity (SUA), using the criteria described in (Valdez et al. [44], Table 2). From 1024 recorded channels, we isolated 881 clusters of SUA. There were 625 clusters of SUA in the non-excluded sessions for the primary results reported here. In agreement with prior studies [27,45,46], we used the number of sorted action potentials in an interval spanning 200–1000 ms after stimulus onset as our dependent measure of neuronal responses to each stimulus. Per consensus in the field, we considered recordings from the same channel in separate sessions to be independent observations. To further check that this treatment is correct, we identified all clusters of SUA recorded on the same recording channel in multiple sessions which had average firing rates differing by less than 50% between sessions and waveform shapes that retained the same sign (positive or negative peak) between sessions. There were 19 such clusters in total. After eliminating these from the dataset, we obtained the same results as reported here with minor variations.

Multinomial Logistic Regression. In order to determine whether neuronal firing in the response interval (200–1000 ms after stimulus onset) changed from background firing in a manner that indicates the race of face presented (stimulus race), we used logistic regression with the ratio of the response firing rate on each trial compared to the firing in a background interval of the same length prior to stimulus onset (1000–200 ms before stimulus onset; [44,47,48]). Our sole predictor (the number of action potentials within the

200–1000 ms window) was assigned a coefficient, indicating the degree of the predictor's correlation with the predicted outcomes. Statistically significant changes in coefficient values from 0 (i.e., excitatory or inhibitory differences in firing rates, measured by t-tests for nonzero coefficients) could signal different odds of any race category being present, versus a reference level (background firing rate), where odds ratios reflect the probability of a category is present, given a unit change in the predictor [49]. This is a simplified version of the point-process framework that Truccolo et al. [50] proposed.

After determining the coefficients for neurons, we compared the numbers of neurons in each brain area that significantly predicted race categories to those expected by chance under a binomial distribution, using a Benjamini–Hochberg adjustment for false discovery rate [51]. All statistical tests used a significance criterion of 0.05 and were performed using the R statistical package [52].

Changes from Background Firing. The multinomial logistic regression tests inherently compare the response to background firing; however, to increase confidence that the observed neuronal responses reflected stimulus-driven activity, rather than noise, we also used a hypothesis test to determine how probable the observed responses would be under a combined null hypothesis. This null hypothesis was that firing rates were equivalent for all race categories (equivalent to a one-way ANOVA) and were also equal to their respective background firing rates [53]. We applied a resampling technique [54] wherein we computed the empirical distribution of a test statistic equal to the sum of the squared deviations of the mean responses for each race category from the mean background firing rate. We then computed the probability of obtaining the observed test values. This test indicates how unlikely it would be to observe the responses, divided among race categories if neural firing rates were insensitive to such information.

Information To Decode Race Categories. In order to further examine the ability to predict stimulus race based on neural firing, we computed the mutual information between neural response counts (in the 200–1000 ms interval) and stimulus races for all trials, in each brain region [52, infotheo package, mutinformation function using the entropy measure].

3. Results

The primary questions we sought to answer in this initial study were (1) whether individual neurons would preferentially respond to certain race categories, (2) whether such responding would vary according to brain regions, (3) whether stronger responses would be observed for either black or white faces, and (4) whether neurons would appear coded for specific races or would be more generally sensitive to changes across trials.

Neuronal Firing Rates Predict Race Categories. We first applied multinomial logistic regression models to predict, based on neural firing rates, when patients were viewing specific race categories. In this primary analysis, we report results only from single-unit activity (SUA), or well-isolated neurons, bilaterally by brain area among 11 Caucasian participants. Appendix B shows results across all enrolled volunteers (including two African-American patients) and also results from multi-unit activity (MUA).

Figure 3 shows the percentage of neurons in each brain area that significantly predicted each race category. These results are derived across all recordings, calculated using the number of well-isolated neurons in each brain structure as denominators, thus no error bars appear in Figure 3. In the amygdala, ACC, and hippocampus, there were significant numbers of neurons that predicted the depicted race categories (binomial test, $p < 0.05$). In the vmPFC, the percentage did not reliably differ from zero ($p > 0.05$). The percentages for each brain area and racial grouping are listed in Table 1.

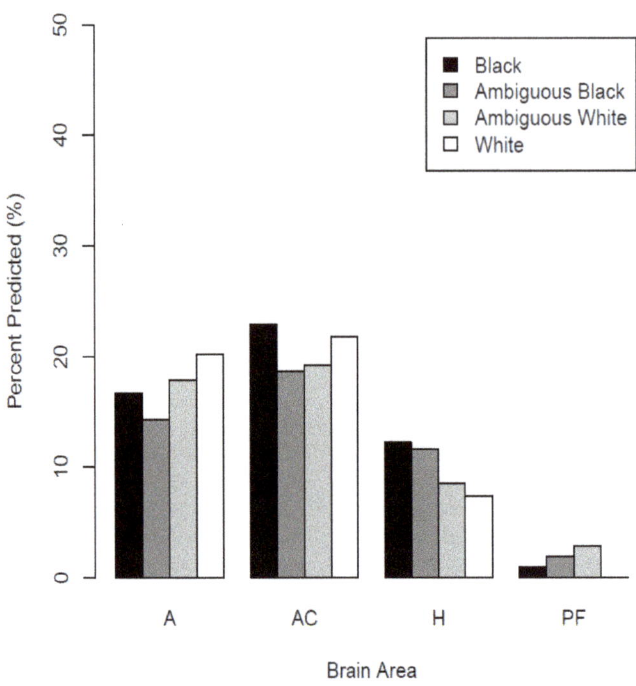

Figure 3. Percentages of neurons (SUA) significant for a test of the effect of stimulus race category on firing rate (relative to background firing) by brain area. The *x*-axis shows brain areas. The total neurons recorded were: Amygdala (n = 168), ACC (n = 188), Hippocampus (n = 164), and the percentages stated are the corresponding percentages of these totals in each brain area. Note that there was no significant percentage of race-predictive neurons in the vmPFC out of the total (n = 105).

Table 1. Percentages of neurons with a significant response to stimulus race category (relative to background firing rates) for each brain area. Asterisks denote significant proportions ($p < 0.05$). Percentages are the number of neurons with a significant test in each area divided by the total number in the area multiplied by 100.

	Number of Neurons	Black	Ambiguous Black	Ambiguous White	White
Amygdala	168	28 (17%) *	24 (14%) *	30 (18%) *	34 (20%) *
ACC	188	43 (23%) *	35 (19%) *	36 (19%) *	42 (22%) *
Hippocampus	164	20 (12%) *	19 (12%) *	14 (9%)	12 (7%)
vmPFC	105	1 (1%)	2 (2%)	3 (3%)	0 (0%)

When doing so, there were no apparent differences in the percentages of neurons predicting racial category in the amygdala (24% vs. 22%) or ACC (26% vs. 28%). However, there was a more notable difference in the percentages of hippocampal neurons predicting racial category (12% vs. 19%).

To illustrate the magnitude of a significant response to one race category, relative to background firing rate, Figure 4 shows an elevated median response from one left-amygdala neuron in response to Black and Ambiguous Black faces. Overall, while response magnitudes were low in terms of absolute spikes per trial, they were comparable to prior human single-neuron studies [27,44,45,55].

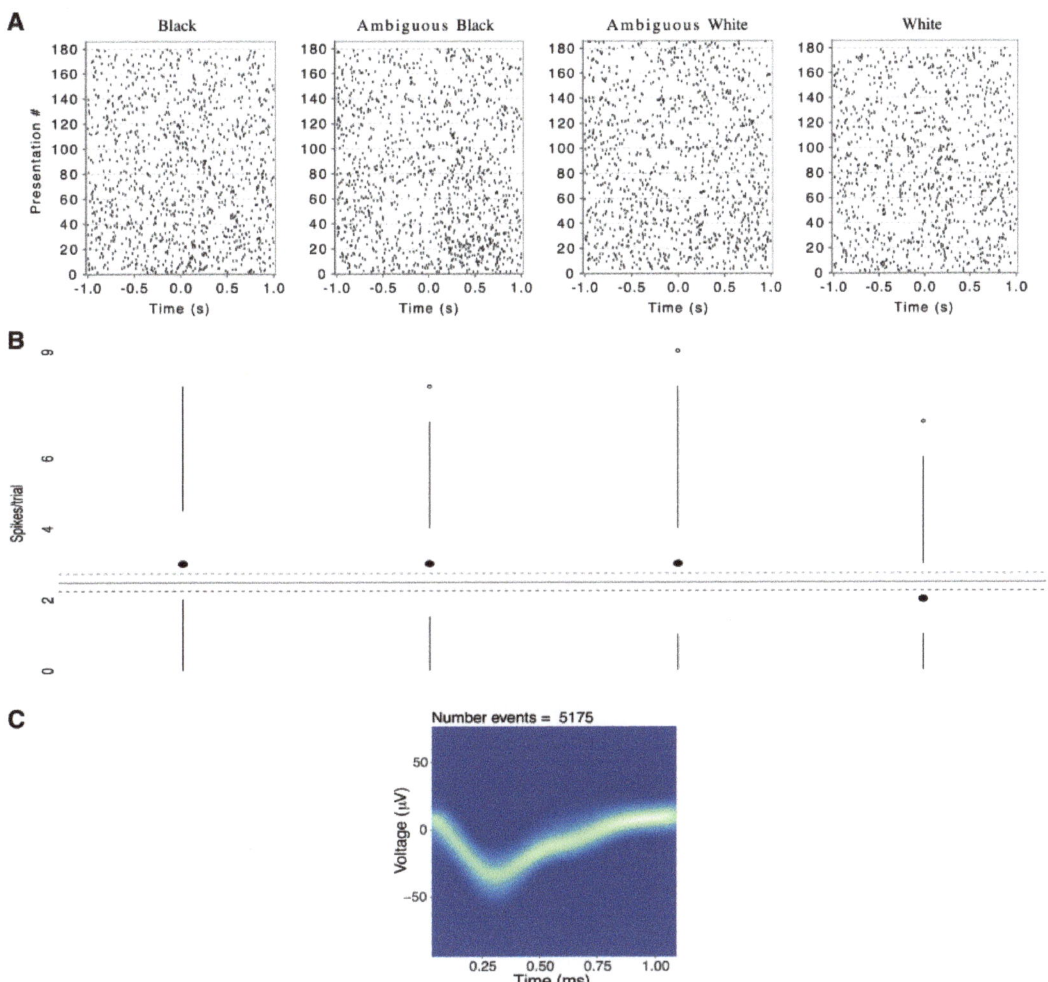

Figure 4. Responses of a single neuron in the left amygdala of one subject, predictive of the black and ambiguous black stimulus categories. (**A**) Raster plots of firing for each presentation of a face from each race category. Time 0 is the onset of stimulus presentation. (**B**) Modified boxplot of firing rates for all presentations of a face from each race category. The center dots show the median response per race category. Vertical lines extend from $\pm(1.58 \ast \mathrm{IQR})/\mathrm{sqrt}(n)$, where IQR = interquartile range, n = number of observations. This is equivalent to a 95% confidence interval for differences between medians. Small circles show responses outside that range. The solid grey line shows the mean of background firing; dashed grey lines represent a 95% confidence interval for the median of background firing. (**C**) Density plot of the extracellular potential waveforms for each putative action potential from this neuron. Threshold crossing at 0.27 ms.

Changes from Background Test (CBT). As illustrated by one neuron in Figure 4, even when neural firing predicted race categories, the response differences were often subtle. This may reflect our experimental procedure, which required focused attention to emotional expressions, whereas race varied as an irrelevant background dimension. To further test that the observed responses were not due to random fluctuations of background activity, we used a bootstrapping procedure as described in Methods [53].

Table 2 shows there were significant differences between responses and background counts in more neurons than expected by chance ($p < 0.05$) in all brain areas. When these significant neural responses were further subdivided by race categories, the results were virtually identical to those in Figure 3, suggesting that the logistic regression and bootstrapping analyses produced a very high agreement.

Table 2. Neuronal responses per brain area that differed from background firing.

Area	Total # of Neurons	# Significant	Binomial *p*-Value	Adjusted *p*-Value
A	168	85 (34%)	8.9×10^{-47}	1.6×10^{-35}
ACC	188	74 (39%)	5.3×10^{-46}	2.1×10^{-45}
H	164	44 (27%)	2.6×10^{-20}	3.5×10^{-20}
vmPFC	105	16 (15%)	6.8×10^{-5}	6.8×10^{-5}

A = amygdala, ACC = anterior cingulate cortex, H = hippocampus, vmPFC = ventromedial prefrontal cortex.

Number of Race Categories Encoded. To expand upon the results in Figure 3, we next tested whether race-responsive neurons were selective (e.g., preferentially responding to one specific race) or were more broadly tuned. For each brain area, we counted neurons that significantly coded zero, one, two, three, or four race categories in the multinomial logistic regression analysis. Coding four race categories, for example, suggests general responsiveness to all presented images. The results are shown in Figure 5. In the amygdala, 5% of neurons encoded a single race category, while 18% of neurons encoded two or more categories. In the ACC, 12% of neurons encoded a single category, while 19% encoded two or more categories. In the hippocampus, 9% encoded a single category, while 12% encoded two or more categories. As shown, roughly 60% of all recorded neurons showed no reliable firing-rate changes based on stimulus race.

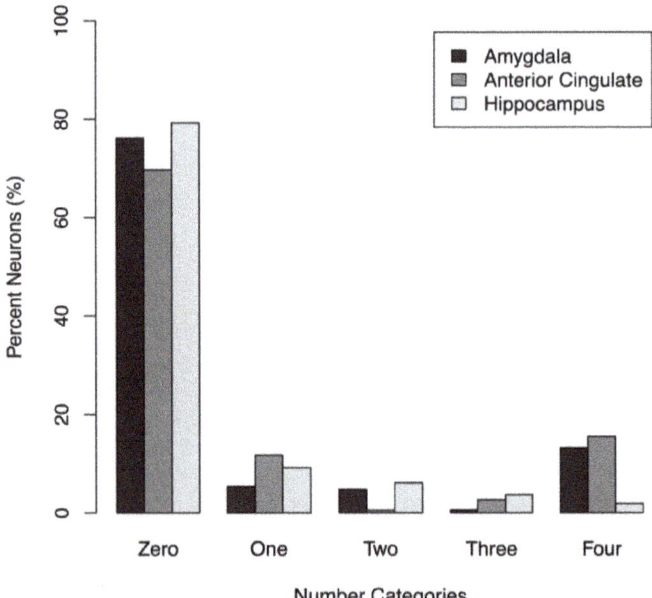

Figure 5. Percentages of neurons that significantly coded different numbers of race categories. The *x*-axis shows the number of race categories observed per neuron; bar shading indicates brain areas. Amygdala (*n* = 168), ACC (*n* = 188), Hippocampus (*n* = 164). There were no significant proportions of race-predictive neurons in the vmPFC (*n* = 105).

Calculation of information to decode race categories. As shown in Figure 5, neural responses to race categories were rarely one-to-one, we found robust differences in the average amounts of available information across brain areas (ANOVA, $p = 0.007$). Table 3 shows the amount of information available per brain area, along with the neurons per area that would be required to distinguish among all four categories (i.e., two bits of information). As shown, slightly fewer neurons are required in the amygdala (relative to other brain areas) to successfully decode stimulus races, whereas considerably more neurons are required in the ventromedial prefrontal cortex. The latter result is not surprising, as the vmPFC showed relatively few race-sensitive neurons.

Table 3. Average information for decoding race in single-neuron firing in each brain area.

Brain Area	Mutual Information (Bits)	# Neurons to Decode 2 Bits
A	0.022	90.1
ACC	0.021	94.4
H	0.021	94.5
vmPFC	0.156	128.0

A = amygdala, ACC = anterior cingulate cortex, H = hippocampus, vmPFC = ventromedial prefrontal cortex.

As previously reported, there were robust effects of the depicted emotions [27], which directly occupied participants' attention.

4. Discussion

The present study examined single-neuron responses during perception of Black and White faces, collected during a task wherein facial emotions were classified, and the race was varied as an orthogonal, irrelevant dimension. Despite the attention being directed toward facial expressions, we observed race-selective neural firing rates in the amygdala, ACC, and hippocampus. Regression models showed notable percentages of neurons encoded multiple race categories in the amygdala and ACC, but not the hippocampus. Separate tests verified that race-predictive neurons had firing rates that reliably differed from background firing upon stimulus presentation. The tuning of race-predictive neurons in the amygdala and ACC appears rather broad, characterized by responses to one or more race categories, with low lifetime sparsity [56]. This type of sparsity is defined as the fraction of stimuli that evoke statistically reliable neuronal responses [31]. Although our main analyses included only Caucasian participants viewing White and Black faces, the results imply a distributed neural representation for racial categories. This is generally consistent with Valdez et al.'s study [44] showing distributed coding (with similarly broad tuning and low lifetime sparsity) for stimuli from several categories, including faces, during a similar visual discrimination task. Tuning in the hippocampus showed lower lifetime sparsity, with a greater fraction of neurons responding to zero or just one race category.

Our results showed that similar percentages of neurons in the amygdala and ACC predicted the presence of racial groups, a result at odds with prior reports in the imaging literature. fMRI studies have suggested different hemodynamic responses when people perceive in-group vs out-group members [57–60]. In the amygdala, for example, there is greater activation when subjects briefly view racial out-group faces, while in the prefrontal cortex, there is greater activation when the same subjects view racial in-group faces [14]. Racial out-group-related increases in amygdala activity also occur when subjects perform social categorization based on perceived face ages [61]. Such differential activation regarding racial in-group versus out-group also occurs in the ACC across various tasks [1].

By contrast, we did not find neurons in the amygdala or ACC that were disproportionately selective for either racial group presented. This seems to indicate a lack of correlation between single-neuron coding and race-driven changes in regional blood flow. Such a disparity may arise because we did not record local field potentials (LFPs), a peri-synaptic activity thought to be comprised of excitatory and inhibitory postsynaptic potentials and dendritic afterhyperpolarizations, which have a known correlation with blood oxygen level-dependent (BOLD) signal changes [62]. The disparity could also reflect the narrow spatial focus of a human single-neuron recording or it could confirm a reported dissociation between BOLD signals, spiking activity, and LFPs [63]. Task differences may also help explain the different neural responses. Cunningham et al. [14] observed race-selective BOLD differences in the amygdala only when faces were shown very briefly (30 ms), not with longer exposures (525 ms). In the current study, faces were shown for 1000 ms, well beyond the race-selective "window" suggested by the Cunningham et al. results. Additionally, neural responses to race features were expected to be somewhat attenuated in the present study, as attention was task-oriented toward the orthogonal emotional dimension. Clearly, more data from these different methodologies would allow stronger conclusions about neural responses to racial features, as well as verifying the out-group selectivity we observed in hippocampal neurons.

The patients in this experiment have epilepsy, which may broadly affect their cognition. Additionally, we specifically recorded brain areas that were identified as potentially containing seizure onset zones. Although prior research suggests that epilepsy does not affect neural firing patterns in response to visual stimuli [64], we re-analyzed our recordings for responses to racial categories, excluding any areas that were identified as potential seizure onset zones. The results, presented in Appendix B, confirmed the existence of broadly distributed race representation, as in Figure 3 and Table 1, and also showed a significant fraction of neurons responding to the ambiguous in-group in the hippocampus. As noted earlier, our participants were limited to patients receiving microwire evaluation, preventing us from achieving a balanced design with respect to the patients' races. With only two African-American participants, we could not properly test a full 2 × 2 design, crossing subject, and stimulus races. We, therefore, excluded those two participants in the results presented above but did analyze their results, as summarized in Appendix B. The analyses showed no effects related to the participant's race. Future studies to examine this question in greater detail could include a larger number of African-American participants or could analyze data averaged over many neurons.

Another future question building on these results is the relative contributions of skin tone versus facial morphology to the changes in neural firing rates reported here. The stimuli used combined changes in skin tone and facial morphology when varying race. Since we previously described changes in firing rates in these brain areas depending on the brightness and contrast of images [65], it would be informative to vary skin tone and facial morphology and determine how each contributes to variations of firing rate with race. In the current study, however, we expected neural responses to racial variations to be relatively attenuated, as attention was selectively directed to facial affect. Given prior research in face perception [28,66,67], it is likely that skin color is the most diagnostic stimulus dimension during rapid face classification. For our purposes, spiking rates suggested that MTL neurons track humans' implicit classification of faces by race, despite such variations being task irrelevant. Even if the neural responses we observed were driven by skin color variations, they illustrate how the brain responds to socially relevant facial features, regardless of selective attention being focused elsewhere (see also [68] for classic research on dimensional interactions during stimulus processing).

Since the recording methods employed in this study did not allow the determination of sub-nuclei within the various brain areas, future studies could employ more accurate localization to discern potential functional differences within separate nuclei of the amygdala. For example, such a study could examine differences in neuronal responses in the basolateral and lateral amygdala when processing fearful and reward-related stimuli [69].

Although our results did not reveal strong selectivity for either racial group in brain areas critical for social judgments and decision making, they do constitute further evidence for neural representations of race that carry several implications. First, the results support the hypothesis that race is processed as a basic perceptual feature [67], triggering robust neural responses even when attention is focused on an orthogonal stimulus dimension (in this case, facial affect). As noted, prominent theories of the other-race effect posit that the human brain classifies racial outgroup members categorically, only further processing them as individuals with extended effort [28]. Such differences in immediate perception are not universal, for example being absent when people perform match-to-sample tasks with in-group and out-group faces [70,71]. Notably, the match-to-sample task requires focused attention on precise facial features. When task performance does not demand face individuation, neural responses (such as repetition suppression in the FFA) suggest more categorical processing for out-group faces. Second, regarding specific brain structures, neural firing in the amygdala links stimulus perception to judgments [72,73]. As such, race responsiveness may represent a correlate of increased vigilance [74] (e.g., in dorsal subregions) and is evidence of sustained stimulus evaluation [75]. Race-selective neural firing in the ACC may signal preconscious race classification [18,19]. Activity in the ACC also likely reflected subjects' sustained error monitoring during the experimental task.

The present study suggests that the perception of racial features has an identifiable and mostly distributed (rather than highly selective) neural basis. Other interesting lines of future research include whether race-specific neural firing changes after perceptual training [76], when faces are known individuals [77], or by examining perception in cultures that place less emphasis on dichotomized racial boundaries [78].

Author Contributions: A.B.V. conducted experiments, analyzed data, and wrote the paper. M.H.P. conducted validation experiments and edited the paper. P.N.S. designed the experiments, conducted them, supervised data analysis, and wrote the paper. S.D.G. designed experiments, designed stimuli and procedures, and edited the paper. D.M.T. supervised clinical work and edited the paper. All authors have read and agreed to the published version of the manuscript.

Funding: This research was funded by NIH Grant 1R21DC009871-0, the Barrow Neurological Foundation, the Arizona Biomedical Research Commission #09084092, and the Veritas Fund.

Institutional Review Board Statement: This research protocol was approved by the Institutional Review Board of Saint Joseph's Hospital and Medical Center in October 2005, August 2007, May 2009, November 2010, October 2012, and February 2013.

Informed Consent Statement: Informed consent was obtained from all subjects involved in the study.

Data Availability Statement: The data underlying this report are available at the Open Science Foundation at https://osf.io/nf7s8/ (accessed on 29 May 2022) as well as by request to info@neurtex.org (accessed on 29 May 2022).

Acknowledgments: We thank the patients at the Barrow Neurological Institute who volunteered for these experiments, Kris A. Smith for performing neurosurgery, and E. Cabrales for technical assistance.

Conflicts of Interest: The authors declare no conflict of interest.

Significance Statement: Identifying the members of one's social group is critical to social interactions. Several brain areas, such as the amygdala, are critical to this identification. Prior imaging studies have examined the hemodynamic correlates of race discriminations. Here we examined the activity of single neurons in the brains of epilepsy patients to determine how neurons fire as human subjects view face images belonging to different races. We report that neuronal firing rates in the amygdala, hippocampus, and prefrontal cortex all predicted race categories. These results suggest that race-selective neurons are broadly represented across brain regions and that such neurons respond to race-specific cues even when attention is directed toward different stimulus dimensions.

Appendix A. Patient Details

Table A1. Characteristics and clinically identified seizure foci for all subjects (*n* = 14). Note the three subjects who were excluded from the primary analysis.

Age	Sex	Handedness	Ethnicity/Race	Brain Areas with Seizure Foci	Included in Primary Analysis
38	Female	Right	Caucasian White	Left amygdala, left hippocampus	No
54	Female	Right	Caucasian White	Left amygdala, right amygdala, right hippocampus	Yes
42	Male	Right	Caucasian White	Left amygdala, left hippocampus	Yes
35	Male	Left	Hispanic White	Right amygdala, right hippocampus	Yes
41	Male	Right	African-American Black	Left hippocampus, left frontal lobe	No
21	Female	Right	Caucasian White	Left amygdala, left hippocampus	Yes
24	Female	Right	Caucasian White	Right amygdala, right hippocampus, right frontal lobe	Yes
23	Male	Right	Caucasian White	Right amygdala, right hippocampus	Yes
54	Female	Right	Caucasian White	Left hippocampus	Yes
56	Female	Right	Caucasian White	Right occipital lobe	Yes
40	Female	Right	Hispanic White	Right amygdala, right hippocampus	Yes
54	Female	Right	Caucasian White	Right hippocampus	Yes
20	Male	Right	Caucasian White	Right temporal lobe	Yes
53	Female	Right	African-American Black	Right amygdala, right hippocampus, left amygdala, left hippocampus	No

Appendix B. Extended Analyses

Appendix B contains tables and statistical notations excluding identified seizure onset zones, and analyses covering the entire sample excluding one patient whose session contained a technical error (the first patient in Table A1). For completeness, we also report tables for multi-unit analyses (MUA), which largely mirror those for the single-units (SUA) reported in the main text.

Table A2. Percentages of neurons significant for race category (relative to background firing) for each brain area, excluding all recordings from identified seizure onset zones. Values with asterisks denote significant proportions in binomial tests (*p* < 0.05), after correction for multiple comparisons.

	# Neurons	Black	Ambiguous Black	Ambiguous White	White
Amygdala	101	26 (26%) *	22 (22%) *	28 (27%) *	30 (30%) *
ACC	188	43 (23%) *	35 (19%) *	36 (19%) *	41 (22%) *
Hippocampus	84	14 (17%) *	12 (14%) *	9 (11%) *	7 (8%)
vmPFC	105	1 (1%)	2 (2%)	3 (3%)	0 (0%)

Table A3. Neurons significant for specific race categories (relative to background firing) by side and brain area (SUA), across all patients ($n = 13$), as indicated by multinomial logistic regression models. Values with asterisks denote significant proportions in binomial tests ($p < 0.05$), after correction for multiple comparisons.

Area	Total # of Neurons	Sig. for White	Sig. for Black	Sig. for Ambiguous White	Sig. for Ambiguous Black
LA	128	30 (23%) *	27 (21%) *	30 (23%) *	24 (19%) *
RA	89	7 (8%)	3 (3%)	3 (3%)	4 (4%)
LACC	164	24 (15%) *	28 (17%) *	20 (12%) *	21 (13%) *
RACC	92	20 (22%) *	17 (18%) *	18 (20%) *	17 (18%) *
LH	143	10 (7%)	12 (8%)	7 (5%)	15 (10%) *
RH	99	6 (6%)	10 (10%)	10 (10%)	7 (7%)
LvmPFC	87	1 (1%)	1 (1%)	4 (5%)	0 (0%)
RvmPFC	79	0 (0%)	0 (0%)	0 (0%)	2 (3%)

A = amygdala, ACC = anterior cingulate cortex, H = hippocampus, PFC = ventromedial prefrontal cortex.

Table A4. Significant neuronal responses (relative to background firing) by side and brain area (SUA) among all subjects ($n = 13$), as indicated by the changes from background test.

Side/Area	Total # of Neurons	# Significant	Adjusted Binomial *p*-Value
LA	128	56 (44%)	2.6×10^{-37}
RA	89	17 (19%)	2.4×10^{-6}
LACC	164	56 (34%)	8.5×10^{-31}
RACC	92	29 (32%)	1.6×10^{-15}
LH	143	29 (20%)	1.8×10^{-10}
RH	99	27 (27%)	6.0×10^{-13}
LvmPFC	87	16 (18%)	6.9×10^{-6}
RvmPFC	79	7 (9%)	1.0×10^{-1}

A = amygdala, ACC = anterior cingulate cortex, H = hippocampus, vmPFC = ventromedial prefrontal cortex.

Table A5. Clusters (MUA) significant for specific race categories (relative to background firing) by side and brain area across all patients ($n = 13$), as indicated by multinomial logistic regression models. Values with asterisks denote significant proportions in binomial tests ($p < 0.05$), after correction for multiple comparisons.

Area	Total # of Clusters	Sig. for White	Sig. for Black	Sig. for Ambiguous White	Sig. for Ambiguous Black
LA	248	34 (14%) *	46 (19%) *	29 (12%) *	37 (15%) *
RA	173	5 (3%)	5 (3%)	1 (1%)	7 (4%)
LACC	207	43 (21%) *	31 (15%) *	28 (14%) *	22 (11%) *
RACC	213	32 (15%) *	30 (14%) *	34 (16%) *	34 (16%) *
LH	226	13 (6%)	10 (4%)	5 (2%)	6 (3%)
RH	198	7 (4%)	7 (4%)	6 (3%)	6 (3%)
LvmPFC	217	11 (5%)	13 (6%)	11 (5%)	10 (5%)
RvmPFC	197	4 (2%)	1 (1%)	1 (1%)	2 (1%)

A = amygdala, ACC = anterior cingulate cortex, H = hippocampus, vmPFC = ventromedial prefrontal cortex.

Table A6. Reliable cluster responses (MUA) by side and brain area across all patients (n = 13), as indicated by the change from background test.

Side/Area	Total # of Clusters	# Significant	Adjusted Binomial p-Value
LA	248	77 (31%)	$1.3 \times 10-38$
RA	173	32 (18%)	$2.6 \times 10-10$
LACC	207	74 (36%)	$1.2 \times 10-41$
RACC	213	58 (27%)	$3.4 \times 10-26$
LH	226	38 (17%)	$1.3 \times 10-10$
RH	198	21 (11%)	$1.7 \times 10-3$
LvmPFC	217	29 (13%)	$2.2 \times 10-6$
RvmPFC	197	12 (6%)	$2.8 \times 10-1$

A = amygdala, ACC = anterior cingulate cortex, H = hippocampus, vmPFC = ventromedial prefrontal cortex.

Representational Similarity Analysis. As a final step, we conducted representational similarity analyses (RSA) [79–81] to help visualize any joint effects of stimulus race and affect. To perform RSA, responses for each neuron to each image were first normalized to the average background firing rate of that neuron in the interval from 1000–200 ms prior to stimulus presentation (this interval was chosen to mirror that used for computing responses) by subtracting the mean background spike count and dividing by the standard deviation of the background spike counts. Next, Pearson's correlations between the normalized responses for all image pairs were computed. For pairs consisting of a stimulus with itself, the variance of the observations, rather than the correlation (which would always equal one), was used. Finally, these values were plotted in two different orderings. Of primary interest (see Figure A1), pairs were first ordered by affect and race, such that the outer larger blocks showed effects of race, averaged across affects. Figure A1 shows the average correlations of neural responses in the ACC, ordered first by race categories and then by emotions for faces in the stimulus set 1. A strong response to race would manifest as stronger correlations in major blocks along the diagonal, relative to the off-diagonal blocks, as shown. This figure also shows a tendency for stronger correlations when White-White or Black-Black face pairs were compared. Both appear as a pattern of less purple in the four central major blocks. Similar (although less visually apparent) patterns were observed for the hippocampus and amygdala, and for the stimulus set 2.

To statistically assess the apparent pattern of stronger correlations in the main diagonal blocks, we used the technique from [82,83], a permutation test [54] based on the Wilcoxon rank-sum test statistic [33]. For each brain area, we computed the rank-sum statistic comparing all correlation values (in both stimulus sets) in the off-main versus the on-main diagonal blocks (excluding the variances). We then randomly permuted the race labels of the second image per pair and re-computed the rank-sum statistics. This permutation was performed 1000 times; the rank of the original (non-permuted) test statistic was used to obtain the corresponding p-value. The pattern of stronger correlations in the main diagonal blocks was significant in all four brain areas, indicating that race-responsive neurons were reasonably consistent in their response profiles. Admittedly, the statistical evidence is challenging to visually appreciate in Figure A1. Since race variations were orthogonal our experimental task, we posit that race-sensitive neural responses were attenuated.

Figure A1. Representational similarity analysis of neural firing in the anterior cingulate cortex to face pairs, grouped first by race, then by affect. Each pixel represents the Pearson's correlation coefficient for image pairs, with race and affect determined by the pixel's position along the x and y axes (except for pixels with identical images in the pair, wherein the variance is shown). Stronger correlations are observed for the white-white and black-black pairs.

Appendix C. Photographs

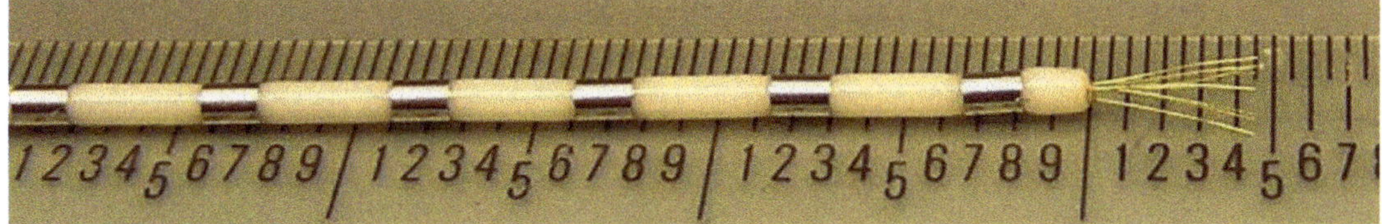

Figure A2. Image of clinical macroelectrode and microwire electrodes to scale. Numbers on the ruler indicate mm of length.

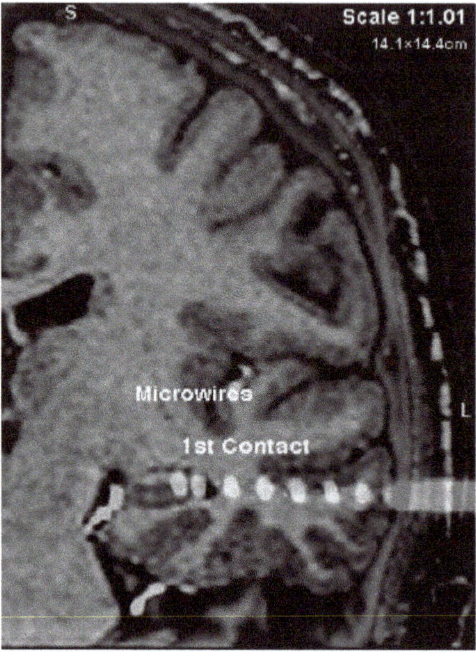

Figure A3. CT scan of the left hemisphere with macroelectrode and microwires implanted in the hippocampus.

References

1. Molenberghs, P.; Louis, W.R. Insights from fMRI Studies into Ingroup Bias. *Front. Psychol.* **2018**, *9*, 1868. [CrossRef] [PubMed]
2. Amodio, D.M. The neuroscience of prejudice and stereotyping. *Nat. Rev. Neurosci.* **2014**, *15*, 670–682. [CrossRef] [PubMed]
3. Cikara, M.; Van Bavel, J.J. The Neuroscience of Intergroup Relations: An Integrative Review. *Perspect. Psychol. Sci.* **2014**, *9*, 245–274. [CrossRef] [PubMed]
4. Han, S. Neurocognitive Basis of Racial Ingroup Bias in Empathy. *Trends Cogn. Sci.* **2018**, *22*, 400–421. [CrossRef]
5. Mattan, B.D.; Wei, K.Y.; Cloutier, J.; Kubota, J.T. The social neuroscience of race-based and status-based prejudice. *Curr. Opin. Psychol.* **2018**, *24*, 27–34. [CrossRef]
6. Golby, A.J.; Gabrieli, J.D.E.; Chiao, J.Y.; Eberhardt, J.L. Differential responses in the fusiform region to same-race and other-race faces. *Nat. Neurosci.* **2001**, *4*, 845–850. [CrossRef]
7. Adolphs, R. The neurobiology of social cognition. *Curr. Opin. Neurobiol.* **2001**, *11*, 231–239. [CrossRef]
8. Dubois, S.; Rossion, B.; Schiltz, C.; Bodart, J.; Michel, C.; Bruyer, R.; Crommelinck, M. Effect of Familiarity on the Processing of Human Faces. *NeuroImage* **1999**, *9*, 278–289. [CrossRef]
9. George, N.; Driver, J.; Dolan, R.J. Seen gaze-direction modulates fusiform activity and its coupling with other brain areas during face processing. *Neuroimage* **2001**, *13*, 1102–1112. [CrossRef]
10. Kawashima, R.; Sugiura, M.; Kato, T.; Nakamura, A.; Hatano, K.; Ito, K.; Fukuda, H.; Kojima, S.; Nakamura, K. The human amygdala plays an important role in gaze monitoring: A PET study. *Brain* **1999**, *122*, 779–783. [CrossRef]
11. Whalen, P.J.; Shin, L.M.; McInerney, S.C.; Fischer, H.; Wright, C.I.; Rauch, S.L. A functional MRI study of human amygdala responses to facial expressions of fear versus anger. *Emotion* **2001**, *1*, 70–83. [CrossRef] [PubMed]
12. Wicker, B.; Perrett, D.I.; Baron-Cohen, S.; Decety, J. Being the target of another's emotion: A PET study. *Neuropsychologia* **2003**, *41*, 139–146. [CrossRef]
13. Yang, T.T.; Menon, V.; Eliez, S.; Blasey, C.; White, C.D.; Reid, A.J.; Gotlib, I.H.; Reiss, A.L. Amygdalar activation associated with positive and negative facial expressions. *NeuroReport* **2002**, *13*, 1737–1741. [CrossRef] [PubMed]
14. Cunningham, W.A.; Johnson, M.K.; Raye, C.L.; Gatenby, J.C.; Gore, J.C.; Banaji, M.R. Separable Neural Components in the Processing of Black and White Faces. *Psychol. Sci.* **2004**, *15*, 806–813. [CrossRef]
15. Hart, A.J.; Whalen, P.J.; Shin, L.M.; McInerney, S.C.; Fischer, H.; Rauch, S.L. Differential response in the human amygdala to racial outgroup vs. ingroup face stimuli. *NeuroReport* **2000**, *11*, 2351–2354. [CrossRef]
16. Mah, L.W.Y.; Arnold, M.C.; Grafman, J. Deficits in social knowledge following damage to ventromedial prefrontal cortex. *J. Neuropsychiatry Clin. Neurosci.* **2005**, *17*, 66–74. [CrossRef]

17. Herzmann, G.; Curran, T. Neural Correlates of the In-Group Memory Advantage on the Encoding and Recognition of Faces. *PLoS ONE* **2013**, *8*, e82797. [CrossRef]
18. Hughes, B.L.; Ambady, N.; Zaki, J. Trusting outgroup, but not ingroup members, requires control: Neural and behavioral evidence. *Soc. Cogn. Affect. Neurosci.* **2017**, *12*, 372–381. [CrossRef]
19. Katsumi, Y.; Dolcos, S. Neural Correlates of Racial Ingroup Bias in Observing Computer-Animated Social Encounters. *Front. Hum. Neurosci.* **2018**, *11*, 632. [CrossRef]
20. Xu, X.; Zuo, X.; Wang, X.; Han, S. Do You Feel My Pain? Racial Group Membership Modulates Empathic Neural Responses. *J. Neurosci.* **2009**, *29*, 8525–8529. [CrossRef]
21. Kawasaki, H.; Adolphs, R.; Oya, H.; Kovach, C.; Damasio, H.; Kaufman, O.; Howard, M. Analysis of Single-Unit Responses to Emotional Scenes in Human Ventromedial Prefrontal Cortex. *J. Cogn. Neurosci.* **2005**, *17*, 1509–1518. [CrossRef] [PubMed]
22. Kawasaki, H.; Adolphs, R.; Kaufman, O.; Damasio, H.; Damasio, A.R.; Granner, M.; Bakken, H.; Hori, T.; Howard, M.A. Single-neuron responses to emotional visual stimuli recorded in human ventral prefrontal cortex. *Nat. Neurosci.* **2001**, *4*, 15–16. [CrossRef] [PubMed]
23. Wang, S.; Tudusciuc, O.; Mamelak, A.N.; Ross, I.B.; Adolphs, R.; Rutishauser, U. Neurons in the human amygdala selective for perceived emotion. *Proc. Natl. Acad. Sci. USA* **2014**, *111*, E3110–E3119. [CrossRef] [PubMed]
24. Wang, S.; Yu, R.; Tyszka, J.M.; Zhen, S.; Kovach, C.; Sun, S.; Huang, Y.; Hurlemann, R.; Ross, I.B.; Chung, J.M.; et al. The human amygdala parametrically encodes the intensity of specific facial emotions and their categorical ambiguity. *Nat. Commun.* **2017**, *8*, 14821. [CrossRef]
25. Heeger, D.J.; Ress, D. What does fMRI tell us about neuronal activity? *Nat. Rev. Neurosci.* **2002**, *3*, 142–151. [CrossRef]
26. Phelps, E.A.; O'Connor, K.J.; Cunningham, W.A.; Funayama, E.S.; Gatenby, J.C.; Gore, J.C.; Banaji, M.R. Performance on Indirect Measures of Race Evaluation Predicts Amygdala Activation. *J. Cogn. Neurosci.* **2000**, *12*, 729–738. [CrossRef]
27. Newhoff, M.; Treiman, D.M.; Smith, K.A.; Steinmetz, P.N. Gender differences in human single neuron responses to emotional faces. *Front. Hum. Neurosci.* **2015**, *9*, 499. [CrossRef]
28. Hugenberg, K.; Young, S.G.; Bernstein, M.J.; Sacco, D.F. The categorization-individuation model: An integrative account of the other-race recognition deficit. *Psychol. Rev.* **2010**, *117*, 1168–1187. [CrossRef]
29. Ekstrom, A. How and when the fMRI BOLD signal relates to underlying neural activity: The danger in dissociation. *Brain Res. Rev.* **2010**, *62*, 233–244. [CrossRef]
30. Nir, Y.; Fisch, L.; Mukamel, R.; Gelbard-Sagiv, H.; Arieli, A.; Fried, I.; Malach, R. Coupling between Neuronal Firing Rate, Gamma LFP, and BOLD fMRI Is Related to Interneuronal Correlations. *Curr. Biol.* **2007**, *17*, 1275–1285. [CrossRef]
31. Bowers, J.S. On the biological plausibility of grandmother cells: Implications for neural network theories in psychology and neuroscience. *Psychol. Rev.* **2009**, *116*, 220–251. [CrossRef] [PubMed]
32. Goldinger, S.D.; He, Y.; Papesh, M.H. Deficits in cross-race face learning: Insights from eye movements and pupillometry. *J. Exp. Psychol. Learn. Mem. Cogn.* **2009**, *35*, 1105–1122. [CrossRef] [PubMed]
33. Lindgren, B. *Statistical Theory*; Chapman & Hall: New York, NY, USA, 1993.
34. Dymond, A.M.; Babb, T.L.; Kaechele, L.E.; Crandall, P.H. Design considerations for the use of fine and ultrafine depth brain electrodes in man. *Biomed. Sci. Instrum.* **1972**, *9*, 1–5. [PubMed]
35. Fried, I.; Wilson, C.L.; Maidment, N.T.; Engel, J.; Behnke, E.; Fields, T.A.; Macdonald, K.A.; Morrow, J.W.; Ackerson, L.C. Cerebral microdialysis combined with single-neuron and electroencephalographic recording in neurosurgical patients. Technical note. *J. Neurosurg.* **1999**, *91*, 697–705. [CrossRef]
36. Nowell, M.; Rodionov, R.; Diehl, B.; Wehner, T.; Zombori, G.; Kinghorn, J.; Ourselin, S.; Duncan, J.; Miserocchi, A.; McEvoy, A. A Novel Method for Implementation of Frameless StereoEEG in Epilepsy Surgery. *Oper. Neurosurg.* **2014**, *10*, 525–534. [CrossRef]
37. Roessler, K.; Sommer, B.; Merkel, A.; Rampp, S.; Gollwitzer, S.; Hamer, H.M.; Buchfelder, M. A Frameless Stereotactic Implantation Technique for Depth Electrodes in Refractory Epilepsy Using Intraoperative Magnetic Resonance Imaging. *World Neurosurg.* **2016**, *94*, 206–210. [CrossRef]
38. Valdez, A.B.; Hickman, E.N.; Treiman, D.M.; Smith, K.A.; Steinmetz, P.N. A statistical method for predicting seizure onset zones from human single-neuron recordings. *J. Neural Eng.* **2013**, *10*, 016001. [CrossRef]
39. Singular Inversions, Inc. (2004). FaceGenModeller (Version 3.1.4) [Computer software]. Toronto, ON, Canada. Available online: http://www.FaceGen.com (accessed on 29 May 2022). Retrieved October, 2008.
40. Minear, M.; Park, D.C. A lifespan database of adult facial stimuli. *Behav. Res. Methods Instrum. Comput.* **2004**, *36*, 630–633. [CrossRef]
41. Phillips, P.; Wechsler, H.; Huang, J.; Rauss, P.J. The FERET database and evaluation procedure for face-recognition algorithms. *Image Vis. Comput.* **1998**, *16*, 295–306. [CrossRef]
42. Celeux, G.; Govaert, G. A classification EM algorithm for clustering and two stochastic versions. *Comput. Stat. Data Anal.* **1992**, *14*, 315–332. [CrossRef]
43. Celeux, G.; Govaert, G. Gaussian parsimonious clustering models. *Pattern Recognit.* **1995**, *28*, 781–793. [CrossRef]
44. Valdez, A.B.; Papesh, M.H.; Treiman, D.M.; Smith, K.A.; Goldinger, S.D.; Steinmetz, P.N. Distributed Representation of Visual Objects by Single Neurons in the Human Brain. *J. Neurosci.* **2015**, *35*, 5180–5186. [CrossRef] [PubMed]
45. Viskontas, I.V.; Knowlton, B.J.; Steinmetz, P.N.; Fried, I. Differences in Mnemonic Processing by Neurons in the Human Hippocampus and Parahippocampal Regions. *J. Cogn. Neurosci.* **2006**, *18*, 1654–1662. [CrossRef] [PubMed]

46. Wixted, J.T.; Goldinger, S.D.; Squire, L.R.; Kuhn, J.R.; Papesh, M.H.; Smith, K.A.; Treiman, D.M.; Steinmetz, P.N. Coding of episodic memory in the human hippocampus. *Proc. Natl. Acad. Sci. USA* **2018**, *115*, 1093–1098. [CrossRef] [PubMed]

47. Dobson, A.J.; Barnett, A. *An Introduction to Generalized Linear Models*; Chapman & Hall CRC Press: Boca Raton, FL, USA, 2002. [CrossRef]

48. Maindonald, J.; Braun, J. *Data Analysis and Graphics Using R: An Example-Based Approach*; Cambridge University Press: New York, NY, USA, 2003.

49. Hosmer, D.W.; Lemeshow, S. *Applied Logistic Regression*; Wiley: Hoboken, NJ, USA, 2000.

50. Truccolo, W.; Eden, U.T.; Fellows, M.R.; Donoghue, J.P.; Brown, E.N. A point process framework for relating neural spiking activity to spiking history, neural ensemble, and extrinsic covariate effects. *J. Neurophysiol.* **2005**, *93*, 1074–1089. [CrossRef]

51. Benjamini, Y.; Hochberg, Y. Controlling the False Discovery Rate: A Practical and Powerful Approach to Multiple Testing. *J. R. Stat. Soc. Ser. B* **1995**, *57*, 289–300. [CrossRef]

52. Team, R.C. *R: A Language and Environment for Statistical Computing*; R Foundation for Statistical Computing: Vienna, Austria, 2019.

53. Steinmetz, P.N.; Thorp, C. Testing for effects of different stimuli on neuronal firing relative to background activity. *J. Neural Eng.* **2013**, *10*, 056019. [CrossRef]

54. Efron, B.; Tibshirani, R.J. *An Introduction to the Bootstrap (Chapman & Hall/CRC Monographs on Statistics & Applied Probability)*; Chapman and Hall/CRC: Boca Raton, FL, USA, 1994.

55. Kreiman, G.; Koch, C.; Fried, I. Category-specific visual responses of single neurons in the human medial temporal lobe. *Nat. Neurosci.* **2000**, *3*, 946–953. [CrossRef]

56. Rolls, E.T. The representation of information about faces in the temporal and frontal lobes. *Neuropsychologia* **2007**, *45*, 124–143. [CrossRef]

57. Chiao, J.Y.; Iidaka, T.; Gordon, H.L.; Nogawa, J.; Bar, M.; Aminoff, E.; Sadato, N.; Ambady, N. Cultural Specificity in Amygdala Response to Fear Faces. *J. Cogn. Neurosci.* **2008**, *20*, 2167–2174. [CrossRef]

58. Freeman, J.B.; Schiller, D.; Rule, N.O.; Ambady, N. The neural origins of superficial and individuated judgments about ingroup and outgroup members. *Hum. Brain Mapp.* **2010**, *31*, 150–159. [CrossRef] [PubMed]

59. Liu, Y.; Lin, W.; Xu, P.; Zhang, D.; Luo, Y. Neural basis of disgust perception in racial prejudice. *Hum. Brain Mapp.* **2015**, *36*, 5275–5286. [CrossRef] [PubMed]

60. Van Bavel, J.J.; Packer, D.J.; Cunningham, W.A. Modulation of the Fusiform Face Area following Minimal Exposure to Motivationally Relevant Faces: Evidence of In-group Enhancement (Not Out-group Disregard). *J. Cogn. Neurosci.* **2011**, *23*, 3343–3354. [CrossRef] [PubMed]

61. Wheeler, M.E.; Fiske, S.T. Controlling racial prejudice: Social-cognitive goals affect amygdala and stereotype activation. *Psychol. Sci.* **2005**, *16*, 56–63. [CrossRef]

62. Logothetis, N.K. The Underpinnings of the BOLD Functional Magnetic Resonance Imaging Signal. *J. Neurosci.* **2003**, *23*, 3963–3971. [CrossRef]

63. Angenstein, F.; Kammerer, E.; Scheich, H. The BOLD response in the rat hippocampus depends rather on local processing of signals than on the input or output activity. A combined functional MRI and electrophysiological study. *J. Neurosci.* **2009**, *29*, 2428–2439. [CrossRef]

64. Mormann, F.; Dubois, J.; Kornblith, S.; Milosavljevic, M.; Cerf, M.; Ison, M.; Tsuchiya, N.; Kraskov, A.; Quiroga, R.Q.; Adolphs, R.; et al. A category-specific response to animals in the right human amygdala. *Nat. Neurosci.* **2011**, *14*, 1247–1249. [CrossRef]

65. Steinmetz, P.N.; Cabrales, E.; Wilson, M.S.; Baker, C.P.; Thorp, C.K.; Smith, K.A.; Treiman, D.M. Neurons in the human hippocampus and amygdala respond to both low- and high-level image properties. *J. Neurophysiol.* **2011**, *105*, 2874–2884. [CrossRef]

66. Bernstein, M.J.; Young, S.G.; Hugenberg, K. The cross-category effect—Mere social categorization is sufficient to elicit an own-group bias in face recognition. *Psychol. Sci.* **2007**, *18*, 706–712. [CrossRef]

67. Levin, D.T. Race as a visual feature: Using visual search and perceptual discrimination tasks to understand face categories and the cross-race recognition deficit. *J. Exp. Psychol. Gen.* **2000**, *129*, 559–574. [CrossRef]

68. Garner, W. Interaction of stimulus dimensions in concept and choice processes. *Cogn. Psychol.* **1976**, *8*, 98–123. [CrossRef]

69. Janak, P.H.; Tye, K.M. From circuits to behaviour in the amygdala. *Nature* **2015**, *517*, 284–292. [CrossRef] [PubMed]

70. Papesh, M.H.; Goldinger, S.D. Deficits in other-race face recognition: No evidence for encoding-based effects. *Can. J. Exp. Psychol.* **2009**, *63*, 253–262. [CrossRef] [PubMed]

71. Yaros, J.L.; Salama, D.A.; Delisle, D.; Larson, M.S.; Miranda, B.A.; Yassa, M.A. A Memory Computational Basis for the Other-Race Effect. *Sci. Rep.* **2019**, *9*, 19399. [CrossRef]

72. Adolphs, R. Recognizing Emotion from Facial Expressions: Psychological and Neurological Mechanisms. *Behav. Cogn. Neurosci. Rev.* **2002**, *1*, 21–62. [CrossRef]

73. Amaral, D.G.; Insausti, R. Retrograde transport of D-[^{3}H]-aspartate injected into the monkey amygdaloid complex. *Exp. Brain Res.* **1992**, *88*, 375–388. [CrossRef]

74. Whalen, P.J. Fear, vigilance, and ambiguity: Initial neuroimaging studies of the human amygdala. *Curr. Dir. Psychol. Sci.* **1998**, *7*, 177–188. [CrossRef]

75. Wright, C.I.; Fischer, H.; Whalen, P.J.; McInerney, S.C.; Shin, L.M.; Rauch, S.L. Differential prefrontal cortex and amygdala habituation to repeatedly presented emotional stimuli. *NeuroReport* **2001**, *12*, 379–383. [CrossRef]
76. Lebrecht, S.; Pierce, L.J.; Tarr, M.J.; Tanaka, J.W. Perceptual Other-Race Training Reduces Implicit Racial Bias. *PLoS ONE* **2009**, *4*, e4215. [CrossRef]
77. McGugin, R.W.; Tanaka, J.W.; Lebrecht, S.; Tarr, M.J.; Gauthier, I. Race-Specific Perceptual Discrimination Improvement Following Short Individuation Training with Faces. *Cogn. Sci.* **2011**, *35*, 330–347. [CrossRef]
78. Telles, E.E. *Race in Another America: The Significance of Skin Color in Brazil*; Princeton University Press: Princeton, NJ, USA, 2004.
79. Kiani, R.; Esteky, H.; Mirpour, K.; Tanaka, K. Object Category Structure in Response Patterns of Neuronal Population in Monkey Inferior Temporal Cortex. *J. Neurophysiol.* **2007**, *97*, 4296–4309. [CrossRef] [PubMed]
80. Kriegeskorte, N.; Mur, M.; Bandettini, P.A. Representational similarity analysis—Connecting the branches of systems neuroscience. *Front. Syst. Neurosci.* **2008**, *2*, 4. [CrossRef] [PubMed]
81. Reber, T.P.; Bausch, M.; Mackay, S.; Boström, J.; Elger, C.E.; Mormann, F. Representation of abstract semantic knowledge in populations of human single neurons in the medial temporal lobe. *PLoS Biol.* **2019**, *17*, e3000290. [CrossRef] [PubMed]
82. Mormann, F.; Niediek, J.; Tudusciuc, O.; Quesada, C.M.; Coenen, V.A.; Elger, C.E.; Adolphs, R. Neurons in the human amygdala encode face identity, but not gaze direction. *Nat. Neurosci.* **2015**, *18*, 1568–1570. [CrossRef]
83. Chambers, J.M.; Cleveland, W.S.; Kleiner, B.; Tukey, P.A. *Graphical Methods for Data Analysis*; Wadsworth & Brooks: Los Angeles, CA, USA, 1983. [CrossRef]

Review

Handwriting in Autism Spectrum Disorder: A Literature Review

Henriette C. Handle [1,*], Marcus Feldin [1] and Artur Pilacinski [1,2,3]

1   Faculty of Psychology, University of Warsaw, 00-183 Warsaw, Poland
2   Medical Faculty, Ruhr University Bochum, 44892 Bochum, Germany
3   Faculty of Psychology and Educational Sciences, University of Coimbra, 3001-802 Coimbra, Portugal
*   Correspondence: h.handle@student.uw.edu.pl

Abstract: Handwriting is linked to a variety of systems in the human brain and has been likewise demonstrated to be affected by a variety of neurological and developmental disorders. In this paper we provide a narrative review of recent findings regarding the quantitative evaluation of handwriting product in people with autism spectrum disorder. We summarize the experimental approaches and variables measured by most representative studies, such as handwriting speed and quality. We highlight the key issues such as small sample sizes resulting in underpowered designs. Lastly, we draw conclusions and delineate potential research directions, such as the use of machine learning to evaluate multivariate components of handwriting.

Keywords: autism; handwriting; motion tracking; developmental neuroscience

Citation: Handle, H.C.; Feldin, M.; Pilacinski, A. Handwriting in Autism Spectrum Disorder: A Literature Review. *NeuroSci* **2022**, 3, 558–565. https://doi.org/10.3390/neurosci3040040

Academic Editors: Lucilla Parnetti, Federico Paolini Paoletti and Xavier Gallart-Palau

Received: 23 September 2022
Accepted: 19 October 2022
Published: 21 October 2022

Publisher's Note: MDPI stays neutral with regard to jurisdictional claims in published maps and institutional affiliations.

1. Introduction

Autism spectrum disorder (ASD) is a neurodevelopmental disorder which encompasses a broad range of complex developmental and neurobiological disabilities [1].The *Diagnostic and Statistical Manual of Mental Disorders* (DSM-IV) specifies impaired communication and social interaction, restricted interests, repetitive patterns in behavior, deficits in developing and maintaining relationships as well as impaired sensory information processing. In terms of severity the disorder spans from low to high functioning, both in terms of intelligence level and symptoms [2]. Handwriting is known to be challenging for many individuals on the spectrum due to difficulties with fine motor skills, and according to Cartmill and colleagues this difficulty is the main reason why as many as 86% of children diagnosed with ASD are referred to therapy services to improve their handwriting and fine motor skills [3].

Handwriting is typically assessed both through its handwritten product and handwriting process itself, such as speed of production. The handwritten product can be measured with regard to "readability" and "legibility" [4]. The development of computerized software and digital tablets has enabled quantitative measuring of these processes, whereas in the past, the handwritten product would be compared to global evaluation scales and normative samples. Today, researchers use more analytically based computerized evaluations which consider specific aspects of the "readability" and "legibility" criteria [5]

The criteria for the handwritten product are typically: form of the letter, sizing, spacing, line-straightness and consistency and are increasingly becoming standard as a consensus is being reached among the researchers who develop these analytical writing scales [5]. While examining the handwritten product of interest, the process in itself is also an area of interest as it may offer further insight about the writer's handwriting characteristics and potential challenges [4]. The process encompasses, for example, velocity, acceleration, direction, and changes in force [4].

With this literature review, we sought to examine the most current research available on the topic of handwriting quality and handwriting speed within the population of individuals with the Autism Spectrum Disorder.

2. Literature Review Summary

We reviewed six studies focusing on the topic of handwriting (e.g., handwriting quality, handwriting speed) in those with autism spectrum disorder. We selected the studies by browsing the Pubmed database with keywords "handwriting" and "autism". The studies were conducted in a similar manner, with no randomization and with comparisons between test and control groups. The lack of randomization occurred due to the inclusion criteria where the test group was required to be diagnosed with autism spectrum disorder. Autism spectrum disorder will be referred to as ASD throughout this text. The studies are listed and summarized in Table 1 below.

Table 1. A summary of the reviewed studies examining handwriting in ASD in terms of sample sizes, primary outcomes and study design.

Study	No. of Participants	Outcome Domains	Study Design
Fuentes, Mostofsky and Bastian, 2009	14 ASD 14 Control	Handwriting quality	Non-randomized, between-groups with control group
Johnson, Papadopoulos, Fielding, Phillips and Rinehart, 2011	11 HF-ASD 11 Asperger 11 Control	Handwriting quality Handwriting speed	Non-randomized, between-groups with control group
Hellinckx, Roeyers and Van Waelvelde, 2013	70 ASD 61 Control	Handwriting quality Handwriting speed	Non-randomized, between-groups with control group
Rosenblum, Simhon and Gal, 2016	30 HF-ASD 30 Control	Handwriting quality	Non-randomized, between-groups with control group
Li-Tsang, Li, Ho, Lau and Leung, 2018	15 ASD 174 Control	Handwriting quality Handwriting speed	Non-randomized, between-groups with control group
Godde, Tsao, Gepner and Tardif, 2018	21 ASD 42 Control	Handwriting quality Handwriting speed Predictors	Non-randomized, between-groups with control group

3. Children with Autism Show Specific Handwriting Impairments

In a study by Fuentes, Mostofsky and Bastian [6], the authors sought to explore specific aspects of handwriting in which children with autism show difficulties, an area of research which until that date remained largely unexamined. A total of 28 participants, 14 with ASD and 14 typically developing controls, were first administered the WISC-IV (Wechsler Intelligence Scale for Children-IV) where full-scale IQs greater than 80 was observed in all but two subjects who showed marked discrepancies. The Perceptual Reasoning Index (the PRI) was used as the primary intelligence measure since the study involved nonverbal, perceptually based, motor tasks. Subjects were then administered the Minnesota Handwriting Assessment where they were asked to copy several words onto a provided solid line, making their letters the same size as the sample and using their best handwriting (Fuentes, Mostofsky and Bastian, 2009). The sample was then scored on each letter individually based on five categories, namely: alignment, legibility, size, form and spacing. Motor skills were assessed using the Revised Physical and Neurological Examination for Subtle (Motor) Signs (PANESS) [6]. This measure comprises several categories such as balance and timed movements, heel, and toe walking, hopping on one foot and finger apposition. The subjects were also asked to undergo the Block Design test, which is a subtest of the WISC-IV that assesses visuospatial abilities. In this test, the subjects were asked to reconstruct several advanced designs by way of assembling a set of blocks which are components of a larger pattern.

The results showed that, although the children diagnosed with ASD performed worse in the quality of forming letters, neither of the groups demonstrated differences in relation to alignment, size, and spacing. No significant differences were found between groups in terms of age, Block Design, or PRI score. However, consistent with previous studies, the control group performed better overall on the PANESS inventory, specifically in the gait/stances and timed movements subcategories. Stepwise multiple regressions analysis revealed that the PANESS timed movements scores were the strongest predictors when it comes to handwriting performance in the ASD group [6].

4. Handwriting in Children with Autism and Asperger's Disorder

A study by Johnson and colleagues aimed to compare and investigate the handwriting profile of children diagnosed with either high-functioning autism spectrum disorder or Asperger's [7]. The study included three groups with 11 subjects in each group: high-functioning ASD, Asperger's, and a control group. The participants were first asked to fill out the Wechsler Intelligence Scale for children-IV and were subsequently matched according to age and Perceptual Reasoning Index (the PRI) score. The participants were then asked to perform three different writing tasks. Under three different conditions, they would be asked to write a cursive letter "l" on a digital tablet, in different sizes, with five trials per condition. With use of special software, the authors of the study were able to extract kinematic and temporal features, e.g., height, length, width, duration, and pen pressure. The participants then completed a speed subtest from the Handwriting Performance Test to assess change in handwriting over time, as well as handwriting speed. Here they were tasked with writing the words "cat and dog" as many times as they could on a specific line for a duration of two minutes. The height, spacing and width of each of these words, as well as the whole phrase, was then measured and scored [7].

Perhaps the most important finding from this study was that once the participants of the clinical groups were faced with the lack of visual cues, their sizing of the letters increased significantly, which suggests that handwriting size is motored by contextual and visual guides [7]. Decreased space and increased variability in spacing were found between words on the "cat and dog" task in the clinical groups, which could contribute to their under-average handwriting legibility. Overall, few differences were found between the two clinical groups, with the control group performing better in all tasks. Although these two conditions are clinically distinct from one another, there is significant overlap between the neurobiological and clinical symptomatology. This study demonstrated support to the revisions of DSM-V, in which ASD and AD have been merged into the "autism spectrum disorder" category. The study also confirmed the finding that children with these neurobiological conditions perform better when provided with visual cues as guidelines, which serves as important knowledge for those working within educational and/or therapeutic settings as they develop strategies for improving handwriting in these populations [7].

5. Predictors of Handwriting in Children with Autism Spectrum Disorder

The study conducted by Hellinckx, Roeyers and Van Waelvelde aimed to investigate several factors of handwriting quality and speed in children with autism spectrum disorder [8]. The 131 participants completed IQ tests, with FSIQ and WISC-III; 70 of the children were previously diagnosed with ASD and 61 children were typically developing. To measure handwriting quality and speed, the Dutch Systematic Screening of Handwriting tool was used. This tool detects graphomotor disorders in children. Participants were instructed to copy a text as fast and as neatly as they could onto an unruled paper for five minutes or until five sentences were completed. The sentences are used to measure quality through fluency of letter formation, fluency in connecting letters, letter height, regularity of letter height, spaces between words, and spatial alignment of sentences. Handwriting speed was measured by counting the number of letters written in five minutes. Scores below the 5th percentile indicate graphomotor disorder. The children were additionally tested with the M-ABC-2, which measures motor functioning and identifies movement difficulties, as well as the VMI which measures visual-motor integration skills. They were also instructed to complete the One Minute Reading Test which measures the amount of correctly read words for one minute [8].

The results of the study indicate that children with ASD perform poorer on all measures. Children with ASD had lower handwriting quality compared to the control group, as they had difficulties with connecting letters, wrote less fluently, and had irregularities in height and spatial alignment. Additional findings include that handwriting quality is better in higher ages, as well as boys having poorer quality of handwriting than girls. No difference was found between left- and right-handed writing. In the population with ASD, handwriting quality could be predicted with age, gender and the VMI scores. Additionally, handwriting speed was correlated with age. The study successfully identified three predictors of handwriting quality and speed: age, gender, and VMI results. The study also shows that participants with ASD have poorer handwriting skills, as the quality and speed of their handwriting is poorer than the neurotypical participants. However, better coordination between visual input and finger movement resulted in higher quality output in those with ASD, indicating that improving these skills may result in better performance on handwriting quality and handwriting speed tests [8].

6. Unique Handwriting Performance Characteristics of Children with High-Functioning Autism Spectrum Disorder

The aim of the study by Rosenblum, Simhon and Gal was to compare the product characteristics and handwriting process of children who have been diagnosed with high-functioning autism spectrum disorder (HF-ASD) with a control group consisting of typically developing children, to find the best way of differentiation between the two groups [1]). The participants were 60 children between 9 and 12 years old; 30 who were diagnosed with HF-ASD, and 30 who were age- and gender-matched controls. They were asked to perform three different writing tasks on a digital tablet, which used the computerized handwriting evaluation system (ComPET). The tasks consisted of the participants writing their name and surnames, copying a paragraph, and writing a story based on a picture they were provided with—so-called free-style writing [1]. Upon completion, the participants' paragraph copying result was then assessed with the use of the Hebrew Handwriting Evaluation (HHE), which is a standardized test that comes with a rating of global legibility (i.e., the overall clarity of handwriting), the number of letters erased and/or overwritten, the number of unrecognizable letters and spatial arrangement [9]. The computerized handwriting evaluation system (ComPET) measured pen tilt, temporal measures in seconds, on-paper stroke duration, in-air stroke duration, stroke width and height, as well as pen pressure.

Significant differences were found across all tasks in relation to stroke time on-paper; however, mean stroke in-air was significantly different for the name task and the free-style writing tasks, but not for the paragraph copying task. This suggests that when a child diagnosed with HF-ASD is provided with a visual guideline such as an image or a paragraph, they invest the same amount of time planning the next pen stroke as the typically developed children do. When that image or paragraph is removed, the children diagnosed with HF-ASD need more time to produce, reconstruct and plan their handwriting. This could be explained by challenges with visual perception in individuals with HF-ASD [1]. Overall, the results showed significantly higher scores for the control group. Perhaps more importantly though, the results showed that both pen stroke duration in-air and on-paper helped predict the speed of handwriting, which could suggest that a child diagnosed with HF-ASD invests a lot of energy in the mechanical process related to producing handwriting [1].

7. The Relationship between Sensorimotor and Handwriting Performance in Chinese Adolescents with Autism Spectrum Disorder

The study conducted by Li-Tsang, Li, Ho, Lau and Leung aims to explore the connection between sensorimotor control and handwriting problems [10]. The control group consisted of 174 typically developing adolescents, and the test group consisted of 15 participants with diagnosed ASD. To measure handwriting, the participants completed the Computerized Handwriting Speed Test System (CHSTS-2). In this measure, the participants are asked to copy 130 Chinese characters and 120 English words on A4 paper connected to a tablet with an electronic pen. The test provides information on ground time (pen on paper), airtime (pen off paper), handwriting speed (character per minute), standard deviation of writing time per character, pen pressure, SD of pen pressure, and readability (correctness of words and number of words recognized by the system). The participants were also tested on motor skills, visual perceptual skills, visual-motor integration, and eye movements. These were tested through the measures the Bruininks-Oseretsky Test of Motor Proficiency (BOT), the Motor-Free Visual Perception Test (MVPT), the Beery-Buktenica Developmental Test of Visual-Motor Integration (VMI), and the Developmental Eye Movement test (DEM), respectively [10].

The results of the study show that participants with ASD had more ground and airtime, wrote more slowly, and had larger variations in writing speed than the neurotypical participants. They also showed less stability in their handwriting as can be seen through larger writing speed and pen pressure variations. Poorer manual dexterity was associated with ASD. In Chinese handwriting, poor manual dexterity increased ground time and reduced writing speed. In English handwriting, poor manual dexterity increased airtime and writing speed variation. Manual dexterity was found to be a significant mediator between ASD and Chinese handwriting, whereas ground time and writing speed were found to be a significant mediator between ASD and English handwriting. In terms of readability, participants with ASD show comparable results to the neurotypical participants, which is argued as resulting from specialized training in these aspects in childhood. This study indicates that handwriting results are consistent in both Chinese and English handwriting tasks, which suggests that the handwriting is influenced by ASD rather than the language skills of the participants. Manual dexterity is the main aspect found to affect handwriting results, which indicates that manual dexterity training might improve handwriting in adolescents with ASD [10].

8. Characteristics of Handwriting Quality and Speed in Adults with Autism Spectrum Disorders

The study conducted by Godde, Tsao, Gepner and Tardif aimed to explore features of handwriting and the role of perceptual-motor skills in adults with autism spectrum disorder [11]. The 63 participants were divided into three groups, the first being adults with ASD and the other two being control groups. The participants with ASD completed the Raven's Standard Matrices test with visuospatial reasoning, nonverbal intelligence tests and inductive-logical reasoning to identify the developmental age of the participants. The control groups consisted of non-ASD adults who matched the chronological age of the test group, and 21 non-ASD children who matched the developmental age of the test group. Handwriting quality and speed of the participants were measured with the Concise Evaluation Scale for Children's Handwriting (BHK). The participants are asked to copy a text for five minutes, or until the first five sentences are completed. These first five sentences are larger than the following sentences. The test results display scores on letter size, left margin widening, word alignment, word spacing, chaotic writing, irregularities in joining strokes, collision of letters, letter size, height of letters, letter distortion, ambiguous letter forms, correct letter forms, and unsteady writing. The participants were also assessed on their perceptual-motor skills through the Developmental Neuropsychological Assessment (NEPSY-1). Finger dexterity was measured through a finger-tapping task, fine motor coordination was measured through imitating hand or finger positions from a model, graphomotor skills were measured through a figure copy task, and visuomotor integration was measured by drawing a line inside a path as quickly as possible without leaving the track [11]. The results of the study show that the participants with ASD performed poorer than both control groups. No differences between the control groups were found. Differences included poorer word alignment, ambiguous letter forms, marginal effects in left margin widening, insufficient word spacing, and inconsistent letter size. In terms of writing speed, the participants with ASD wrote more slowly than the chronological age group, but no difference was found between the participants with ASD and the developmental age group. In the participants with ASD, handwriting speed was significantly influenced by finger dexterity, graphomotor skills, and visuomotor integration. Developmental age was the best predictor for handwriting quality. The findings of the study highlight that handwriting difficulties in those with ASD persist throughout adulthood. Training skills such as finger dexterity, graphomotor skills, and visuomotor integration might improve handwriting quality and speed of those with ASD in adulthood [11].

9. Discussion

In each of the reviewed studies, researchers used similar designs, comparing handwriting capacities of participants diagnosed with ASD to control groups comprising neurotypical participants. This allows for comparisons between the results of the studies. In extension, it appears that most of these studies agree with each other. The studies show that the participants with ASD perform poorer on measures of handwriting quality and handwriting speed compared to typically developed participants. Four separate studies indicate that motor skills play a large part in handwriting in ASD participants, highlighting that a better score on factors such as manual dexterity, graphomotor skills, and general motor skills might aid ASD participants to perform better in terms of handwriting quality and speed. Moreover, it can be used as predictor of handwriting scores. Additionally, four of the studies found results supporting that visual input, visual perception, and visuomotor integration is important and might be a predictor for handwriting quality and speed in those with ASD. Improving motor skills, as well as visual perception and visuomotor integration, can be important factors in improving handwriting quality and speed in people diagnosed with ASD, and allowing people diagnosed with ASD to have visual input while performing tasks like the ones in the research can also improve the output.

Most of the studies focused on children diagnosed with ASD, however two of the studies focused on adolescents or adults diagnosed with ASD. These studies were added to the literature review to analyze whether the difficulties ASD participants have in the studies including children would persist throughout adulthood. Both studies that focused on adolescents or adults maintained that difficulties do persist. The first study focusing on adolescents found manual dexterity to be the main aspect that affects handwriting results. The second study found that not only manual dexterity, but also graphomotor skills and visuomotor integration are important aspects that influence handwriting in ASD participants in adulthood as well as in childhood. The persistence of these difficulties implies that improving training or support in these areas in childhood can improve handwriting outcomes in adults diagnosed with ASD.

It is worth noting that in chronological order, the studies in this field that focus on the topic of handwriting output in participants diagnosed with ASD improve with time. The earlier studies are less specific and focus on a broader horizon, while the later studies become narrower and more specific in learning which factors truly affect handwriting output. This implies that we have become more knowledgeable on the topic, and that we move closer and closer to learning about skills that can truly aid the population on the spectrum in improving handwriting and other variables that are affected by the same skills. An example of that may be such skills as playing instruments or drawing.

On a slightly critical note, one seemingly problematic factor which some of the authors have pointed out is the fact that very few of these studies use the same tools and inventories for measuring and scoring variables [4,5,12]. While different digital software recording, e.g., pen pressure or milliseconds of on-paper stroke duration, may not logically differentiate significantly between their produced results, they will, if programmed correctly, record the same result which one can use to compare between studies with minimal effort. Different tasks, and protocols, in the way they are provided to the participant, may present larger differences and biases in their measurements when attempting to compare such studies.

In the cited studies, the sample sizes are consistently small. Except for one study by Hellinckx, Roeyers and Van Waelvelde [8] with 70 ASD participants, none of the examined studies included more than 30 participants diagnosed with ASD, with an average of just 28.6 participants in our examined studies. One could argue that these numbers are too small to generalize to larger populations of individuals on the spectrum. The statistical significance of the results the researchers have found might be faulty or unreliable, due to the increased margin of error. At the same time, one also must ask how realistic and plausible larger sample sizes really are, taking into consideration the size of the population of interest.

Lastly, recent studies on ASD and the use of tablet devices identified distinct patterns of forces and gesture kinematics when children with ASD used tablets while playing serious games [13]. The authors concluded that children with autism showed generally higher contact forces and faster movements when playing the games. While these games did not directly employ handwriting, such combined analysis of kinematics and dynamics of movement might in the future yield a more complex and sensitive way of testing for ASD and other disorders affecting both cognition and motor control.

10. Conclusions

Autism spectrum disorder affects several skills pertaining to handwriting outcome, such as manual dexterity, graphomotor skills, general motor skills, visual input, visuomotor integration and visual perception. Additionally, visual cues or guidelines are shown to assist those with ASD in better handwriting outcomes. Some of the difficulties pertaining to these skills persist throughout adulthood in those with ASD. A possible recommendation for educators, parents, therapists, and rehabilitators would be to include specific training of these skills in ASD childhood, as well as using visual cues to assist in learning. These skills likely apply to other life skills or hobbies and improving these skills might increase the quality of life for those with ASD. Using visual cues with those with ASD is not an uncommon practice, but studies in this review support the continuation of this aspect.

Unfortunately, despite becoming more knowledgeable about the topic throughout the years, it seems there should be more of an agreement on which measures to use as somewhat of a standard when studying the topic. Using many different measures might lead to many different results that are not comparable to each other, therefore using the same measures would allow for better comparisons. The studies in this review have quite small sample sizes, likely due to accessibility of participants with ASD or not wanting to participate. Small sample sizes bring quite a few disadvantages. The results of these studies might not be applicable to the general population, and the statistical significance of the results may be faulty. A recommendation for further research would be to increase sample sizes, as well as improve and agree on a measurement procedure. Additionally, it could be helpful to the generalizability of these studies to attempt to randomize several test groups and compare these to control groups instead of only having one non-randomized test group. Further research should continue to be narrowed to specific aspects that might affect handwriting output in the autism spectrum disorder population.

Author Contributions: Conceptualization: H.C.H., M.F. and A.P.; Writing (original draft): H.C.H. and M.F.; Writing (review and editing): H.C.H., M.F. and A.P. All authors have read and agreed to the published version of the manuscript.

Funding: This research was funded by FCT project (PTDC/PSI-GER/30745/2017).

Institutional Review Board Statement: Not applicable.

Informed Consent Statement: Not applicable.

Data Availability Statement: Not applicable.

Conflicts of Interest: The authors declare no conflict of interest.

References

1. Rosenblum, S.; Simhon, H.; Gal, E. Unique handwriting performance characteristics of children with high-functioning autism spectrum disorder. *Res. Autism Spect. Disord.* **2016**, *23*, 235–244. [CrossRef]
2. American Psychiatric Association. Anxiety Disorders. In *Diagnostic and Statistical Manual of Mental Disorders*, 5th ed.; American Psychiatric Association: Washington, DC, USA, 2013. [CrossRef]
3. Cartmill, L.; Rodger, S.; Ziviani, J. Handwriting of eight-year-old children with autism spectrum disorder: An exploration. *J. Occup. Ther. Sch. Early Interv.* **2009**, *2*, 103–118. [CrossRef]
4. Rosenblum, S.; Parush, S.; Weiss, P.L. The in air phenomenon: Temporal and spatial correlates of the handwriting process. *Percept. Mot. Ski.* **2003**, *96*, 933–954. [CrossRef] [PubMed]
5. Rosenblum, S.; Weiss, P.L.; Parush, S. Product and process evaluation of handwriting difficulties: A review. *Educ. Psychol. Rev.* **2003**, *15*, 41–81. [CrossRef]
6. Fuentes, C.; Mostofsky, S.; Bastian, A. Children with autism show specific handwriting impairments. *Neurology* **2009**, *73*, 1532–1537. [CrossRef] [PubMed]
7. Johnson, B.; Papadopoulos, N.; Fielding, J.; Phillips, J.; Rinehart, N. Handwriting in Children with Autism and Asperger's Disorder. 2011. Available online: https://www.researchgate.net/publication/233855638_Handwriting_in_children_with_autism_and_Asperger%27s_disorder (accessed on 21 September 2022).
8. Hellinckx, T.; Roeyers, H.; Van Waelvelde, H. Predictors of handwriting in children with Autism Spectrum Disorder. *Res. Autism Spectr. Disord.* **2013**, *7*, 176–186. [CrossRef]
9. Devash, L.; Levi, M.; Traub, R.; Shapiro, M. Reliability and Validity of the Hebrew Handwriting Evaluation. Master's Thesis, The Hebrew University of Jerusalem, Jerusalem, Israel, 1995.
10. Li-Tsang, C.; Li, T.; Ho, C.; Lau, M.; Leung, H. The Relationship Between Sensorimotor and Handwriting Performance in Chinese Adolescents with Autism Spectrum Disorder. *J. Autism Dev. Disord.* **2018**, *48*, 3093–3100. [CrossRef] [PubMed]
11. Godde, A.; Tsao, R.; Gepner, B.; Tardif, C. Characteristics of handwriting quality and speed in adults with autism spectrum disorders. *Res. Autism Spectr. Disord.* **2018**, *46*, 19–28. [CrossRef]
12. Rosenblum, S.; Aassy Margieh, J.; Engel-Yeger, B. A handwriting features of children with developmental coordination disorder—Results of triangular evaluation. *Res. Dev. Disabil.* **2013**, *34*, 4134–4141. [CrossRef] [PubMed]
13. Anzulewicz, A.; Sobota, K.; Delafield-Butt, J.T. Toward the Autism Motor Signature: Gesture patterns during smart tablet gameplay identify children with autism. *Sci. Rep.* **2016**, *6*, 31107. [CrossRef]

NeuroSci

MDPI

Article

The Effect of Doxapram on Proprioceptive Neurons: Invertebrate Model

Bethany J. Ison , Maya O. Abul-Khoudoud, Sufia Ahmed, Abraham W. Alhamdani, Clair Ashley, Patrick C. Bidros, Constance O. Bledsoe, Kayli E. Bolton , Jerone G. Capili, Jamie N. Henning , Madison Moon, Panhavuth Phe, Samuel B. Stonecipher, Hannah N. Tanner, Logan T. Turner, Isabelle N. Taylor, Mikaela L. Wagers, Aaron K. West and Robin L. Cooper *

Department of Biology, University of Kentucky, Lexington, KY 40506, USA
* Correspondence: rlcoop1@email.uky.edu

Abstract: The resting membrane potential enables neurons to rapidly initiate and conduct electrical signals. K2p channels are key in maintaining this membrane potential and electrical excitability. They direct the resting membrane potential toward the K^+ equilibrium potential. Doxapram is a known blocker for a subset of K2p channels that are pH sensitive. We assessed the effects of 0.1 and 5 mM doxapram on the neural activity within the propodite-dactylopodite (PD) proprioceptive sensory organ in the walking legs of blue crabs (*Callinectes sapidus*). Results indicate that 0.1 mM doxapram enhances excitation, while the higher concentration 5 mM may over-excite the neurons and promote a sustained absolute refractory period until the compound is removed. The effect of 5 mM doxapram mimics the effect of 40 mM K^+ exposure. Verapamil, another known K2p channel blocker as well as an L-type Ca^{2+} channel blocker, reduces neural activity at both 0.1 and 5 mM. Verapamil may block stretch activated channels in sensory endings, in addition to reducing the amplitude of the compound action potential with whole nerve preparations. These findings are notable as they demonstrate that doxapram has acute effects on neurons of crustaceans, suggesting a targeted K2p channel. The actions of verapamil are complex due to the potential of affecting multiple ion channels in this preparation. Crustacean neurons can aid in understanding the mechanisms of action of various pharmacological agents as more information is gained.

Keywords: crab; K2p channels; proprioception; sensory

Citation: Ison, B.J.; Abul-Khoudoud, M.O.; Ahmed, S.; Alhamdani, A.W.; Ashley, C.; Bidros, P.C.; Bledsoe, C.O.; Bolton, K.E.; Capili, J.G.; Henning, J.N.; et al. The Effect of Doxapram on Proprioceptive Neurons: Invertebrate Model. *NeuroSci* **2022**, *3*, 566–588. https://doi.org/10.3390/neurosci3040041

Academic Editor: Chul-Kyu Park

Received: 15 September 2022
Accepted: 20 October 2022
Published: 23 October 2022

Publisher's Note: MDPI stays neutral with regard to jurisdictional claims in published maps and institutional affiliations.

1. Introduction

The resting membrane potentials of most cells are maintained by K^+ leak channels, also known as K2p channels (two-pore domain K^+ channels). They were first found in yeast and have since been identified in the genomes of various organisms from plants to humans [1–4]. These channels are active in an open state, allowing K^+ ions to flux, driving the resting membrane potential toward the K^+ equilibrium potential.

K2p channels are classified based on their structures and sensitivities. There are 15 known types of K2p channels in mammals. They are grouped into six subtypes with varying pharmacological profiles and distribution in various tissues. The channels show subtype selectivity to volatile anesthetics, pH, membrane tension, endocannabinoids, hypoxia, heat, and G protein-coupled receptor agonists [5–7]. Some compounds, such as chloroform, halothane, and isoflurane [8], stimulate a subtype of K2p channels resulting in the hyperpolarization of cells. The subtype K2P13 is activated by arachidonic acid [9]. Some channels (i.e., K2P18) can be activated by volatile anesthetics (e.g., isoflurane, sevoflurane, halothane, desflurane) but are also reduced in function by local anesthetics (e.g., including bupivacaine, tetracaine, ropivacaine, mepivacaine, lidocaine [7]), leading to the depolarization of cells.

Doxapram, also known by the trade names Stimulex and Respiram, is an inhibitor of two K2p channels: TASK1 and TASK3, which are in the TASK (TWIK-Related Acid-Sensitive K+) channel subtype [6,10]. Since these channels are present in carotid bodies, doxapram can be used clinically to stimulate associated neurons to enhance respiratory drive when inducing therapeutic hypothermia [11–13] or for treatments of respiratory disorders and apnea in infants [14,15].

The actions of pharmacological agents like doxapram have not been fully examined for potential effects on other cellular functions. Agonists and antagonists of K2p channels may impact many cell types yet to be investigated. Clinically, this is important to understand, as potential undesirable effects may arise by action on other cell types [16–18]. Experimental animal models can aid in screening novel pharmacological agents targeting subtypes of K2p channels. Verapamil, commonly known as a Ca^{2+} channel blocker, was recently shown to inhibit a K2p subtypes known as $K_{2P}18.1$. Learning how K2p channels function and what drives expression of their subtypes may help to understand why the expression profiles vary in disease states, such as cancer [18] and forms of epilepsy [19]. In addition, addressing pharmacological agents used in mammals in other organisms aids in understanding their mechanisms of action.

Invertebrate models have assisted in the advancement of neurophysiology. They are inexpensive and easy to obtain, but also remain stable in simple saline for conducting experiments at room temperature. These models also have regions of neurons (i.e., soma and axons) large enough for intracellular recordings. Squid, crustaceans, and insects are commonly used to address fundamental concepts in physiology and in particular neurobiological topics [20–25].

The blue crab (*Callinectes sapidus*) has recently been used to investigate the effects of extra- and intra-cellular pH on neuronal excitability of sensory neurons [26]. Various aspects of the neural circuitry of the crab, *Cancer borealis*, were shown to respond differentially to extreme differences in extracellular pH. Responses mediated by pH may function via K2p channels expressed in neurons of crab models. Such responses have yet to be examined in genomic analysis or physiologically for acid-sensitive K2p channels, such as the doxapram-sensitive TASK channels. This study was exploratory to determine if sensory neurons respond to doxapram and how the compound alters the activity profile.

2. Materials and methods

2.1. Animals

The preparations used for this study were wild-caught blue crabs (*Callinectes sapidus*) from August and October 2022. They were purchased at a local supermarket in Lexington, KY, USA, which originated from a distribution center in Atlanta, GA, USA. Crabs were all female adults in the range of 10–15 cm in carapace width (from point to point). They were maintained in a seawater aquarium for several days prior to use to assess their health. Upon examination and muscle dissection, some specimens had cysts, while others had missing limbs. All crabs used were alive and active upon autotomizing a leg for experimentation.

2.2. Chemicals and Dissection

The physiological saline used to obtain recordings and maintain preparations consisted of (in mM): 470 NaCl, 7.9 KCl, 15.0 $CaCl_2 \cdot 2H_2O$, 6.98 $MgCl_2 \cdot 6H_2O$, 11.0 dextrose, HEPES acid and 5 HEPES base normalized to pH 7.5. Doxapram powder was added directly to the saline to obtain 0.1 mM or 5 mM concentrations and placed on a vortex (high setting) for 5 min. The saline solution remained opaque for the 5 mM solution. Verapamil hydrochloride was directly dissolved in the crab saline at 0.1 and 5 mM. All compounds were obtained from Sigma-Aldrich (St. Louis, MO, USA).

The dissection to expose the propodite-dactylopodite (PD) organ in the walking legs and to record from the transected PD nerve with a suction electrode is detailed in Pankau et al. [27], and a detailed movie is provided in Majeed et al. [28]. Crabs were induced to autotomize the first or second walking leg by using forceps to apply gentle pressure at the coxa of the limbs. Forceps were used to gently pull the PD nerve away from the primary leg nerve under a dissecting microscope. The proximal aspect of the leg was then secured with insect dissection pins in a Sylgard-lined dish filled with crab saline, and the isolated PD nerve was placed in a suction electrode to measure its activity.

Procedures similar to the PD nerve isolation and compound action potential (CAP) recording techniques described by Pankau et al. [27] were followed. In brief, the PD nerve was isolated by dissecting away the main leg nerve. At the distal end of the PD nerve, a stimulating electrode was placed to induce CAPs. The proximal portion of the nerve extending from the propodite region was placed in a recording electrode to measure the induced CAPs. A ground wire was placed along the side of the dish. Following electrode placement, both the distal and proximal ends of the nerve were packed with petroleum jelly to increase contact between the electrode and nerve surface, therefore maximizing the amplitude of the extracellular signals recorded. The nerve was stimulated until the voltage producing maximum CAP amplitude and duration was identified. With the maximum CAP amplitude and duration serving as a point of reference, any change in the amplitude or area under the trace of the CAPs could be clearly identified upon exposure to the various solutions.

2.3. Data Collection

Treatment groups consisted of two concentrations (0.1 mM and 5 mM) of doxapram or verapamil. For each concentration, six preparations were first exposed to a saline control, then to doxapram or verapamil, then incubated in doxapram or verapamil for 5 min and the activity was assessed. Preparations were incubated for another 15 min (20 min total) and assessed again for activity, after which two saline washes occurred. The preparations were retested to see if changes in neural activity due to the pharmacological compounds were reversible. To confirm the experimental findings, participants in a university course tested five more preparations with 5 mM doxapram. Experiments were also conducted in six different preparations where the saline bath was exchanged with 20 mM KCl followed with 40 mM KCl and a saline wash out.

For each condition, the dactyl (most distal segment of the leg) was moved from a flexed position of 90 degrees to full extension in 1 s and then held in the extended position for at least 9 more seconds. The joint was then returned to the 90-degree bend (Figure 1) All 10 s of data recordings were used during analysis. Throughout each trial, a dissecting pin was used as a stop point to ensure consistency in the maximum range of extension obtained.

2.4. Analysis of the Neural Activity

Spikes were counted in the 10-s window during the movement and time held in the extended position via the software in Chart 7 and Chart 8 (ADI Instruments) (Figure 1). A sine wave fit of the trace was selected, and standard deviation was set to ensure proper signal detection. The amplitudes of spikes had a wide range from large to small; the choice of which spikes to count was determined by eye and was approximately the amplitude of two times the baseline noise.

The number of spikes for each trace was determined by one individual to maintain consistency in analysis. The software does produce errors in determining the signals from noise. Therefore, counts were confirmed by eye and adjusted as needed by modifying the standard deviation within the software. Software issues were detailed in a previous report [29]. To examine reproducibility in the analysis, the university students were provided samples of the same data sets to compare to the original analysis.

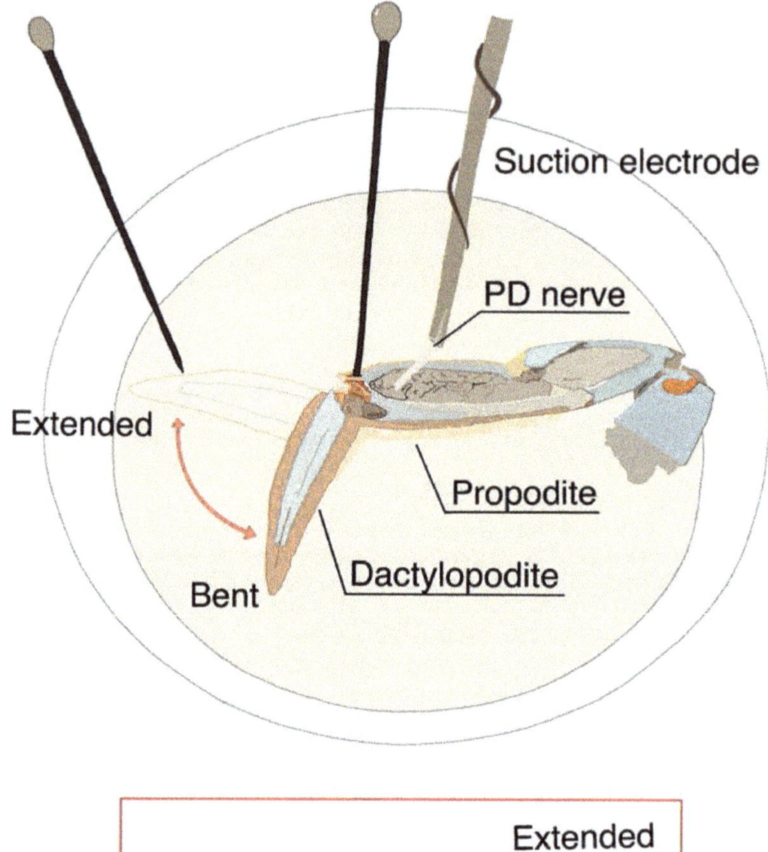

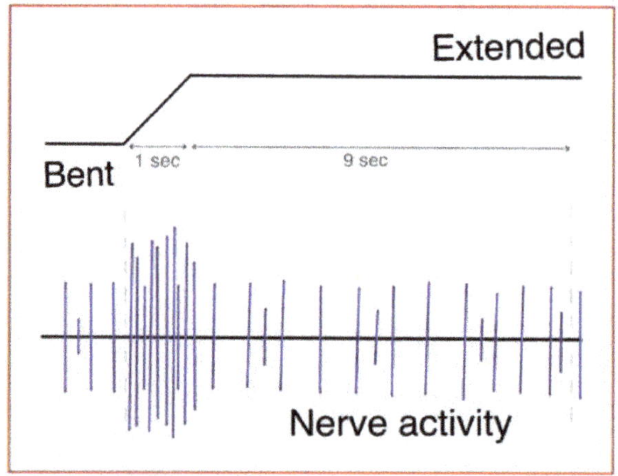

Figure 1. The first or second walking leg of the crab was used to expose the PD organ and associated nerve to various compounds. The joint was initially bent at 90 degrees, then extended out straight within 1 s, and then held for at least another 9 s. The entire 10 s was then used for analysis in the number of spikes that occurred while bathed in different solutions.

2.5. Statistical Methods

A paired T-test or a Wilcoxon Signed Rank Test and a significance level of 0.05 was used in all studies. This study was also conducted, in part, with participants in a university senior-level neurophysiology course as part of an ACURE (Authentic Course-based Undergraduate Research Experience) [26–29]. This helped confirm the experimental findings and the reproducibility of the data analysis.

3. Results

3.1. Effects of Doxapram on Neural Activity

To examine the effect of doxapram at 5 mM on neural activity during movements of the PD joint, the number of spikes were counted (Figure 2). A representative preparation is illustrated in Figure 2 with the initial saline exposure (Figure 2A) followed by a bath exchange to 5 mM doxapram (Figure 2B). The bath was flushed and allowed to incubate for 5 min before repeating three more joint movements (Figure 2C). Afterward, the bath was exchanged with fresh saline which was flushed around the preparation. Three more trials were then repeated (Figure 2D).

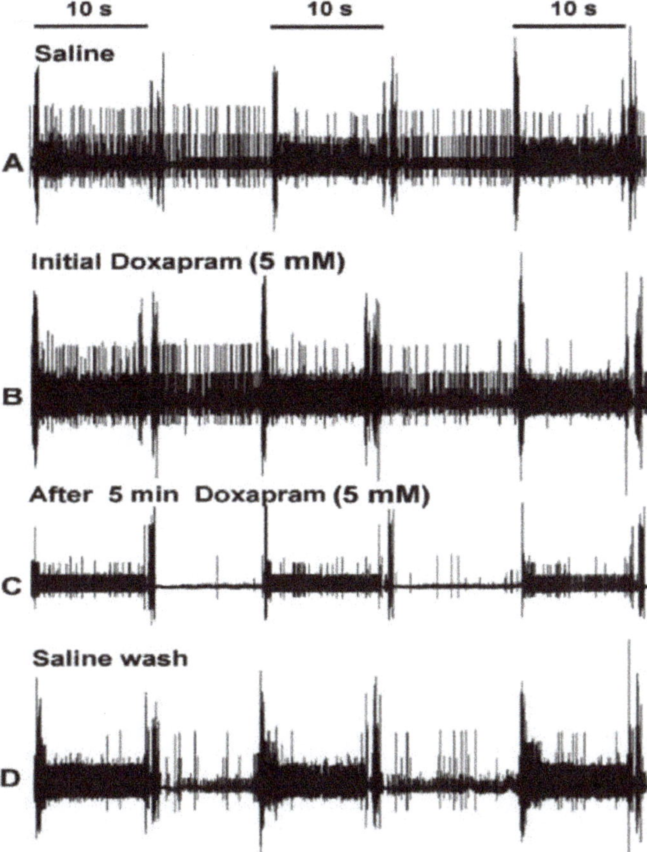

Figure 2. Representation of the effects of doxapram at 5 mM on neural activity for the proprioceptive neurons in the crab PD organ. (**A**) The activity of the nerve in saline with the three movements of the joint (1 s for the movement to an extended position and 9 s or more for being held in a static position of joint extension). (**B**) After the bath is exchanged to doxapram, the joint is then moved again three times. (**C**) Flushing the doxapram solution around the preparation and allowing it to incubate for 5 min. After 5 min, three more movements are made. (**D**) The bath is exchanged two times with fresh saline and the joint movements are repeated. Only the initial 10 s are used for analysis.

The activity for the three trials in each bathing condition from the 6 walking legs and 1 large cheliped is shown in Figure 3A. The data obtained from the large cheliped is shown in red to separate the responses from the walking legs. However, the overall results fall within the range obtained for the walking legs. The mean number of spikes obtained in each of the three trials for each condition is shown in Figure 3B. There is a significant decrease in activity with incubating the preparation for 5 min in doxapram (5 mM) ($p < 0.5$ Paired T-Test, N = 7).

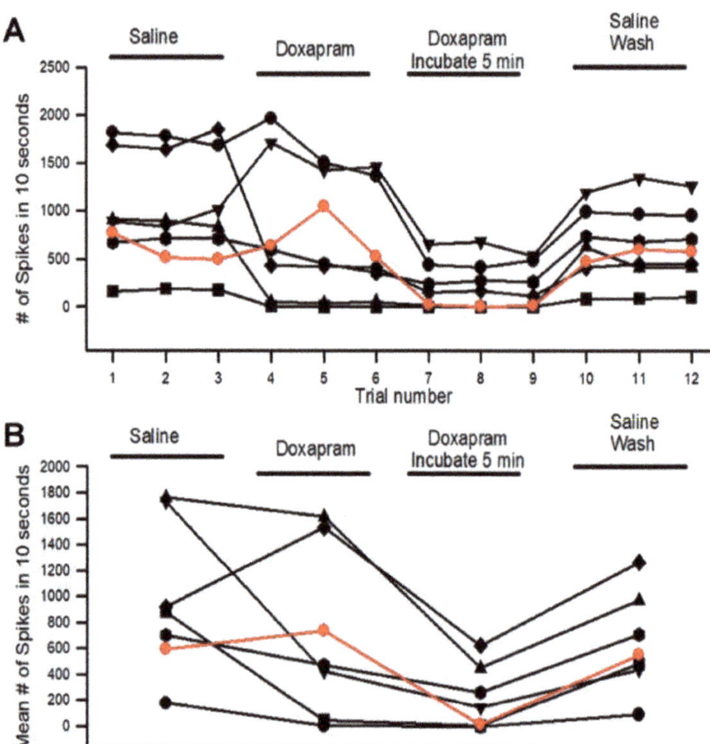

Figure 3. The acute effect of doxapram (5 mM) on neural activity of the PD organ. (**A**) The number of spikes measured in the 10-s window from the beginning movement of the joint starting from a bent position (90 degrees) to fully extended within 1 s and held in an extended position for the next 9 s. This paradigm is repeated three times for each condition. Each line represents a different preparation of an PD organ. Three trials were undertaken with saline, three trials were done immediately after switching the bath to doxapram (5 mM), and they were examined again after incubation for 5 min. The final exchange was to rinse the preparation twice with fresh saline and then move the joint three more times. Each movement was separated by at least 10 s while the joint was held in a bent position. (**B**) The number of spikes in each of the three trails was averaged and graphed in the same manner as in (**A**), which allows an easier view of the overall effects. The red colored trace represents a PD preparation from a chela of the large claw.

To examine reproducibility in the effect of doxapram at 5 mM independently of the data presented above, 17 participants in a university senior-level neurophysiology course repeated the experiments in five groups. The results of the classroom experiments are shown in the Appendix A.

In addition, data sets were provided to the class to analyze the number of spikes. The conditions of the data were blind to the participants. One trained individual analyzed all data sets in this study for all experiments. The results in reproducibility in data analysis are presented in the Appendix A.

Since the number of spikes in three preparations first demonstrated an increase in activity before decreasing with the initial exposure, it was assumed that the initial exposure might take some time for the full effect. One reason the preparations were incubated for 5 min while flushing the bath was to allow the preparations to be well exposed. The blocking of K2p channels would depolarize the neurons and potentially inactivate the ability of the nerve to have repetitive firing. Thus, a lower concentration was used at 0.1 mM with flushing the bath well around the preparation as well as allowing 5 min of incubation time as performed for the 5 mM concentration. The neural activity from a representative preparation is shown in Figure 4. The second trial in each condition (A-Saline; B-Doxapram; C-After 5 min of incubation in doxapram; D- Saline wash) is shown. Illustrating the enlarged second trial of the three for each condition allows one to readily see the differences in the number of individual spikes for each condition.

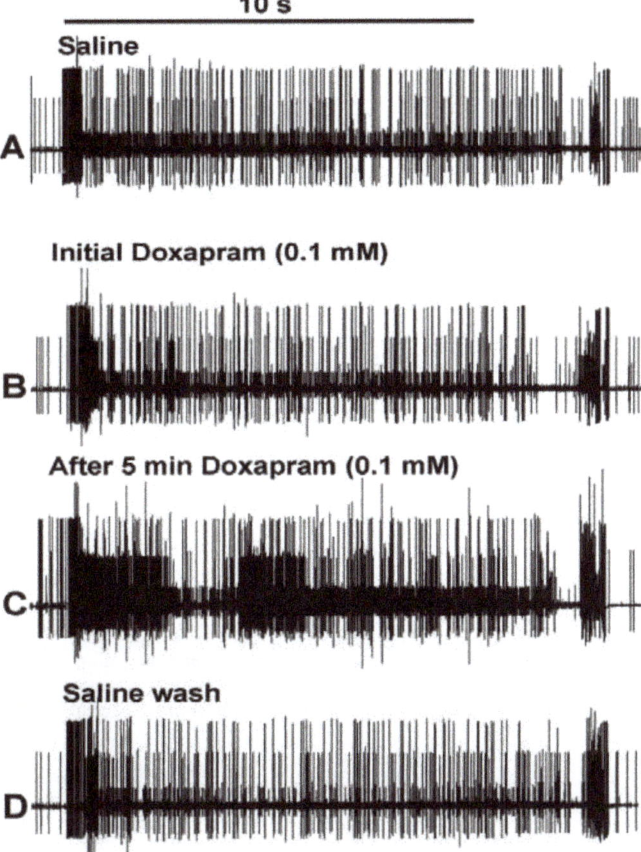

Figure 4. Representative responses to the effects of 0.1 doxapram exposure. The second trial of the three movements for each condition: (**A**) initial saline, (**B**) initial exposure to doxapram, (**C**) 5 min of incubation to doxapram, and (**D**) saline wash out. The number of spikes within the initial 10 s is used for quantification.

The number of spikes varied among the preparations (Figure 5A), but the overall trend showed an increase in the number of spikes after the 5-min incubation. This was readily observed in the mean of the three trials for each condition (Figure 5: $p < 0.05$; paired T-Test; N = 6).

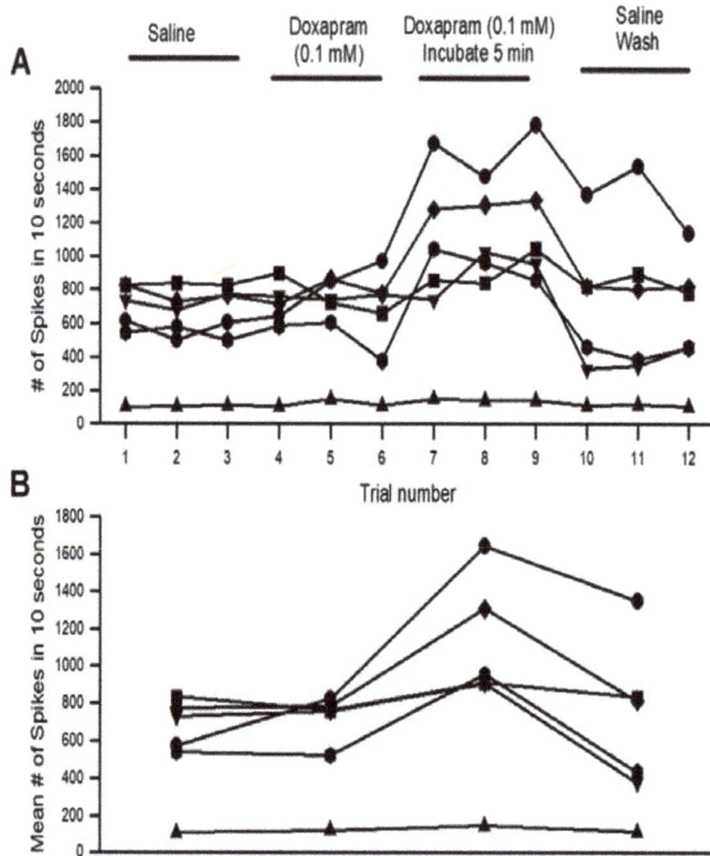

Figure 5. The acute effect of doxapram (0.1 mM) on neural activity of the PD organ. (**A**) The number of spikes measured in the 10 s window from the beginning movement of the joint starting from a bent position (90 degrees) to fully extended within 1 s and held in an extended position for the next 9 s. This paradigm is repeated three times for each condition. Each line represents a different preparation of a PD organ. Three trials were undertaken with saline, three trials were done immediately after switching the bath to doxapram (0.1 mM), and they were examined again after incubation for 5 min. The final exchange was to rinse the preparation twice with fresh saline and then move the joint three more times. Each movement was separated by at least 10 s while the joint was held in a bent position. (**B**) The number of spikes in each of the three trials was averaged and graphed in the same manner as in (**A**), which allows an easier view of the overall effects.

3.2. Effects of Raised Extracellular K^+ on Neural Activity

To address the possibility that doxapram depolarized the neurons to an unexcitable state by depolarizing the membrane and inactivating the voltage-gated sodium channels, the PD organ was exposed to a higher-than-normal concentration of extracellular K^+. The concentration of KCl of 40 mM rapidly depressed the neural activity of the PD nerve during the joint movements (Figure 6). Upon exchanging the bathing media back to saline, the neural activity was able to return (Figures 6D and 7).

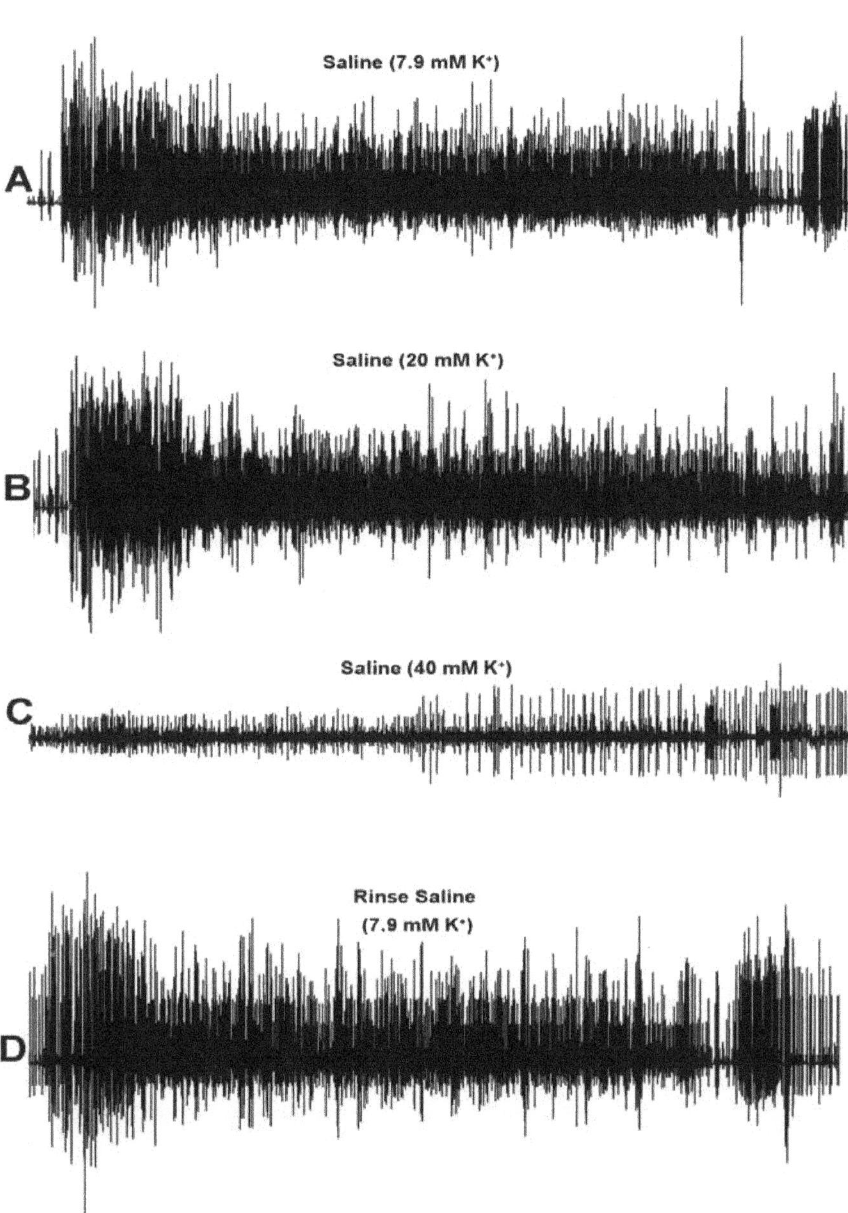

Figure 6. The effect of raised extracellular K^+ on the neural activity for a representative PD organ. The activity of the PD nerve over 10 s when the joint from a 90-degree angle is fully extended within 1 sec and held for 9 s. The activity in saline (**A**) to saline containing 20 mM K^+ (**B**). The 20 mM K^+ did not show a significant change in overall activity. A change to a saline with 40 mM K^+ (**C**) decreased the activity substantially. Some activity is regained with bathing the preparation in fresh saline (**D**).

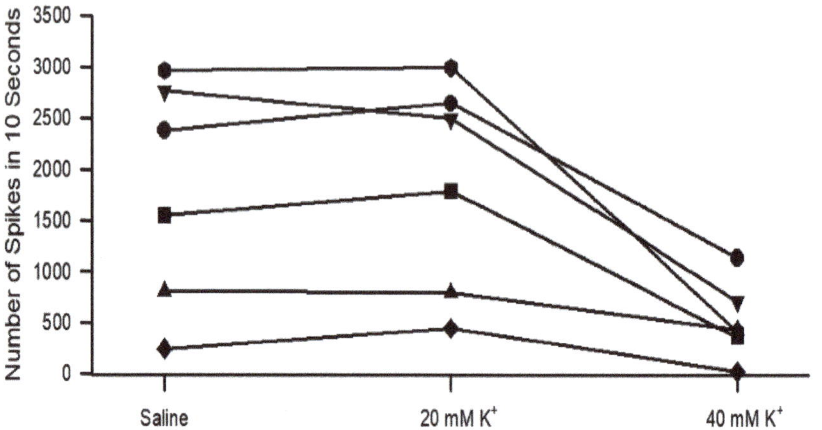

Figure 7. The number of spikes within 10 s when displacing the joint and holding it in a static position for six different preparations. There is no a significant effect for exposure to 20 mM K⁺ but there is a significant decrease in activity when exposed to 40 mM K⁺ ($p < 0.05$ paired T-Test).

3.3. Effects of Verapamil on Neural Activity

Verapamil has recently been described to inhibit a specific subtype of K2p channels (i.e., $K_{2P}18.1$) in mammalian neurons, and is a specific L-Type Ca^{2+} channel blocker in cardiac and neuronal tissues [30,31] For comparison to the effects of doxapram, the same concentrations were examined for verapamil (0.1 and 5 mM). The action of the in-situ PD organ and the excised nerve to examine the effect on the compound action potential (CAP) independent of the sensory endings with the stretch activated ion channels was examined.

A representative preparation for the effect of verapamil at 5 mM on neural activity while moving the PD joint is shown (Figure 8). We used the same paradigm for the exposure with doxapram by bending the joint and extending it three times in each bathing environment (saline, verapamil, saline wash). A total of 10 s were used for the analysis. This consisted of the number of spikes in the first 1 s it took to extend the joint and the subsequent 9 s while the joint was extended. Figure 9 illustrates the number of spikes among the six preparations exposed to 0.1 mM (Figure 9(A1)) and 5 mM (Figure 9(B1)). The activity in saline and initially after changing the bathing media to verapamil, as well as after five and 20 min of incubation, was used for quantification. After vigorous rinsing of the preparations with fresh saline, after exposure to 5 mM verapamil, the activity did not return. After the 0.1 mM exposure and rinsing, the neural activity did partially return. Both 0.1 mM and 5 mM resulted in a significant reduction in neural activity after 20 min of incubation ($p < 0.05$; paired T-Test; N = 6 for each concentration). A more rapid and prominent effect when compared to doxapram at the same concentrations was observed.

The neural activity was rapidly depressed in 5 mM verapamil and over time in 0.1 mM verapamil; thus, the effect on the axonal excitability was examined independent of the sensory endings in the PD organ. The effect on CAPs of the leg nerves were examined for both 0.1 mM and 5 mM. The amplitudes of the evoked CAPs were slightly depressed by exposure to 0.1 mM, but this was not an immediate effect. After 20 min of exposure, some preparations showed little change at 0.1 mM. Since the amplitudes of the CAPs were slightly dampened and appeared to spread over time, the area under the curve was used as an index for the effect of 0.1 mM and 5 mM. Two representative preparations are shown, one for the effect of 0.1 mM (Figure 10(A1–A3)) and one for 5 mM (Figure 10(B1–B3)). The initial amplitude in saline and after 20 min (Figure 10(A2,B2)) as well as after flushing the nerve with fresh saline (Figure 10(A3,B3)) are shown. The areas under the traces for the CAPs of the six preparations for 0.1 mM and the six for 5 mM were used to determine a percentage change from the initial saline. Exposure after 20 min of verapamil and percent change from initial saline to wash out of the verapamil was also examined and was shown to be significantly different ($p < 0.05$; paired T-Test; N = 6 for each concentration). A decrease in the area for the CAPs occurred in both 0.1 and 5 mM, but there was a greater change in area for 5 mM (Figure 11).

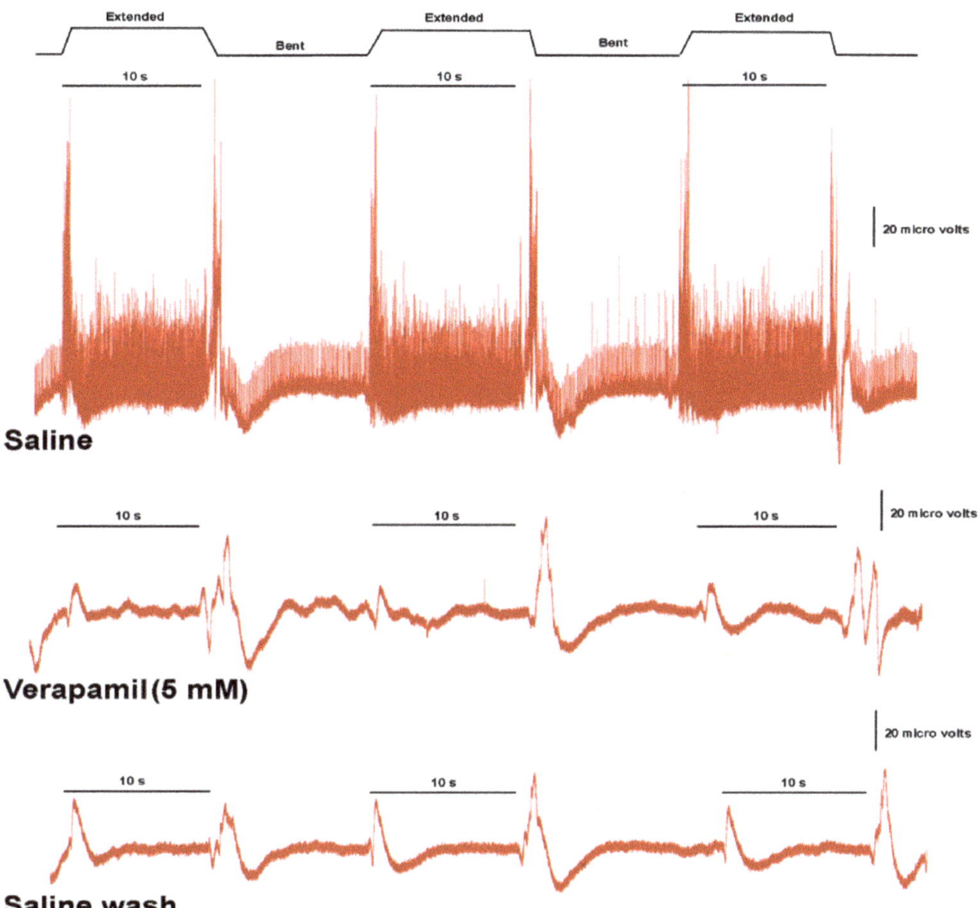

Figure 8. A representative effect of verapamil (5 mM) on neural activity of the PD nerve during the extension and bending of the PD joint. The movement of the joint is shown at the top and the 10 s used for analysis in the number of spikes recorded are shown for each paradigm. Three trials of movement are used initially during saline and when switching the bath to verapamil and following the flushing of the recording chamber with fresh saline.

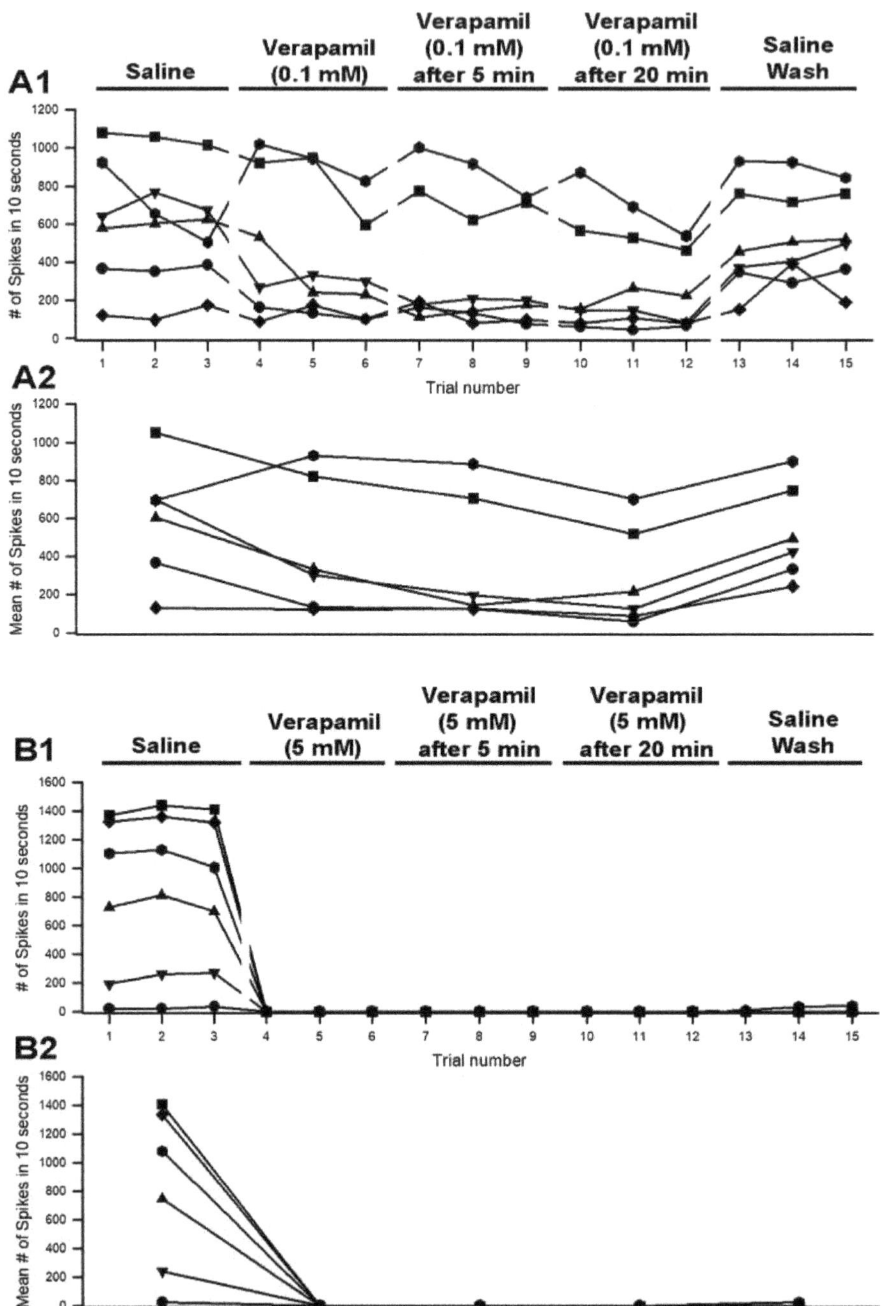

Figure 9. The acute effect of verapamil (0.1 mM and 5 mM) on neural activity of the PD organ. (**A1**) The number of spikes measured in the 10 s window from the beginning movement of the joint starting from a bent position (90 degrees) to fully extended within 1 s and held in an extended position for the next 9 s. This paradigm is repeated three times for each condition. Each line represents a different preparation of a PD organ. Three trials were undertaken with saline, three trials were done

immediately after switching the bath to verapamil (0.1 mM), and they were examined again after incubation for 5 and 20 min. The last step was to rinse the preparation with fresh saline twice and then move the joint three more times. Each movement was separated by at least 10 s while the joint was held in a bent position. (**A2**) The number of spikes in each of the three trials was averaged and graphed in the same manner as in (**A1**), which allows an easier view of the overall effects. (**B1**) The same analysis is shown for the three trials in each condition for exposure to 5 mM verapamil. (**B2**) The average of each of the three trials for each condition is shown. Depression in neural activity was present at 0.1 and 5 mM after 20 min ($p < 0.05$; paired T-Test; N = 6 for each concentration). However, activity was depressed even after the initial exposure to 5 mm ($p < 0.05$; paired T-Test; N = 6).

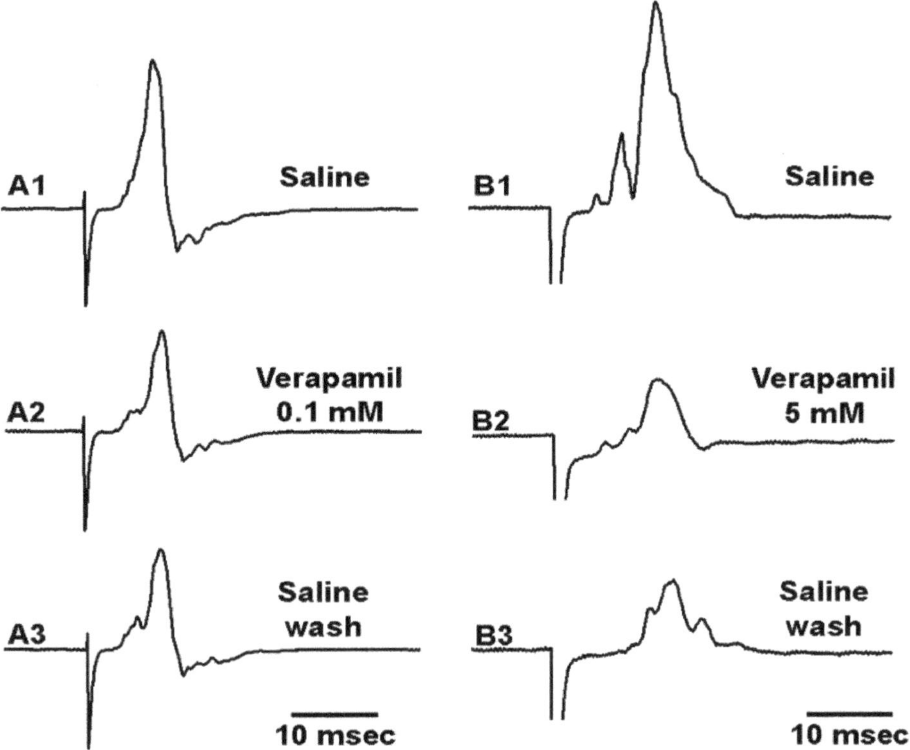

Figure 10. Representative effects of verapamil on the compound action potentials (CAPs) of the walking leg nerve. The CAPs in saline (**A1,B1**) and after 20 min of exposure to 0.1 mM (**A2**) or 5 mM (**B2**) of verapamil showed some depression for both concentrations. (**A3,B3**) Removal of verapamil with fresh saline did not fully recover the amplitude of the CAPs.

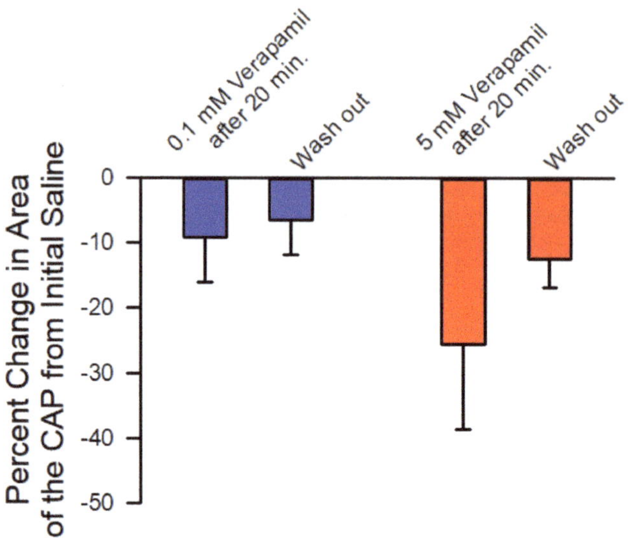

Figure 11. The percent change in the areas of the trace for the compound action potentials (CAPs) in saline and after 20 min of exposure to 0.1 mM or 5 mM of verapamil. The percent change in the area of the CAPs from initial saline to the wash-out is also shown. Some depression still occurred even after washout for both concentrations. ($p < 0.05$; paired T-Test; N = 6 for each concentration from initial saline to after 20 min of exposure to verapamil).

4. Discussion

This study demonstrated that a low concentration of doxapram (0.1 mM) after 5 min of incubation resulted in hyperexcitability of the proprioceptive organ in an excised crab limb. However, a higher concentration (5 mM) after 5 min substantially decreased neural activity. The mechanism of action is likely in blocking of the K2p channels that maintain the resting membrane potential, as well as potentially the stretch activated channels in the sensory endings. It remains to be determined if K2p channels are expressed in these neurons of the crab. If these channels are present, blocking some K2p channels may depolarize the neurons to be closer to the threshold of the voltage-gated Na^+ channels. Thus, the same sensory stimulus would more readily produce action potentials. The higher concentration of doxapram may block more K2p channels, causing depolarization without sufficient repolarization to remove the inactivation of voltage-gated Na^+ channels. This would result in a prolonged refractory period. The proposed mechanism of action is supported by exposure to a high concentration of K^+ which would produce a depolarized state of the neurons and would result in a prolonged state of voltage-gated Na^+ channel inactivation until the high K^+ is removed. Similarly, when doxapram is removed, the neural activity begins to return for the 0.1 mM exposure. This indicates the acute effects without permanent damage to the neurons. The axons were viable with 5 mM exposure to verapamil after 20 min of exposure, despite the PD organ not responding after 20 min of exposure. The stretch activated channels in the sensory endings appeared more sensitive to verapamil than channels in the axon. A working model to explain these observations is illustrated in Figure 12.

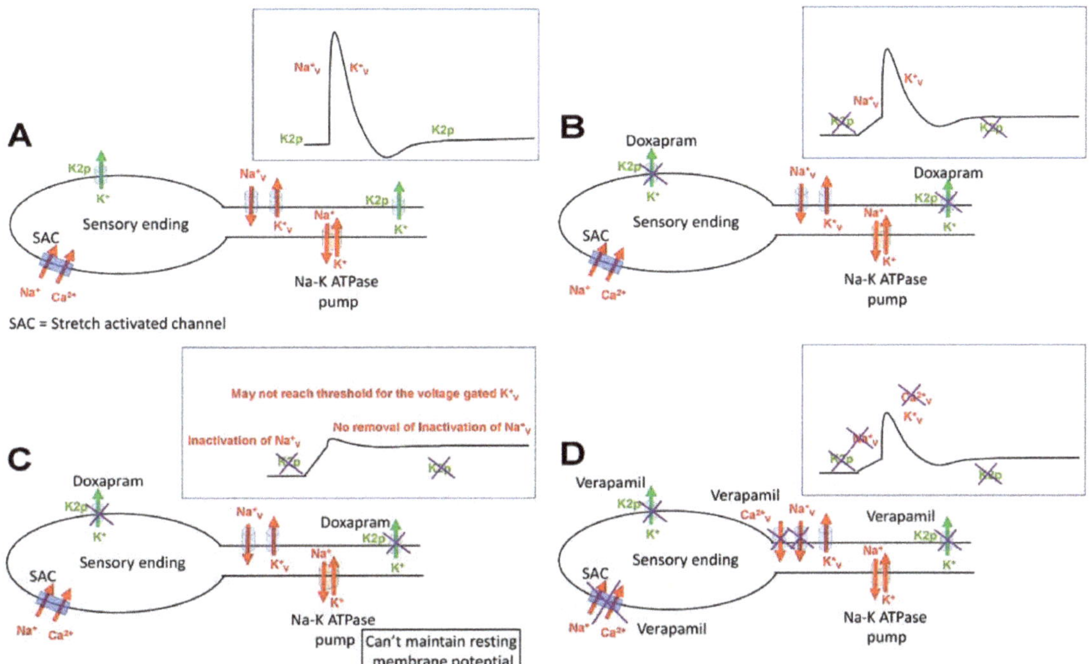

Figure 12. A representative model to explain the observed phenomenon with exposure to doxapram in relation to the neural activity in sensory neurons of the PD organ in a crab preparation. (**A**) A representative sensory neuron which is activated by opening stretch activated ion channels (SACs). The depolarization from activating SACs may reach the threshold to open voltage-gated Na⁺ channels (Na$_v$⁺) and, subsequently, voltage-gated K⁺ channels (K$_v$⁺) to allow action potentials to travel along the nerve (see inset in top right corner). The K2p channels help to maintain the resting membrane potential along with the Na⁺-K⁺ ATP dependent pump. (**B**) In the presence of doxapram, the K2p channels are blocked. (**C**) The effect of blocking the K2p channels depolarizes the neurons. A low level may bring the membrane closer to threshold to activate the Na$_v$⁺ channels and produce more action potentials for the same stimulus. However, if the neuron depolarizes and cannot repolarize, the inactivation of the Na$_v$⁺ channels would result in prolonged absolute refractory periods while exposed to doxapram, not allowing the neuron to be excitable. (**D**) The potential effects of verapamil are illustrated. The SACs are likely blocked rapidly upon exposure to low and high concentrations. On the axon, the voltage gated Ca²⁺ channels (Ca$_v$²⁺) as well as the Na$_v$⁺ may be a target given the compound action potential is slowly depressed over time, independent of the actions on the SACs in the sensory endings.

Since doxapram has an action on crab sensory neurons, this would suggest that a subtype of K2p channels similar to those described in mammals, which are doxapram-sensitive, exist in crabs as well. However, this remains to be confirmed with molecular identification, which is beyond the scope of this initial study. Doxapram is indicated to act on the TASK subtype of K2p channels [6,10]; these are acid-sensitive channels, in which low pH inhibits the function of the channel. It was demonstrated in an earlier study that low pH blocks the neural activity of the crab PD organ [26]. To date, there are no reports as to the types or number of K2p channels represented in the genomes of crustaceans. However, 15 subtypes have been identified in mammal genomes and 11 in the fruit fly *Drosophila* [32,33].It is likely that there are similar subtypes in crustaceans as in insects since they share chordotonal organs of similar anatomical structure used to monitor joint movements, as well as other physiological similarities. It would be of interest to know if multiple subtypes of K2p channels are expressed within a single cell. A high concentration of 5 mM of doxapram did not completely block the neural activity in the crab axons. This would indicate that the membrane potential did not depolarize to zero, and that other non-doxapram sensitive K2p channels are potentially helping to maintain a membrane potential to allow the neurons to remain active. Screening more organisms, from plants to animals, for types of K2p channels would pave the way for pharmacological and physiological studies to examine functional significance.

Invertebrate models can aid in screening novel pharmacological agents once research better understands the similarities and differences among organisms. This study has shown that doxapram, which is used clinically, has an action on crab neurons. It would be of interest to screen other known agonists and antagonists of K2p channels that are used in mammals on this model preparation as well as other invertebrate models such as crayfish (a freshwater crustacean) or *Drosophila* (insects). For example, verapamil (10 μM) is known to inhibit the K2p channel 18.1 [34].

In examining various pharmacological agents on the chordotonal organ, one needs to consider other channels in addition to the K2p channels. Verapamil is an L-type Ca^{2+} channel blocker [30,31]. Since the stretch-activated channels in the sensory endings of the neurons of the PD organ are blocked by Gd^{3+} [35], the channels may be stretch-activated Ca^{2+} channels which could also be blocked by verapamil. The molecular subtype of the stretch-activated channels associated with sensory neurons has yet to be identified in chordotonal organs of crustaceans [36]. The neural activity was silenced right away when bending the PD joint and exposing the preparation to verapamil (5 mM); however, when removing the sensory endings and evoking electrical activity in the nerve, the evoked CAPs took time to show depression in amplitude and upon flushing the exposed nerve the amplitude of the CAPs was able to partially return. The electrical activity was not able to regainwhen the PD organ was intact and initiating neural activity by displacement of the PD organ with joint movement. It would appear that verapamil rapidly blocked stretch-activated channels, and the compound was not able to be removed readily. In addition to verapamil potentially blocking stretch activated channels in the sensory ending, it also appeared that the voltage-gated Na^+ channels in the axon are blocked, since the amplitude of the CAPs slowly decreases in amplitude and can return with flushing the preparation with fresh saline without verapamil. However, if the K2p channels are blocked in the axon, then the nerve may slowly be depolarized over time during the evoked responses. The voltage-gated Na^+ channels would then be in a state of inactivation, because the membrane potential is unable to hyperpolarize enough to remove the inactivation. Flushing away the verapamil allows the membrane potential to return to a negative state to rejuvenated voltage-gated Na^+ channels. A variable needing to be considered is that the action potential in these crab neurons may have a Ca^{2+} component, as with peripheral neurons in crayfish [37,38]. The action potential becomes narrower when the influx of Ca^{2+} is reduced. In central neurons within crayfish, verapamil can alter the shape of the action potential [39]. If verapamil blocks the voltage-gated channels in axons, then the calcium-activated potassium conductance would also be reduced. Calcium-activated potassium channels are known to be present in crayfish neurons [38], If this occurs, the frequency of activity could decrease as the neuron would not reset to baseline as quickly and remove inactivation of the sodium channels. Thus, single channel recordings of the SACs as well as the subtype of ion channels in the axon are needed to better understand the mechanism of action for verapamil on this preparation.

In order to examine reproducibility in the observations and avoid potential bias by a given investigator, five groups of the total 17 students investigated the same procedures as three displacements in saline and 3 more after initial exposure to doxapram (5 mM) and after 5 min of incubation in doxapram followed by two saline flushes of the preparation. As shown in the Appendix A, the overall trends proved consistent. The experiments performed with the course participants may not have been as consistent in the rate of movements as using one individual for all the data sets presented in the Results. University policies indicated that the experiments performed in the classroom had to be conducted within a fume hood due to low airflow in the teaching laboratory.

Since it is now common practice to provide raw data sets with publications, it is important to understand how the data were analyzed, as differences in interpretation could occur. This is particularly relevant when using automated analysis procedures provided by commercial software. It is suggested to use an approach of measuring the spikes which occur outside two times the level of the background noise in the types of recordings performed herein. Issues faced with data analysis are described in the Appendix A. A detailed explanation in analysis with the software used in this study is provided in an earlier study [29]. When following the details provided in Tanner et al., [29] the errors shown can be avoided.

Author Contributions: Conceptualization, R.L.C. and B.J.I.; methodology, B.J.I., M.O.A.-K., S.A., A.W.A., C.A., C.O.B., I.N.T., J.N.H., J.G.C., K.E.B., L.T.T., M.M., P.P., P.C.B., S.B.S., H.N.T., M.L.W., A.K.W. and R.L.C.; software, R.L.C.; validation, B.J.I., M.O.A.-K., S.A., A.W.A., C.A., C.O.B., I.N.T., J.N.H., J.G.C., K.E.B., L.T.T., M.M., P.P., P.C.B., S.B.S., H.N.T., M.L.W., A.K.W. and R.L.C.; formal analysis, B.J.I., M.O.A.-K., S.A., A.W.A., C.A., C.O.B., I.N.T., J.N.H., J.G.C., K.E.B., L.T.T., M.M., P.P., P.C.B., S.B.S., H.N.T., M.L.W., A.K.W. and R.L.C.; investigation, B.J.I., M.O.A.-K., S.A., A.W.A., C.A., C.O.B., I.N.T., J.N.H., J.G.C., K.E.B., L.T.T., M.M., P.P., P.C.B., S.B.S., H.N.T., M.L.W., A.K.W. and R.L.C.; resources, R.L.C.; data curation, R.L.C.; writing—original draft preparation, B.J.I. and R.L.C.; writing—review and editing, B.J.I., M.O.A.-K., S.A., A.W.A., C.A., C.O.B., I.N.T., J.N.H., J.G.C., K.E.B., L.T.T., M.M., P.P., P.C.B., S.B.S., H.N.T., M.L.W., A.K.W. and R.L.C.; visualization, B.J.I., M.O.A.-K., S.A., A.W.A., C.A., C.O.B., I.N.T., J.N.H., J.G.C., K.E.B., L.T.T., M.M., P.P., P.C.B., S.B.S., H.N.T., M.L.W., A.K.W. and R.L.C.; supervision, R.L.C.; project administration, R.L.C.; funding acquisition, R.L.C. All authors have read and agreed to the published version of the manuscript.

Funding: Department of Biology, University of Kentucky. Personal funds (R.L.C.). Chellgren Endowed Funding (R.L.C.). College of Arts and Sciences summer fellowship (M.L.W.).

Institutional Review Board Statement: Not applicable.

Informed Consent Statement: Not applicable.

Data Availability Statement: All data are available in the manuscript and are available upon request.

Acknowledgments: We thank Syed Z. Ali and Eric G. Johnson (University of Kentucky College of Medicine) for useful discussion on the topic of doxapram and clinical procedures.

Conflicts of Interest: The authors declare that they have no conflict of interest.

Appendix A

Appendix A.1. Reproducibility in the Effects of Doxapram with a University Course

Five groups with a total of 17 students investigated the same general procedures as reported in the main study. Each group displaced their crab leg three times in saline, then three times after initial exposure to doxapram (5 mM), then three times after five minutes of incubation in doxapram, and finally three times after two saline washes. The activity profiles for the five preparations performed with the course are shown in Figure A1.

Conditions varied depending on how the joint was moved and the extent of flushing solutions around the dish in which the preparation was placed. Preparations were dissected prior to the course starting. Some were dissected 3 to 4 h before being examined; however, the saline bath was exchanged with fresh saline about every hour prior to experimentation. In addition, due to requirements of the university chemical safety committee, the experiments needed to be performed within a fume hood, which produced some vibration and thus activity of the sensory neurons.

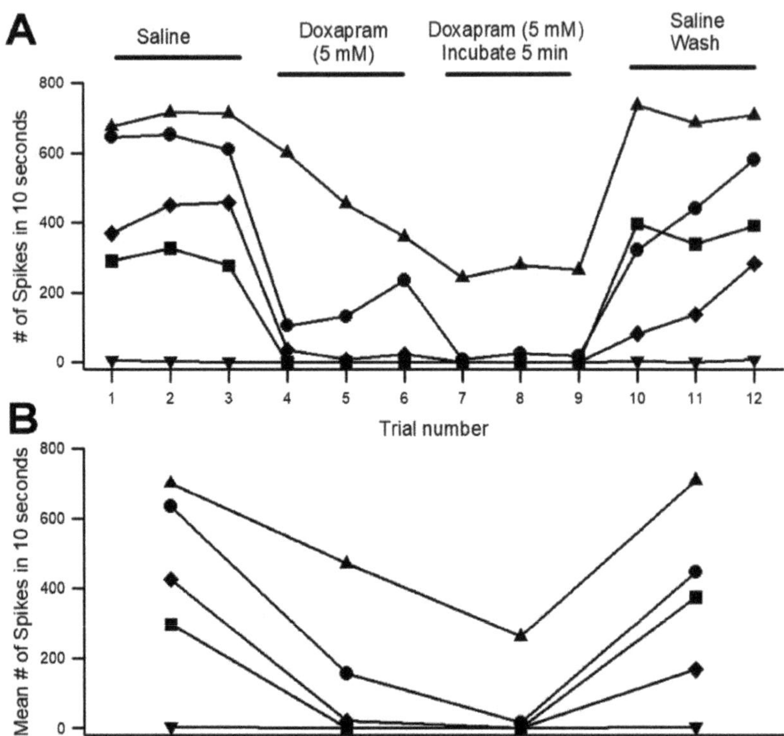

Figure A1. The activity obtained from the PD nerve for five different groups within a university course for the effect of doxapram (5 mM). (**A**) The number of spikes were measured in the same way and by the same person who measured all the traces independently of the student participants for consistency in analysis. (**B**) The number of spikes in each of the three trials was averaged and graphed in the same manner as in (**A**), which allows an easier view of the overall effects.

Appendix A.2. Reproducibility in Analysis of Given Data Sets

Two data sets were given to four different groups of participants without knowing the compound provided or its concentration. Thus, this was a blind study in the analysis. Each group consisted of two participants working together. An explanation of how to count the number of spikes within 10 s of the stimulus was provided. The results were compared to a person deemed as a master in analysis since this person analyzed all the data sets for the whole study. The two given data sets are shown as A (#1) and B (#2) within Figure A2. Spike counts in the A data set were relatively consistent for the four different groups of participants, and the overall trends were similar. The data set shown in Figure A2B proved to be challenging to the course participants using the automated software for analysis. One of the four groups had a similar trend as the master.

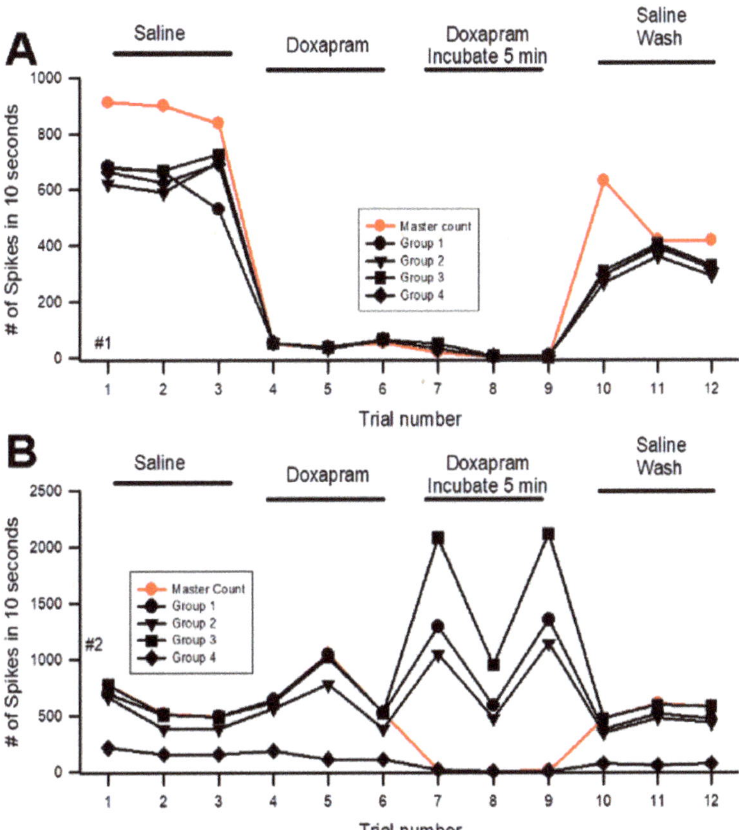

Figure A2. Four groups of different participants, with two people per group, analyzed the same data sets (A #1 and B #2). (**A**) The four groups who analyzed the #1 data set fits well with the analysis from a master in analysis, whose counts are indicated by the red line. The master analyzed all data sets in the study. (**B**) The participants deviated from the master in analysis for three of the four groups in the condition where the doxapram had been incubated for 5 min. The other experimental conditions followed a similar trend as that of the master.

Further investigation and discussion with the class revealed the reasoning for the discrepancy. Upon analysis of the data in the initial three saline trials, the standard deviation of the mean value of the trace was set to a value of 2 (Figure A3(A1)). This setting was ideal to detect the spikes above the background noise. However, since the number of spikes decreased in the set of three trials after the 5-min incubation in doxapram (5 mM) (Figure A3(B1)), the automated software picked up noise, as it was still determining it to be above the mean by two times the standard deviation of the trace. In this case, there were no large spikes, so small deflections in the noise were counted in the automated detection process (Figure A3(A2,B2)). A detailed explanation of these issues in analysis and how to correct for the errors with the software is provided in an earlier study [29]. Since the baseline moves in some cases while moving the joint and the spikes along with the baseline are easily seen visually, but if a set threshold is used to detect spikes, errors will arise as the baseline moves and crosses a threshold.

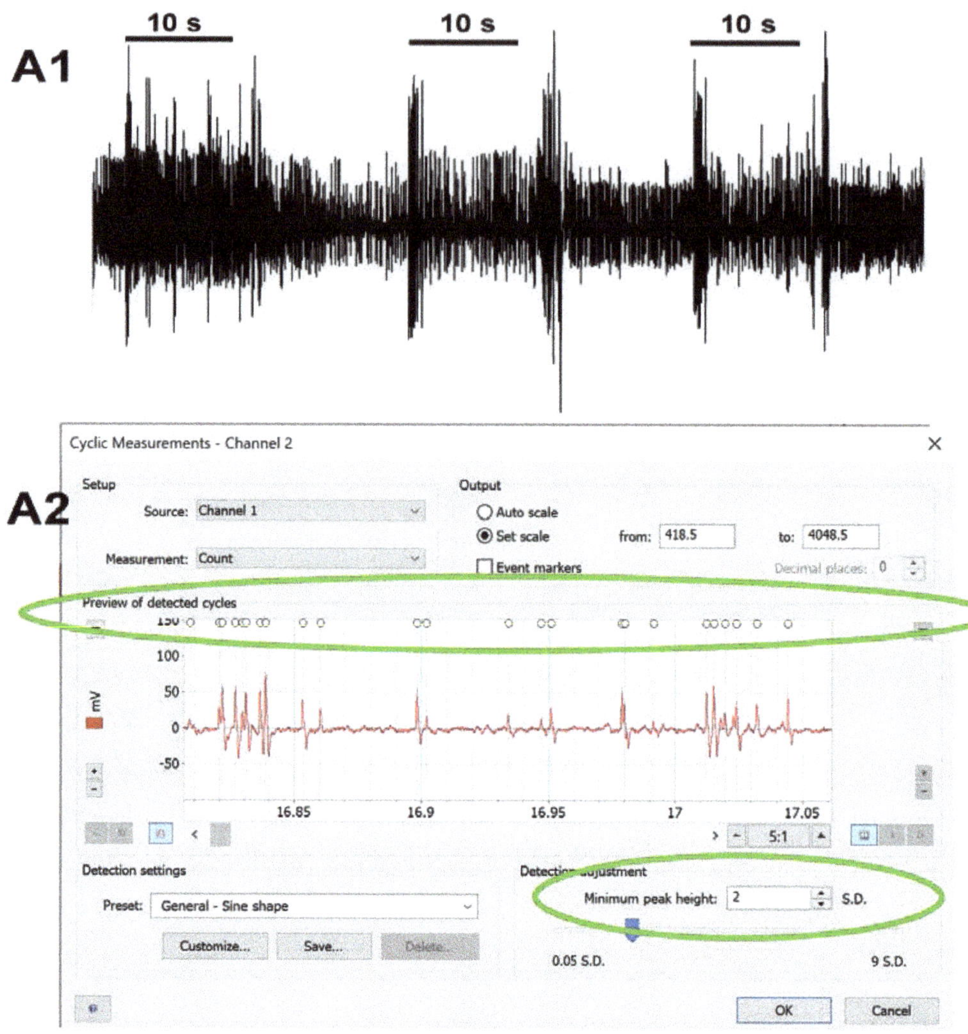

Figure A3. *Cont.*

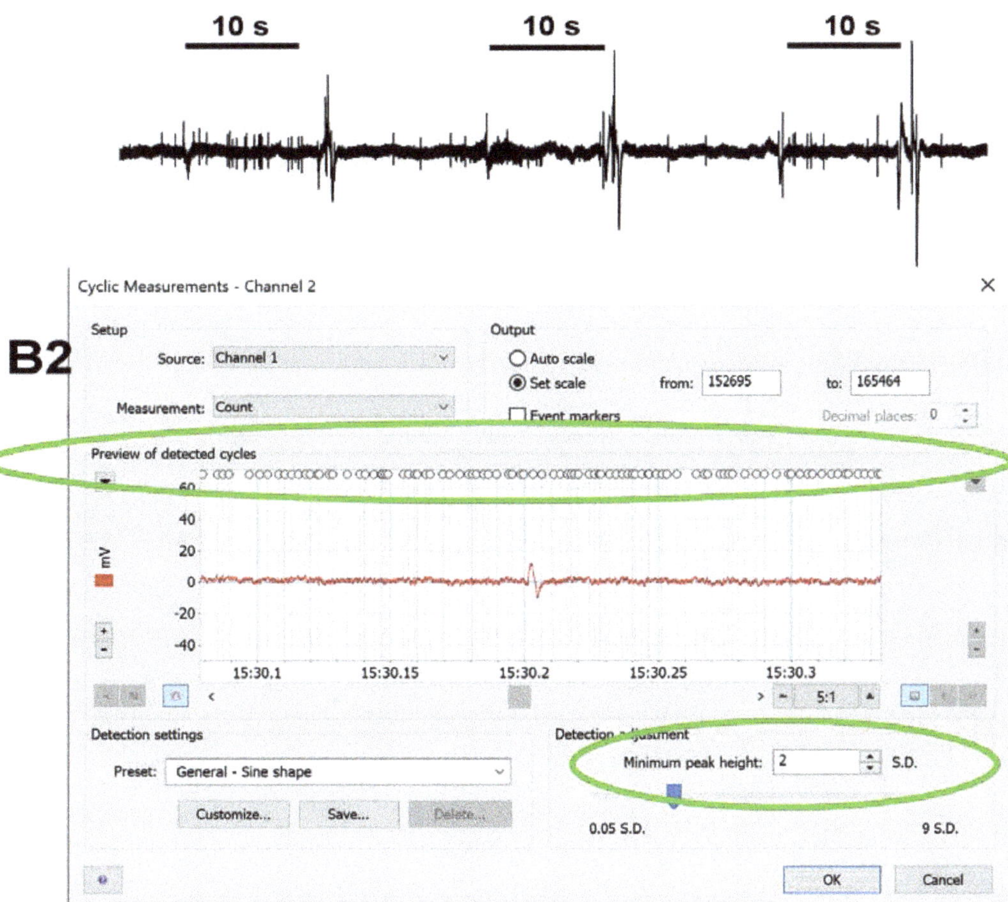

Figure A3. Analysis of data with the same automated measure. Two traces are provided: three trials in saline (**A1**) and the three trials in doxapram after 5 min of incubation (**B1**). The automated analysis is set to 2 SD (standard deviations of the mean in the traces; see small green ellipse in (**A2,B2**)) for both (**A1,B1**), but it produces differences in what is detected, as indicated in the large green ellipse in (**A2,B2**). The open circles inside the ellipse indicate what deflections in the trace are counted as spikes. Since there is little neural activity in (**B1**), the noise of the trace is being detected as spikes, producing an artificially high number of spikes.

References

1. Buckingham, S.D.; Kidd, J.F.; Law, R.J.; Franks, C.J.; Sattelle, D.B. Structure and function of two-pore-domain K+ channels: Contributions from genetic model organisms. *Trends Pharmacol. Sci.* **2005**, *26*, 361–367. [CrossRef] [PubMed]
2. Goldstein, S.A.; Price, L.A.; Rosenthal, D.N.; Pausch, M.H. ORK1, a potassium-selective leak channel with two pore domains cloned from Drosophila melanogaster by expression in Saccharomyces cerevisiae. *Proc. Natl. Acad. Sci. USA* **1996**, *93*, 13256–13261. [CrossRef] [PubMed]
3. Goldstein, S.A.; Wang, K.W.; Ilan, N.; Pausch, M.H. Sequence and function of the two P domain potassium channels: Implications of an emerging superfamily. *J. Mol. Med.* **1998**, *76*, 13–20. [CrossRef] [PubMed]

4. Plant, L.D.; Goldstein, S.A.N. Two-Pore Domain Potassium Channels. In *Handbook of Ion Channels*, 1st ed.; Zheng, J., Trudeau, M.C., Eds.; CRC Press: Boca Raton, FL, USA, 2015; ISBN 9780429193965.

5. Enyedi, P.; Czirják, G. Molecular background of leak K+ currents: Two-pore domain potassium channels. *Physiol. Rev.* **2010**, *90*, 559–605. [CrossRef] [PubMed]

6. Kim, D. Physiology and pharmacology of two-pore domain potassium channels. *Curr. Pharm. Des.* **2005**, *11*, 2717–2736. [CrossRef]

7. Kamuene, J.M.; Xu, Y.; Plant, L.D. The pharmacology of two-pore domain potassium channels. *Handb. Exp. Pharmacol.* **2021**, *267*, 417–443. [PubMed]

8. Patel, A.J.; Honore, E.; Lesage, F.; Fink, M.; Romey, G.; Lazdunski, M. Inhalational anesthetics activate two-pore-domain background K channels. *Nat. Neurosci.* **1999**, *2*, 422–426. [CrossRef]

9. Rajan, S.; Wischmeyer, E.; Karschin, C.; Preisig-Muller, R.; Grzeschik, K.H.; Daut, J.; Karschin, A.; Derst, C. THIK-1 and THIK-2, a novel subfamily of tandem pore domain K+ channels. *J. Biol. Chem.* **2001**, *276*, 7302–7311. [CrossRef]

10. Cotten, J.F.; Keshavaprasad, B.; Laster, M.J.; Eger, E.I., 2nd; Yost, C.S. The ventilatory stimulant doxapram inhibits TASK tandem pore (K2P) potassium channel function but does not affect minimum alveolar anesthetic concentration. *Anesth. Analg.* **2006**, *102*, 779–785. [CrossRef]

11. Komatsu, R.; Sengupta, P.; Cherynak, G.; Wadhwa, A.; Sessler, D.I.; Liu, J.; Hurst, H.E.; Lenhardt, R. Doxapram only slightly reduces the shivering threshold in healthy volunteers. *Anesth. Analg.* **2005**, *101*, 1368–1373. [CrossRef]

12. Yost, C.S. A new look at the respiratory stimulant doxapram. *CNS Drug Rev.* **2006**, *12*, 236–249. [CrossRef] [PubMed]

13. Song, S.S.; Lyden, P.D. Overview of therapeutic hypothermia. *Curr. Treat. Options. Neurol.* **2012**, *14*, 541–548. [CrossRef] [PubMed]

14. Cunningham, K.P.; MacIntyre, D.E.; Mathie, A.; Veale, E.L. Effects of the ventilatory stimulant, doxapram on human TASK-3 (KCNK9, K2P9.1) channels and TASK-1 (KCNK3, K2P3.1) channels. *Acta Physiol.* **2020**, *228*, e13361. [CrossRef] [PubMed]

15. Vliegenthart, R.J.; Ten Hove, C.H.; Onland, W.; van Kaam, A.H. Doxapram treatment for apnea of prematurity: A systematic review. *Neonatology* **2017**, *111*, 162–171. [CrossRef]

16. Baxter, A.D. Side effects of doxapram infusion. *Eur. J. Intensive Care Med.* **1976**, *2*, 87–88. [CrossRef] [PubMed]

17. Fathi, M.; Massoudi, N.; Nooraee, N.; Beheshti Monfared, R. The effects of doxapram on time to tracheal extubation and early recovery in young morbidly obese patients scheduled for bariatric surgery: A randomised controlled trial. *Eur. J. Anaesthesiol.* **2020**, *37*, 457–465. [CrossRef] [PubMed]

18. Lee, L.M.; Müntefering, T.; Budde, T.; Meuth, S.G.; Ruck, T. Pathophysiological role of K_{2P} channels in human diseases. *Cell Physiol. Biochem.* **2021**, *55*, 65–86.

19. Holter, J.; Carter, D.; Leresche, N.; Crunelli, V.; Vincent, P. A TASK3 channel (KCNK9) mutation in a genetic model of absence epilepsy. *J. Mol. Neurosci.* **2005**, *25*, 37–51. [CrossRef]

20. DeGiorgis, J.A.; Jang, M.; Bearer, E.L. The giant axon of the squid: A simple system for axonal transport studies. *Meth. Mol. Biol.* **2022**, *2431*, 3–22.

21. Dow, J.A.T.; Simons, M.; Romero, M.F. Drosophila melanogaster: A simple genetic model of kidney structure, function and disease. *Nat. Rev. Nephrol.* **2022**, *18*, 417–434. [CrossRef]

22. Ecovoiu, A.A.; Ratiu, A.C.; Micheu, M.M.; Chifiriuc, M.C. Inter-species rescue of mutant phenotype-the standard for genetic analysis of human genetic disorders in *Drosophila melanogaster* model. *Int. J. Mol. Sci.* **2022**, *23*, 2613. [CrossRef] [PubMed]

23. Haley, J.A.; Hampton, D.; Marder, E. Two central pattern generators from the crab, *Cancer borealis*, respond robustly and differentially to extreme extracellular pH. *Elife* **2018**, *7*, e41877. [CrossRef] [PubMed]

24. Otopalik, A.G.; Pipkin, J.; Marder, E. Neuronal morphologies built for reliable physiology in a rhythmic motor circuit. *Elife* **2019**, *8*, e41728. [CrossRef] [PubMed]

25. Ugur, B.; Chen, K.; Bellen, H.J. Drosophila tools and assays for the study of human diseases. *Dis. Model. Mech.* **2016**, *9*, 235–244. [CrossRef] [PubMed]

26. Dayaram, V.; Malloy, C.A.; Martha, S.; Alvarez, B.; Chukwudolue, I.; Dabbain, N.; Goleva, S.; Hickey, T.; Ho, A.; King, M.; et al. The effect of CO_2, intracellular pH and extracellular pH on mechanosensory proprioceptor responses in crayfish and crab. *Am. J. Undergrad. Res.* **2017**, *14*, 85–99.

27. Pankau, C.; Nadolski, J.; Tanner, H.; Cryer, C.; Di Girolamo, J.; Haddad, C.; Lanning, M.; Miller, M.; Neely, D.; Wilson, R.; et al. Effects of manganese on physiological processes in Drosophila, crab and crayfish: Cardiac, neural and behavioral assays. *Comp. Biochem. Physiol. C* **2022**, *251*, 109209.

28. Majeed, Z.R.; Titlow, J.; Hartman, H.B.; Cooper, R.L. Proprioception and tension receptors in crab limbs: Student laboratory exercises. *J. Vis. Exp.* **2013**, *80*, e51050. [CrossRef]

29. Tanner, H.N.; Atkins, D.E.; Bosh, K.L.; Breakfield, G.W.; Daniels, S.E.; Devore, M.J.; Fite, H.E.; Guo, L.Z.; Henry, D.K.J.; Kaffenberger, A.K.; et al. Effect of TEA and 4-AP on primary sensory neurons in a crustacean model. *J. Pharmacol. Toxicol.* **2022**, *17*, 14–27. [CrossRef]

30. Triggle, D.J. Calcium channel antagonists: Clinical uses–past, present and future. *Biochem. Pharmacol.* **2007**, *74*, 1–9. [CrossRef]

31. Reuter, H. Calcium channel modulation by neurotransmitters, enzymes and drugs. *Nature* **1983**, *301*, 569–574. [CrossRef]

32. Adams, M.D.; Celniker, S.E.; Holt, R.A.; Evans, C.A.; Gocayne, J.D.; Amanatides, P.G.; Scherer, S.E.; Li, P.W.; Hoskins, R.A.; Galle, R.F.; et al. The genome sequence of Drosophila melanogaster. *Science* **2000**, *287*, 2185–2195. [CrossRef] [PubMed]

33. Littleton, J.T.; Ganetzky, B. Ion channels and synaptic organization: Analysis of the Drosophila genome. *Neuron* **2000**, *26*, 35–43. [CrossRef]

34. Park, H.; Kim, E.J.; Ryu, J.H.; Lee, D.K.; Hong, S.G.; Han, J.; Han, J.; Kang, D. Verapamil inhibits TRESK ($K_{2P}18.1$) current in trigeminal ganglion neurons independently of the blockade of Ca^{2+} influx. *Int. J. Mol. Sci.* **2018**, *19*, 1961. [CrossRef] [PubMed]
35. Dayaram, V.; Malloy, C.A.; Martha, S.; Alvarez, B.; Chukwudolue, I.; Dabbain, N.; Goleva, S.; Hickey, T.; Ho, A.; King, M.; et al. Stretch activated channels in proprioceptive chordotonal organs of crab and crayfish are sensitive to Gd3+ but not amiloride, ruthenium red or low pH. *IMPULSE* **2017**, *14*. Available online: https://impulse.pubpub.org/pub/8xdynrg8 (accessed on 20 August 2022).
36. McCubbin, S.; Jeoung, A.; Waterbury, C.; Cooper, R.L. Pharmacological profiling of stretch activated channels in proprioceptive neuron. *Comp. Biochem. Physiol. C* **2020**, *233*, 108765. [CrossRef]
37. Wojtowicz, J.M.; Atwood, H.L. Presynaptic membrane potential and transmitter release at the crayfish neuromuscular junction. *J. Neurophysiol.* **1984**, *52*, 99–113. [CrossRef]
38. Sivaramakrishnan, S.; Brodwick, M.S.; Bittner, G.D. Presynaptic facilitation at the crayfish neuromuscular junction. Role of calcium-activated potassium conductance. *J. Gen. Physiol.* **1991**, *98*, 1181–1196. [CrossRef]
39. Kuroda, T. The effects of D600 and verapamil on action potential in the X-organ neuron of the crayfish. *Jpn. J. Physiol.* **1976**, *26*, 189–202. [CrossRef]

Article

Deficits in Cerebellum-Dependent Learning and Cerebellar Morphology in Male and Female BTBR Autism Model Mice

Elizabeth A. Kiffmeyer, Jameson A. Cosgrove, Jenna K. Siganos, Heidi E. Bien, Jade E. Vipond, Karisa R. Vogt and Alexander D. Kloth *

Department of Biology, Augustana University, Sioux Falls, SD 57197, USA
* Correspondence: alexander.kloth@augie.edu

Abstract: Recently, there has been increased interest in the role of the cerebellum in autism spectrum disorder (ASD). To better understand the pathophysiological role of the cerebellum in ASD, it is necessary to have a variety of mouse models that have face validity for cerebellar disruption in humans. Here, we add to the literature on the cerebellum in mouse models of autism with the characterization of the cerebellum in the idiopathic BTBR T + Itpr3^{tf}/J (BTBR) inbred mouse strain, which has behavioral phenotypes that are reminiscent of ASD in patients. When we examined both male and female BTBR mice in comparison to C57BL/6J (C57) controls, we noted that both sexes of BTBR mice showed motor coordination deficits characteristic of cerebellar dysfunction, but only the male mice showed differences in delay eyeblink conditioning, a cerebellum-dependent learning task that is known to be disrupted in ASD patients. Both male and female BTBR mice showed considerable expansion of, and abnormal foliation in, the cerebellum vermis—including a significant expansion of specific lobules in the anterior cerebellum. In addition, we found a slight but significant decrease in Purkinje cell density in both male and female BTBR mice, irrespective of the lobule. Finally, there was a marked reduction of Purkinje cell dendritic spine density in both male and female BTBR mice. These findings suggest that, for the most part, the BTBR mouse model phenocopies many of the characteristics of the subpopulation of ASD patients that have a hypertrophic cerebellum. We discuss the significance of strain differences in the cerebellum as well as the importance of this first effort to identify both similarities and differences between male and female BTBR mice with regard to the cerebellum.

Keywords: autism spectrum disorder; mouse model; idiopathic; cerebellum

Citation: Kiffmeyer, E.A.; Cosgrove, J.A.; Siganos, J.K.; Bien, H.E.; Vipond, J.E.; Vogt, K.R.; Kloth, A.D. Deficits in Cerebellum-Dependent Learning and Cerebellar Morphology in Male and Female BTBR Autism Model Mice. *NeuroSci* **2022**, *3*, 624–644. https://doi.org/10.3390/neurosci3040045

Academic Editor: Asher Ornoy

Received: 15 September 2022
Accepted: 4 November 2022
Published: 9 November 2022

Publisher's Note: MDPI stays neutral with regard to jurisdictional claims in published maps and institutional affiliations.

1. Introduction

Autism spectrum disorder (ASD) is a neurodevelopmental disorder marked by socio-communicative deficits, repetitive behaviors, and stereotyped interests [1]. It is commonly associated with several neurological and non-neurological comorbidities, including motor delay and disruption, cognitive delay, epileptic seizures, and gastrointestinal disturbances [1]. It is estimated that 1 in 44 children born today will receive a diagnosis of ASD, with males being 3–4.2 times more likely to receive an ASD diagnosis than females [2–4], though this number might represent substantial underdiagnosis of girls and women with ASD [2,5,6]. Over the last two decades, there has been an avalanche of research addressing the genetics and neural correlates of ASD, with the long-term goals of identifying biomarkers for early diagnosis and discovering effective treatments for all patients.

The cerebellum has emerged as a brain area of intense interest for ASD researchers, sparked by three lines of evidence that have been reviewed widely [7–13]. First, many ASD patients have cerebellar malformation, including abnormal cerebellar volume [14–17], alteration of Purkinje cell shape and density [18–21], or disruption of cerebellar white matter tracts [22]. This malformation is often observed early on in life and is strongly predictive of a later diagnosis of ASD [23,24]; for this reason, it has been hypothesized that the cerebellum may play a key early role in the development of brain areas associated with the core ASD behaviors [13]. Second, the cerebellum is an area of the brain in which many ASD susceptibility genes are highly co-expressed, suggesting that mutations at these loci may disrupt cerebellar function [25]. Third, the cerebellum is the locus for disruptions of motor behavior, which are observed in up to 87% of ASD patients [26,27]. Delay eyeblink conditioning, a form of classical conditioning known to require an intact cerebellum [28–30], is also commonly disrupted in ASD patients, who learn the task more slowly, learn to perform the task at a lower rate, or produce inadequate motor responses associated with the task [31–33]. More recent work has gone beyond the motor role of the cerebellum: this work has focused on malformation or malfunction of specific lobules of the cerebellum that are connected with brain areas associated with core ASD behaviors. These studies suggest a regional specificity to disruptions of cerebellar anatomy, activity, and behavior [11,14,15,19,34–36]. Key questions about the relationship between cerebellum and ASD remain—including the exact role of the specific cerebellar lobules in the development of the disorder and the degree to which these findings apply equally to male and female patients—but are beginning to be addressed through preclinical studies, including those employing animal models.

A large fraction of the work on the cerebellum in ASD animal models has focused on rodents modeling single, high-confidence susceptibility genes and environmental models of maternal infection and toxin exposure [37,38]. This work has identified features of the cerebellar pathophysiology that are highly penetrant across ASD cases, uncovering the causative role of region-specific cerebellar function in the development of ASD, an has provided an important proving ground for novel therapeutics that may be used in patients [36,39–50]. In addition, these studies have begun to identify high-confidence targets for rescue in therapeutics studies. For example, it has recently been suggested that delay eyeblink conditioning, which is affected in ASD model mice [45], may be a target that has a one-to-one correspondence between preclinical models and patients [51]. While this work has been critical for understanding the role of the cerebellum in ASD, it has been narrowly focused on models that represent a remarkably small fraction of syndromic or environmental cases in ASD [52]. It remains to be seen how broadly these findings apply to idiopathic cases of ASD, which represent most patients and capture the complex genetic and environmental etiology of the disease. To answer this question, it is important to examine the cerebellum in idiopathic rodent models of ASD.

One commonly used idiopathic mouse model of ASD is the BTBR T + Itpr3^{tf}/J (BTBR) mouse [53,54]. This inbred mouse line displays many phenotypes that are analogous to the core disruptions seen with ASD, including disrupted social behavior; disrupted ultrasonic vocalization; deficient performance in cognitive tasks; and repetitive species-specific behaviors such as excessive grooming and disrupted marble burying [55–60]. Few studies have determined whether these phenotypes occur equally in male and female mice [61,62]. Studies that have examined the cerebellum in BTBR mice have found hyperplasia [63,64], disrupted gene expression and epigenetic regulation [65], and signs of immune dysfunction and oxidative stress [65,66]. Only one recent study has examined the BTBR cerebellum on a lobular level, discovering altered neuronal signaling associated with social behavior in lobules IV/V [67]. Despite these findings, there are some significant open questions about the BTBR mouse model, including the degree to which cerebellum-related behavior is dysfunctional, whether disruptions to Purkinje cell density and morphology are present, whether cerebellar effects differ based on lobule, and whether these findings are present in both males and females. Addressing these questions will be important in establishing the BTBR strain as a valid mouse model for exploring the role of the cerebellum in ASD.

In the present study, we examine cerebellum-specific motor learning in the BTBR mice to see if the strain phenocopies what has been observed in ASD patients. We also investigate anatomical and morphological alterations in the adult BTBR mice and determine whether these alterations depend on which lobule of the cerebellar vermis is affected. Importantly, we examine, for the first time, whether sex is an important biological variable in cerebellar dysfunction in BTBR mice.

2. Materials and Methods

2.1. Animals

Male and female BTBR T + Itpr3^{tf}/J (BTBR) mice were bred at Augustana University using breeding pairs obtained from Jackson Laboratories, Bar Harbor, Maine (stock no. 002282; RRID:MGI:2160299). Male and female C57BL/6J (C57) mice were bred at Augustana University using breeding pairs obtained from Jackson Laboratories, Bar Harbor, Maine (stock no. 000664; RRID:IMSR_JAX:000664). Mice were between 8 and 16 weeks old in all experiments. Sample sizes for each experiment–consistent with prior experiments on the cerebellum in ASD model mice [45]—are shown in Table 1, listed by experiment and figure.

Table 1. Sample size for all experiments, listed by sex, strain, experiment, and figure number. Shaded cells are indicated for figures that only report data from a single sex. *, sample size reported as number of cells/number of mice.

Experiment (Figure)	Male Mice		Female Mice	
	C57	BTBR	C57	BTBR
Rotarod (Figure 1A,B)	8	8	8	8
Eyeblink conditioning (Figure 1C,D)	13	12	10	11
Brain weight (Figure 2A)	10	12	-	-
Vermal anatomy on Nissl-stained tissue (Figure 2C–I)	8	8	-	-
Brain weight (Figure 3A)	-	-	7	10
Vermal anatomy on Nissl-stained tissue (Figure 3C–I)	-	-	6	6
Purkinje cell density (Figure 4B,C)	8	7	6	6
Golgi–Cox Purkinje cell analysis (Figure 5B–D,F–H) *	18/6	20/7	18/5	22/6
Golgi–Cox Spine density analysis (Figure 5E,I) *	27/10	25/10	15/6	15/5

All mice were housed on a 12 h light-dark cycle (7 a.m.–7 p.m.) in open-top mouse cages (Ancare, Bellmore, NY, USA) in groups of 2–5 littermates per cage. Animals had ad libitum access to food and water during this period. All procedures were conducted in accordance with protocols approved by the Augustana University Institutional Animal Care and Use Committee.

2.2. Accelerating Rotarod

Testing on the accelerating rotarod, which measures motor function and motor coordination [68], was carried out as previously described [69]. Briefly, mice were tested on two separate days, with three trials delivered on the first day and two trials delivered 48 h later. Each day began with 30 min of habituation in a brightly lit room. During each trial, mice were placed using a wooden dowel into one of four lanes of a rod rotating at a constant speed of 4 rpm. Once the trial began, the rotarod accelerated to a speed of 40 rpm over 5 min. The trial for each mouse ended when the mouse fell off the rotarod, completed two complete somersaults around the rotarod, or reached the end of the 5 min trial; end-of-trial times were recorded. Subsequent trials on the same day started 10 min later. The rotarod was cleaned with 70% ethanol between trials.

2.3. Surgery

Surgery was conducted according to previously published protocols [45]. Briefly, behavioral mice had a custom titanium head plate surgically attached to their skulls. During surgery, each mouse was anesthetized with isoflurane (1–2% in oxygen, 1 L/min, for 15–25 min) and mounted in a stereotaxic head holder (David Kopf Instruments, Tujunga, CA, USA). The scalp was shaved and cleaned, and an incision was made down the midline of the scalp. The skull was cleaned, and the margin of the incision was held open using cyanoacrylate glue. The center of the head plate was positioned over bregma and attached to the skull with Metabond dental adhesive (Parkell, Edgewood, NY, USA). Following surgery, the mice were monitored for at least 24 h as they recovered from the surgery.

2.4. Eyeblink Conditioning

Eyeblink conditioning experiments were conducted according to previously published protocols (Supplementary Figure S1A) [45]. Briefly, eyeblink conditioning consisted of 3 sessions of habituation followed by 12 sessions of training, with each session taking place in a sound-proof, light-proof box [70]. During each session, animals were head-fixed to a metal support structure and atop a freely rotating foam wheel (constructed from EVA Bumps Foam Roller, 6″ diameter, Bean Products, Chicago, IL, USA). Following the 3 sessions of habituation, animals sat stably and calmly above the wheel, locomoting freely on occasion, without any struggling, as in prior experiments using this technique [45,71,72]. This position allowed a platform for delivering unconditioned (US) and conditioned (CS) stimuli to the animal in a controlled manner. The US (airpuff, 30–40 psi) could be delivered to the cornea through a P1000 pipette tip. The intensity and timing of the puff were controlled by a Picospritzer III (Parker Hannefin, Lakeview, MN, RRID:SCR_018152) connected to a compressed air tank. The position of the needle was adjusted each day for each mouse to ensure that a complete eyeblink was induced by the airpuff. The conditioned stimulus (CS; ultraviolet LED) was delivered to the contralateral eye. Eyelid deflection was monitored using a PSEye Camera run by custom Python software (RRID:SCR_008394) [71]. This same software automatically initiated the trials and delivered the US and the CS via a digital-analog conversion board (National Instruments, Austin, TX, USA). No measurements were taken from the foam wheel.

The animals were allowed to habituate to this apparatus for at least 120 min total over the course of 3 days. Over this time period, the animals demonstrated that they could run freely on the wheel without struggling. Following habituation, acquisition took place over 12 training sessions (1 session/day, 6 days/week), during which the animals received 22 blocks of 10 trials each. The CS (ultraviolet light, 280 ms) was paired with an aversive US (airpuff to the cornea, 30–40 psi, 30 ms, co-terminating with the CS). Each block consisted of 9 paired US-CS trials and 1 unpaired CS trial, arranged pseudorandomly within the block. Each trial was separated by a randomly assigned interval of at least 12 s.

Videos were then analyzed offline using a custom MATLAB (Mathworks, Natick, MA, RRID:SCR_001622) script with experimenter supervision (Supplementary Figure S1B–D) using a method similar to that previously published [73]. Regions of interest containing the eye receiving the corneal airpuff (contralateral to the eye receiving the CS) and part of the animals' faces were smoothed, thresholded, and binarized. Then, the number of white pixels—corresponding to total eyelid closure—was tracked across every frame of the video. For each US-CS trial, data within 1500 ms of the recorded US onset was normalized to the range between the signal minimum during the 280 ms period following CS onset and the signal maximum during the 500 ms period following US onset. Consistent with the prior literature [45], a successful conditioned response (CR) occurred on a US-CS trial if the normalized signal exceeded 0.15 between 100 and 250 ms following CS onset; a trial was excluded if the normalized signal exceeded 0.15 prior to this period. Data are reported as percent CR performance, the percentage of counted trials on which a successful CR occurred. For each unpaired CS trial, the recorded response was normalized to the size of the UR during the previous 9 US-CS trials, then evaluated for the presence of a CR; a CR was counted as present if it exceeded 0.15 between 100 ms and 400 ms after the onset of the CS and remained below 0.05 below 0 ms and 99 ms. Peak time was calculated from smoothed CS curves and averaged across the final three training sessions for each animal.

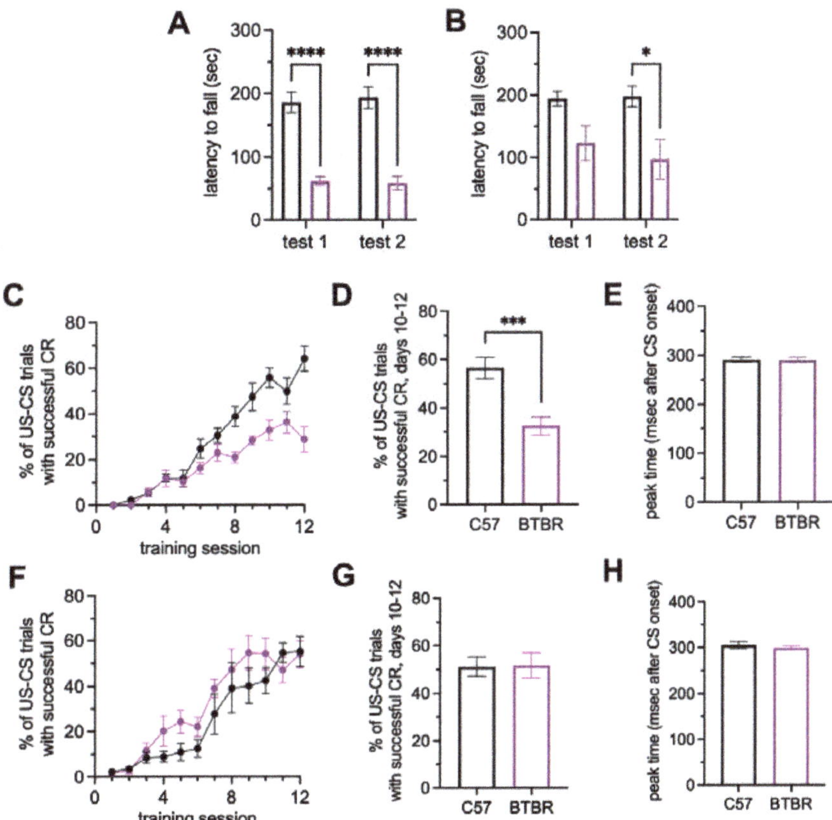

Figure 1. Male BTBR show motor learning and coordination deficits, while female BTBR mice show only motor coordination deficits. (**A**) Male BTBR mice fall earlier than male C57 mice across two days of rotarod testing. (**B**) Female BTBR mice fall earlier than female C57 mice across two days of rotarod testing. (**C–E**) Male BTBR mice lag behind C57 mice in conditioned response performance in the delay eyeblink conditioning task across twelve training days (**C**) with a significant difference at the end of training (**D**). However, there is no difference in response timing on CS trials (**E**). (**F–H**) Female BTBR mice reach comparable levels of conditioned response performance in the delay eyeblink conditioning task across twelve training days (**F**) with no difference in performance at the end of training (**G**). However, there is no difference in response timing on CS trials (**H**). Black, C57B6/J mice; purple, BTBR mice. Error bars denote standard error of the mean. Asterisks denote significant results from two-sample *t*-tests (**A,B,D,F**) or planned comparisons following significant effects in a two-way ANOVA (**C,E**). *, $p < 0.05$; ***, $p < 0.001$; ****, $p < 0.0001$.

2.5. Tissue Processing and Analysis

Tissues from BTBR and C57 mice were used to analyze the cerebellum at the gross anatomical and cellular levels. All experiments were conducted using previously published protocols [45] but will be recapitulated here. For Nissl staining and immunohistochemistry, mice were anesthetized with 0.15 mL ketamine-xylazine (0.12 mL 100 mg/mL ketamine and 0.80 mL mg/mL xylazine diluted 5× in saline), weighed, and transcardially perfused with 4% formalin in pH 7.4 phosphate-buffered saline (PBS). The brain was extracted, weighed, and stored at 4 °C in 4% formalin in PBS for 24 h. Thereafter, brains were stored in 0.1% sodium azide PBS at 4 °C for vibratome sectioning. For Golgi–Cox staining, mice were deeply anesthetized with gaseous isoflurane and decapitated immediately. The brain was removed quickly into ice-cold PBS and processed using the FD Rapid GolgiStain kit (FDNeurotechnologies, Inc., Columbia, MD, USA) according to manufacturer instructions.

For Nissl staining, the cerebellum was sliced sagittally into 50 μm sections with a vibrating microtome (Compresstome, Precisionary Instruments, Greenville, NC, USA; RRID:SCR_018452). Every fourth section slice was mounted onto gelatinized Fisherbrand SuperFrost microscope slides (Thermo Fisher Scientific, Waltham, MA, USA) and allowed to dry overnight before being stained. Other sections were stored in PBS for immunohistochemistry (see below). Standard Nissl stain procedures were used as previously published [45], and the slides were sealed and coverslipped with Permount (Fisher Scientific, Fair Lawn, NJ, USA) before being imaged with 4× objective and 10× eyepiece magnification on a Leica LSI 3000 microscope. Serial images based on Allen Mouse Brain Reference Atlas-referenced sections (RRID:SCR_013286) were taken from vermal (sections 10–11); these locations were approximately 1000–1100 μm apart [74,75]. From these images, we measured the length of the molecular and granule cell layers in each section, the area of the molecular and granule cell layers in each section, and the overall section areas. Thickness was determined using a previously published technique [44]. We also counted the number of lobules in each section. All image analysis took place using ImageJ (National Institutes of Health, Bethesda, MD, RRID:SCR_003070).

For immunohistochemistry, the cerebellum was sliced sagittally into 50 μm sections with a Compresstome and stored in PBS. Sections were immunostained with goat anti-calbindin (1:300) as the primary antibody and anti-goat Alexa Fluor 488 (1:200) as the secondary antibody (Invitrogen, Eugene, OR, USA). Sections were counterstained with 4′,6-diamidino-2-phenylindole (DAPI, 1:1000; Invitrogen, Eugene, OR). The sections were mounted onto gelatinized slides and left to dry (at least 2 h) before being coverslipped with VectaShield (Vector Laboratories, Burlingame, CA, USA). The sections were imaged with 10× objective and 10× eyepiece magnification on a Leica LSI 3000 microscope. Purkinje cell density was measured in medial and lateral sections on a lobular basis by measuring the length of the cell layer and counting the number of calbindin-positive cells in each lobule using ImageJ.

For Golgi–Cox staining, the cerebellum was sliced sagittally in 120 μm sections using a Compresstome. The sections were mounted on slides and dried overnight in darkness before being processed according to the FD Rapid GolgiStain kit instructions. After processing, slides were coverslipped and sealed with Permount. The sections were then imaged with 40× and 100× objectives and a 10× eyepiece on a Leica LSI 3000 microscope. The maximum height of the dendritic arbor and the cross-sectional area of the soma was measured using ImageJ. In addition, Sholl analysis was conducted on images taken at 20× objective and 10× eyepiece magnification to quantify the complexity of the dendritic arbor. Briefly, the number of intersections of the dendritic arbor with concentric circles drawn using ImageJ at 8 μm intervals from the soma was counted [76]. In addition, the dendritic spine density for these cells was quantified from the distal dendrites in an unbiased manner. Each cell was examined with 100× oil-immersion objective and 10× eyepiece magnification, the spines on every seventh branchlet (with a random starting point) were counted, and the length of the branchlet was measured. Density was calculated by dividing by the length of the branchlet.

2.6. Statistics

All histological data was collected by experimenters blinded to the mouse strain. Behavioral data could not be collected by blinded experiments because of the coat color of the mouse; however, the data collected from these experiments were either processed in a semi-supervised manner or statistically analyzed by an experimenter blinded to mouse strain. Statistical tests used in each experiment are summarized in Supplementary Materials Excel File. Eyeblink conditioning data were analyzed using two-way ANOVAs with repeated measures; main strain effects were reported regardless of significance, whereas main session effects (which would indicate learning over time) are significant, and session × strain interactions are not significant unless otherwise indicated. Two-way ANOVA tests with Bonferroni-corrected post hoc comparisons were used for comparing layer and lobule area thickness and Purkinje cell density in the cerebellum; main strain effects were reported regardless of significance. Two-way repeated measures ANOVA tests with Bonferroni-corrected post hoc comparisons were used for data from the Sholl analysis. All pairwise statistical tests were unpaired two-sample t-tests unless otherwise noted. The data were analyzed using Prism (GraphPad Software, San Diego, CA, RRID:SCR_002798). The significance level was $\alpha = 0.05$ unless otherwise noted. All results are depicted as mean $\pm$ standard error of the mean (SEM) unless otherwise noted.

2.7. Code Availability

All code used in data collection and analysis is available upon request.

3. Results

We carried out four sets of experiments to uncover strain differences that depended on sex between C57 and BTBR mice. In describing the results below, we report precise *p*-values; exact statistics can be found in Supplementary Materials Excel File.

In order to identify potential disruptions of motor coordination, we carried out the accelerating rotarod task on BTBR mice and C57 controls over the course of two training days. Male BTBR mice fell off the accelerating rotarod significantly earlier than their C57 controls on both days (Figure 1A; main effect of strain, $p < 0.0001$; differences on both days, $p < 0.0001$), indicating a severe inability to adapt to new motor circumstances. Likewise, female BTBR mice tended to fall off the accelerating rotarod significantly earlier than their C57 controls, particularly on testing day 2 (Figure 1B; main effect of strain, $p = 0.0103$; test day 1, $p = 0.0824$; test day 2, $p = 0.0105$).

We also tested whether BTBR mice were deficient in delay eyeblink conditioning, a motor learning task known to require the cerebellum [28–30,77]. Over the course of 12 training sessions, male BTBR mice lagged significantly behind their C57 counterparts in terms of conditioned response performance (Figure 1C; training session × strain interaction, $p < 0.0001$), particularly on days 8, 10, and 12 (Bonferroni corrected post hoc tests, $p < 0.05$). When we examined the average conditioned response performance rate over the last three days of training, conditioned response rates in male BTBR mice were significantly lower than those of C57 male mice (Figure 1D; $p = 0.0004$). When we examined successful CS-only trials to determine whether the peak time was altered between strains, we found no significant difference (Figure 1E; $p = 0.9198$). Intriguingly, when we performed the same experiment over 12 training sessions in female mice, we found no significant difference in the time course of learning between female BTBR mice and their C57 counterparts (Figure 1F; main effect of session, $p < 0.0001$). A small difference between strains, which failed to reach significance, suggested accelerated learning in female BTBR mice, in stark contrast to the lag in the male BTBR mice (main effect of strain, $p = 0.2531$). Expectedly, when we examined the average conditioned response performance rate over the last three days of training, female BTBR mice performed at statistically equivalent levels as female C57 mice (Figure 1G; $p = 0.4743$). When we examined successful CS-only trials to determine whether the peak time was altered between strains, we found no significant difference (Figure 1H; $p = 0.5155$).

We proceeded to examine whether there were differences in cerebellar anatomy between strains. When we examined overall brain weight in male mice, we found no significant difference between strains (Figure 2A; $p = 0.5091$). We then examined Nissl-stained sagittal midline vermal sections of the cerebellum. We noted qualitatively that sections from male BTBR mice tended to be larger than sections from their C57 counterparts and showed signs of abnormal foliation (Figure 2B). When we quantified these differences, we found that vermal sections from male BTBR mice were indeed hyperplastic in terms of overall area (Figure 2C; $p = 0.0004$), with significant expansion across layers of the cerebellum (Figure 2D; main effect of strain, $p < 0.0001$; main effect of layer, $p < 0.0001$), specifically in the molecular cell layer (MCL; $p = 0.0004$) and granule cell layer (GCL; $p = 0.0098$) but not white matter ($p = 0.2697$). In addition, there was a significant abnormal foliation (Figure 2E; $p < 0.0001$), with the average male BTBR section having four additional folia. We then sought to determine whether the expansion and abnormal foliation were uniform across the cerebellum or whether it varied by lobule. Our analysis confirmed overall expansion across lobules (Figure 2F; main effect of strain, $p < 0.0001$; main effect of the lobule, $p < 0.0001$), while Bonferroni-corrected post hoc tests revealed significant expansion in lobules I/II ($p = 0.0012$), IV/V ($p = 0.0037$), and IX ($p = 0.0396$). Given these results, we tested whether the area occupied by the MCL and GCL varied by lobule. We discovered significant differences between strains for MCL (Figure 2G; main effect of strain, $p < 0.0001$) and GCL (Figure 2H; main effect of strain, $p < 0.001$), with differences appearing largely in the anterior cerebellum. We observed significant differences in both layers in lobules I/II (MCL, $p = 0.0015$; GCL, $p = 0.0076$) and significant differences in GCL in lobules IV/V, VI, and VII. We asked whether the increases we observed were driven by differences in the thickness of the layer rather than an increase in the perimeter of the section and found no significant effect of strain on thickness (Supplementary Figure S2A–C, $p > 0.05$ for main effects of strain). Finally, our analysis confirmed abnormal foliation between strains that depend on lobule (Figure 2I; strain x lobule interaction, $p < 0.0001$); male BTBR mice showed additional folia predominantly in lobules in the anterior cerebellum, including lobules I/II ($p = 0.0076$) and IV/V ($p < 0.0001$), along with lobules VI ($p < 0.0001$) and VII ($p = 0.0004$).

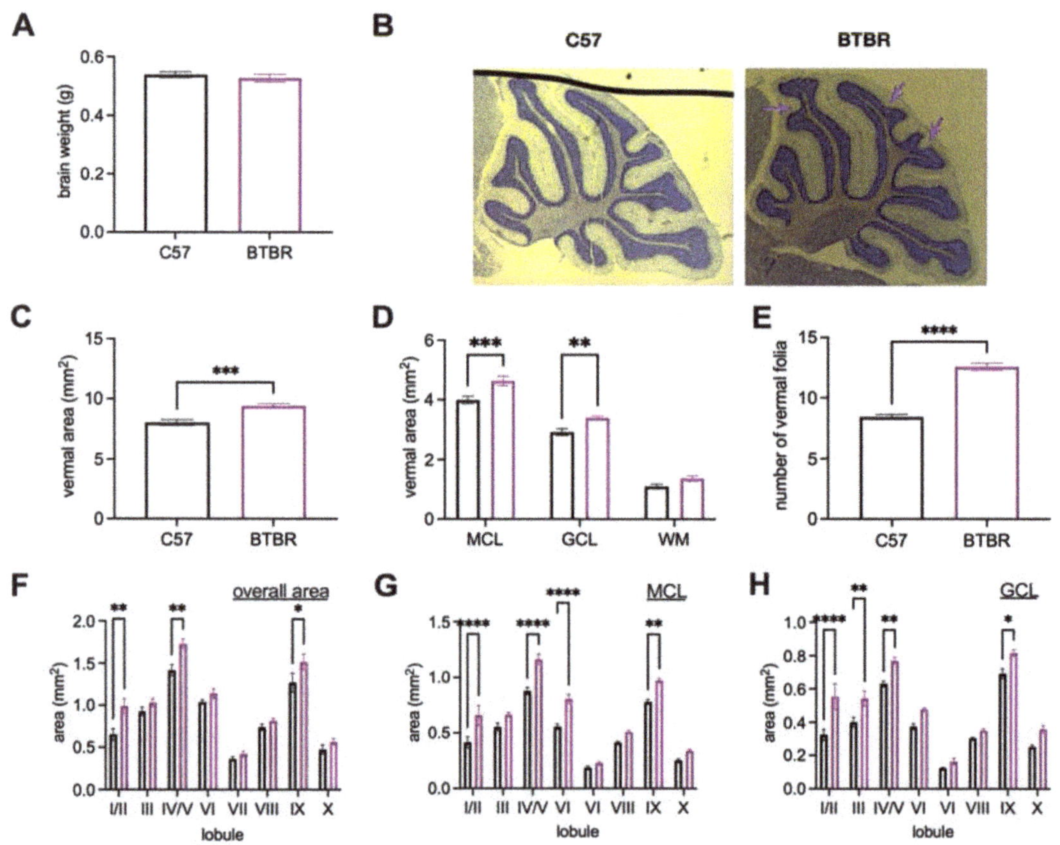

Figure 2. *Cont.*

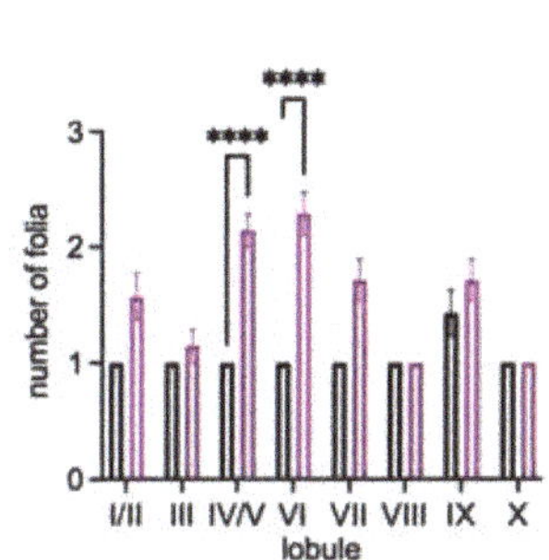

Figure 2. Male BTBR mice show vermal enlargement and foliation that varies by lobule. (**A**) Brain weight is comparable between strains. (**B**) Representative image of gross anatomical differences between C57 (left) and BTBR (right) sagittal vermis sections. Arrows identify additional lobules in the BTBR section. (**C**) Area of the midline vermal section is significantly larger in BTBR mice. (**D**) Molecular cell layer (MCL) and granule cell layer (GCL) are significantly enlarged in the BTBR vermis. (**E**) The number of folia in the vermis is significantly different in BTBR mice. (**F**) Enlargement of vermis area in BTBR mice depends on lobule. (**G**) Enlargement of area of the molecular cell layer in BTBR mice depends on lobule. (**H**) Enlargement of area of the granule cell layer in BTBR mice depends on lobule. (**I**) Abnormal foliation in BTBR mice depends on lobule. Black, C57B6/J mice; purple, BTBR mice. Error bars denote standard error of the mean. Asterisks denote significant results from two-sample *t*-tests (**A–E**) or planned comparisons following a significant two-way ANOVA (**F–I**). *, $p < 0.05$; **, $p < 0.01$; ***, $p < 0.001$; ****, $p < 0.0001$.

In female mice, we first found no significant difference in brain weight between strains (Figure 3A; $p = 0.8221$). As in male BTBR mice, examination of Nissl-stained sagittal midline vermal sections appeared larger and tended to have more folia than sections from their C57 counterparts (Figure 3B). When we quantified these differences, we found that vermal sections were indeed hyperplastic (Figure 3C; $p < 0.0001$), with significant expansion across layers (Figure 3D; strain x layer interaction, $p < 0.0001$). The magnitude of this expansion depended on the layer, with a substantial expansion in the MCL and GCL (Bonferroni-corrected post hoc test, $p < 0.05$). In addition, there was significant abnormal foliation (Figure 3E, $p < 0.0001$), with midline sections from female BTBR mice having, on average, three additional folia than their C57 counterparts. When we examined expansion on a lobule-by-lobule basis, we found that expansion depended on lobule (Figure 3F; strain x lobule interaction, $p = 0.0058$), with Bonferroni-corrected post hoc tests revealing a significant expansion in lobules I/II ($p < 0.0001$), III ($p = 0.0311$), IV/V ($p < 0.0001$), VI ($p = 0.0008$), and IX ($p = 0.0033$). Given these results, we tested whether the area occupied by the MCL and GCL varied by lobule. We discovered significant differences between strains for MCL (Figure 3G; main effect of strain, $p < 0.0001$) and GCL (Figure 3H; main effect of strain, $p < 0.001$), with differences appearing largely in the anterior cerebellum. We observed significant differences in both layers in lobules I/II (MCL, $p < 0.0001$; GCL, $p < 0.0001$) and IV/V (MCL, $p = 0.0090$; GCL, $p = 0.0008$) and other significant differences on one area in lobules III (GCL, $p = 0.0073$) and VI (MCL, $p < 0.0001$). We asked whether the increases we observed were driven by differences in the thickness of the layer rather than an increase in the perimeter of the section and found a significant effect of strain on MCL thickness ($p = 0.0010$) and no significant effect of strain on GCL thickness (Supplementary Figure S2D–F, $p = 0.088$). Finally, our analysis confirmed abnormal foliation between strains that depend on lobule (Figure 3I); female BTBR mice showed additional folia predominantly in lobules in the anterior cerebellum, including lobules I/II ($p < 0.0001$) and IV/V ($p = 0.0008$) as well as lobule VI ($p < 0.0001$).

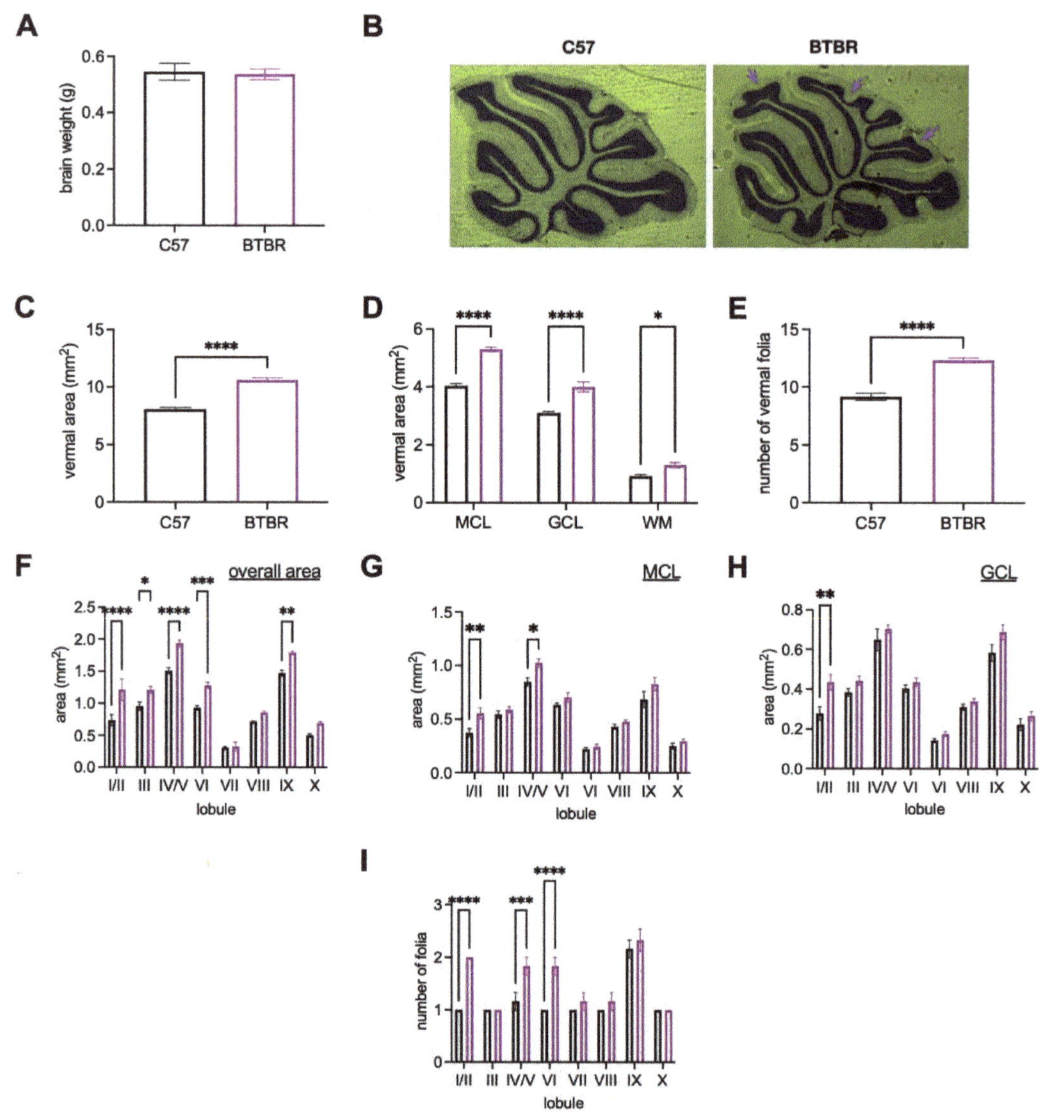

Figure 3. Female BTBR mice show vermal enlargement and foliation that varies by lobule. (**A**) Brain weight is comparable between strains. (**B**) Representative image of gross anatomical differences between C57 (left) and BTBR (right) sagittal vermis sections. Arrows identify additional lobules in the BTBR section. (**C**) Area of the midline vermal section is significantly larger in BTBR mice. (**D**) Molecular cell layer (MCL), granule cell layer (GCL), and white matter areas are all significantly enlarged in the BTBR vermis. (**E**) The number of folia in the vermis is significantly different in BTBR mice. (**F**) Enlargement of vermis area in BTBR mice depends on lobule. (**G**) Enlargement of area of the molecular cell layer in BTBR mice depends on lobule. (**H**) Enlargement of area of the granule cell layer in BTBR mice depends on lobule. (**I**) Abnormal foliation in BTBR mice depends on lobule. Black, C57B6/J mice; purple, BTBR mice. Error bars denote standard error of the mean. Asterisks denote significant results from two-sample t-tests (**A–E**) or planned comparisons following a significant two-way ANOVA (**F–I**). *, $p < 0.05$; **, $p < 0.01$; ***, $p < 0.001$; ****, $p < 0.0001$.

We then asked whether the gross anatomical differences were accompanied by cellular differences commonly observed in the cerebellum of ASD patients and autism mouse models, including altered Purkinje cell density and morphology [19,78–81]. An analysis of the linear density of calbindin-positive neurons in midline vermal sagittal sections of male BTBR mice and their C57 counterparts (Figure 4A) showed significant differences between strains (Figure 4B; main effect of strain, $p = 0.0363$); however, Bonferroni-corrected post hoc tests revealed no significant differences with specific lobules ($p > 0.05$ for all comparisons). We performed a similar analysis of the linear density of calbindin-positive neurons in midline sagittal sections of female BTBR mice and their C57 counterparts. As in the male BTBR and C57 mice, there was a significant difference between female BTBR and C57 mice (Figure 4C; main effect of strain, $p = 0.0094$; main effect of the lobule, $p = 0.0046$). At the same time, Bonferroni-corrected post hoc tests also revealed no significant differences with specific lobules ($p > 0.05$ for all comparisons; one near-significant finding in lobule IX).

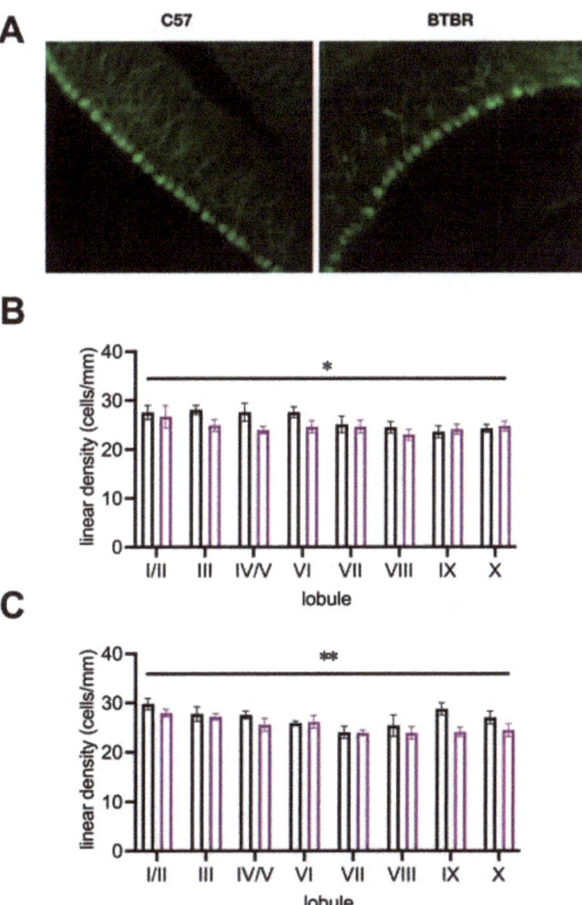

Figure 4. BTBR mice of both sexes have slight, global decreases in vermal Purkinje cell density. (**A**) Representative images of calbindin-stained Purkinje cells in male BTBR and C57 mice. (**B**) Lobule-by-lobule analysis shows a broad decrease in male BTBR mice that is not lobule-specific. (**C**) Lobule-by-lobule analysis shows a road decrease in female BTBR mice that is not lobule-specific. Black, C57B6/J mice; purple, BTBR mice. Error bars denote standard error of the mean. Asterisks denote main effect of strain. *, $p < 0.05$; **, $p < 0.01$.

Purkinje cells were also analyzed in terms of cell body size, dendritic arbor height, dendritic spine density, and branching (Figure 5A). When we examined Golgi-stained cells from male mice, Sholl analysis revealed no significant difference in the complexity of the dendritic arbors of Purkinje cells from BTBR and C57 mice (Figure 5B; main effect of strain, $p = 0.6478$). In addition, we found no significant difference in Purkinje cell body size (Figure 5C; $p = 0.2075$) or Purkinje cell dendritic arbor height (Figure 5D; $p = 0.6305$). When we examined differences in dendritic spines on distal branches of Purkinje cells, we identified a trend toward lower dendritic spine density in male BTBR mice (Figure 5E; $p = 0.1478$). When we examined Golgi-stained cells from female mice, Sholl analysis revealed a significantly more complex dendritic arbor in Purkinje cells from female BTBR mice compared to female C57 mice (Figure 5F; main effect of strain, $p = 0.0159$). In addition, there was a trend toward enlarged cell bodies in Purkinje cells from female BTBR mice (Figure 5G; $p = 0.0652$) but no significant difference in Purkinje cell dendritic arbor height (Figure 5H; $p = 0.2261$). Finally, when we examined differences in dendritic spines on distal branches of the Purkinje cells, we identified a significantly lower dendritic spine density in female BTBR mice (Figure 5I; $p < 0.0001$).

Finally, we examined the male and female datasets side-by-side to identify consistent differences among sex and to determine if there were any instances in which the female BTBR mice differed from both male and female C57 mice (Supplementary Figure S3). In most instances, carrying out a comparison via two-way ANOVA revealed the same statistical differences between BTBR and C57 mice in both strains. We verified significant main effects of strain with no significant effect of sex in rotarod performance (Supplementary Figure S3A,B), the number of vermal folia (Supplementary Figure S3G), overall linear density (Supplementary Figure S3H), and dendritic spine density (Supplementary Figure S3K) (all main effects, $p < 0.05$). Likewise, there was no main effect of strain or sex for brain weight (Supplementary Figure S3D, $p > 0.05$). There was one instance in which there was a main effect of sex and strain with no interaction: dendritic spine density (Supplementary Figure S3K, $p < 0.05$). There was a significant sex x strain interaction for eyeblink conditioning performance, consistent with our prior findings (Supplementary Figure S3C, $p = 0.0082$); in this case, BTBR females performed as well as both C57 males and C57 females. There was also significant sex x strain interactions for vermal area (Supplementary Figure S3F, $p = 0.0077$) and Purkinje cell soma area (Supplementary Figure S3I, $p = 0.0146$), suggesting instances where enlargement occurs for one sex or strain group (in Figure 3F, BTBR females have a larger vermal area than all other groups, $p < 0.05$ for all comparisons; and in Figure 3I, the BTBR females have a larger somatic area, $p < 0.05$ for all comparisons). In one instance where the female BTBR mice were statistically different from the female C57 mice, the female BTBR were not quite significantly different from the C57 males, namely in session 2 rotarod ($p = 0.0975$, Supplementary Figure S3B). In two instances, there was a difference between sex that did not appear for both strains: for BTBR vermal area ($p = 0.0015$, Supplementary Figure S3F) and in eyeblink conditioning performance on the last three days ($p = 0.0225$, Supplementary Figure S3C). Overall, with a few exceptions, these comparisons confirm widespread strain differences with few sex differences, with female BTBR mice performing differently from both sexes of C57 mice in both cases.

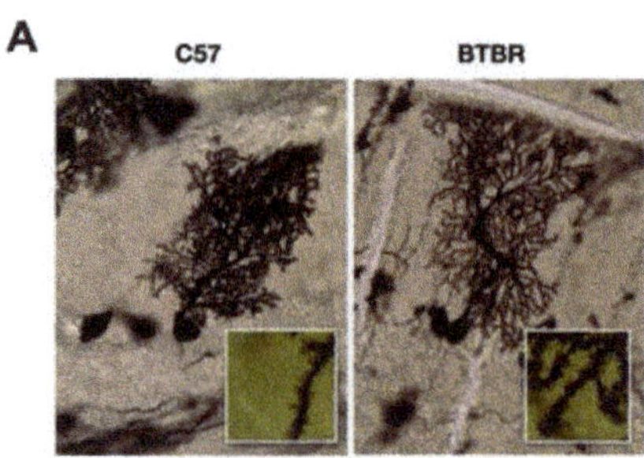

Figure 5. *Cont.*

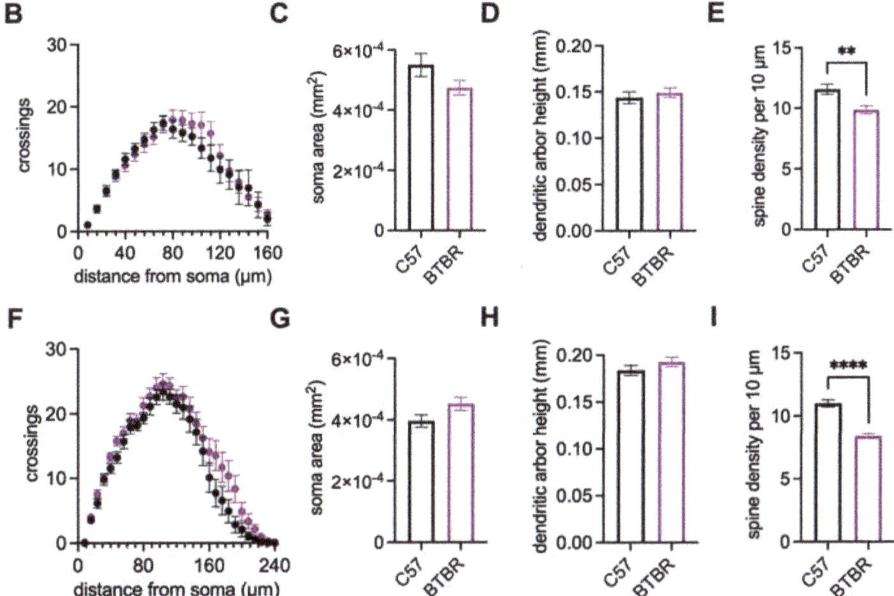

Figure 5. Male and Female BTBR mice show alterations to Purkinje cell dendritic branching and spine density. (**A**) Representative examples of Purkinje cells from BTBR (left) and C57 (right) mice. (**B**) Sholl analysis shows no difference between male BTBR and C57 mice. (**C**) Purkinje cell bodies are similar in area in male BTBR mice. (**D**) Dendritic arbor height is not different between groups of male mice. (**E**) Male BTBR mice have fewer dendritic spines on their distal branches. (**F**) Sholl analysis shows a slight increase in the complexity of dendritic arbors of Purkinje cells from female BTBR mice. (**G**) Purkinje cell bodies are similar in area in female BTBR mice. (**H**) Dendritic arbor height is not different between groups of female mice (**I**) Female BTBR mice have fewer dendritic spines on their distal branches. Black, C57B6/J mice; purple, BTBR mice. Error bars denote standard error of the mean. Asterisks denote significant results from two-sample *t*-tests (**C–E**) or planned comparisons following a significant two-way ANOVA (**B**). **, $p < 0.01$; ****, $p < 0.0001$.

4. Discussion

We set out to characterize cerebellum-specific phenotypes of BTBR T + Itpr3^{tf}/J to determine whether it would be a suitable mouse model for understanding the cerebellar basis of ASD in both sexes. We discovered a high degree of concordance between sexes in our measurements, with a small number of exceptions. BTBR mice tend to show deficits in motor learning, with male mice in particular lagging behind in both tasks we examined. At a gross anatomical level, the BTBR cerebellum is hyperplastic, with significant vermal expansion and abnormal foliation occurring most substantially in lobules and IV/V and VI. Purkinje cells tend to have a lower density across the BTBR vermis than their C57 counterparts, though this decrease is not confined to a single lobule. In addition, there are notable disruptions in the structure of the dendritic arbor: BTBR cells most notably have a significantly lower dendritic spine density than C57 cells.

Our finding of significant motor learning impairments in the BTBR mouse model is consistent with previous literature. One prior work by Xiao and colleagues noted a disruption of rotarod performance in male BTBR mice [82]. The present study confirms that finding, while also adding that female mice have a similar—albeit less severe—deficit. Our finding that male BTBR mice have a deficit in delay eyeblink conditioning, a motor task known to require the cerebellum, is novel but consistent with other ASD mouse models. Prior studies show that delay eyeblink conditioning dysfunction is widespread in ASD mouse models, with deficits in either the ability to acquire delay eyeblink conditioning or to perform the conditioned eyeblink with the correct magnitude or timing [45,47,48,83,84]. The present study adds to this body of the literature. Deficits in eyeblink conditioning tend to cluster with the part of the cerebellar circuit in the eyeblink region that is most likely to be disrupted, setting up future research probing the BTBR cerebellum at the neural circuit level [45]. Delay eyeblink conditioning deficits do arise in ASD patients, with fewer disruptions in the ability to learn and more frequent disruptions in the timing of the conditioned response than is demonstrated here [31–33].

Interestingly, we did not discover the same conditioning deficit in the female mice, and there is some evidence to indicate that female mice acquired conditioning somewhat more quickly than their C57 counterparts. This intriguing finding does mirror a result in the patient literature suggesting faster learning in the delay eyeblink conditioning task [32] (but notice the lack of timing deficits here) and generally mirrors sex differences in the task in the neurotypical population [85]. How might a sex difference in delay eyeblink conditioning arise? Differences in the speed of eyeblink acquisition have been ascribed to the role of the hormonal stress response in learning in female mice [86] or differences in the activity of neurons in the motor areas of the cerebellum [87]. It is possible that sex differences in stress processing [88] or sex differences in the electrophysiology of Purkinje cells [89] in the BTBR mice might account for this difference. Some researchers have suggested that delay eyeblink conditioning could represent a rare phenotype that occurs similarly in patients and model mice; such a finding would provide easily interpretable outcomes for therapeutic studies and provide a clearer path to understanding the cerebellar pathophysiology of autism [51,90]. However, for eyeblink conditioning to be a useful biomarker, much more work will need to be performed to determine how well mouse models, like our BTBR mouse model, map onto a segment of the patient population in males and females.

We discovered that mice of both sexes have vermal hyperplasia and abnormal foliation. The finding that male mice have hyperplasia is consistent with previous studies showing that the cerebellum occupies a larger percentage of brain volume in BTBR mice than it does in C57 mice [63,64]. Our finding that the same feature occurs in female mice is novel. In addition, we are the first group to uncover hyperplasia that is regionally specific, identifying significant enlargement in the anterior cerebellum, lobule VI, and lobule IX. This finding of vermal hyperplasia is certainly at odds with literature that shows that many ASD patients have cerebellar hypoplasia [15,23,91], though there are reports that are more consistent with our findings of regional hyperplasia [14,16,92]. Indeed, an exhaustive study of twenty-six ASD mouse models suggests that malformation of the cerebellum varies widely and may be indicative of multiple subpopulations within the ASD patient population [63]. Our findings may apply more narrowly to one of these subpopulations. In addition, we have discovered abnormal foliation in both male and female mice, confirming one previous study in male mice [82] and notable because the earlier literature had rejected the notion of anatomical abnormality in the cerebellum [93]. This foliation defect is indicative of disruption of the maturation of the cerebellum early in postnatal murine brain development [94]; the observation that early granule cell layer development as well as Purkinje cell migration defects [82] might account for this disrupted foliation is in line with our observation of regional differences in the area of the granule cell layer and molecular cell layer. However, further investigation is required to understand the significance of hyperfoliation of some lobules and not others.

We also discovered a disruption of Purkinje cell density and morphology in both male and female mice. Our finding of a reduced Purkinje cell density is consistent with patient reports of lowered Purkinje cell number [19,21,80,81] and is consistent with a report on reduced Purkinje cell number in juvenile BTBR male mice [82]. The same finding was observed in female mice for the first time. This finding is consistent with the idea that deficits in Purkinje cell migration during late cerebellar development might underlie the anatomical differences in the cerebellum [82]. However, unlike our other measurements, one previous study in BTBR male mice [82], and one notable study showing regional specificity in Purkinje cell density loss [19], we did not find evidence that any one lobule was more disrupted than another. Likewise, we did not find any differences in the size of Purkinje cell bodies as observed in BTBR mice [82] and in patients, and we only observed minor differences in the complexity of the dendritic arbor. Notably, in both male and female BTBR mice, we noted a reduction in the density of dendritic spines, perhaps indicating a reduction of excitatory drive to Purkinje cells that is critical for cerebellar development and cerebellar learning. While this finding is different from that of increased numbers of immature dendritic spines in male juvenile BTBR mice [82], these results from adult mice might indicate an over-pruning process that takes place later in development. However, the significance of dendritic spine density in ASD remains an open question, particularly because of the high variability of the direction and magnitude of dendritic spine dysgenesis across ASD studies [20,95].

Our findings identified lobules I/II, IV/V, VI, and IX as drivers of the differences between the BTBR cerebellum and the C57 cerebellum in both sexes. What is the significance of these lobules in ASD and ASD-related behavior? Dysplasia has been long observed in some lobules but not others [14,15,19], but studies of connectivity have revealed the deeper, nonmotor role of the cerebellum [96]. Such studies have identified the anterior vermis—lobule I through lobule IV/V—as centrally involved in the stereotyped behavior seen in ASD patients because of its functional connectivity with cerebral areas involved in this behavior [34]. Likewise, the posterior vermis—including lobule IX—has been observed to be involved in emotional regulation and social function [34]. Lobule VI has also been identified for its role in stereotyped behavior [97]. Regarding lobule IV/V, a recent study using chemogenetic manipulation in BTBR mice shows a complex role for the lobule in motor function, social behavior, and memory [67]. It is possible that the lobules have a complex relationship with ASD-relevant behavior. However, despite the growing body of evidence illustrating a clear link between cerebellar lobules and specific aspects of ASD behavior, the current study does not attempt to connect the observed regional abnormalities with any individual behavior. Furthermore, this study did not attempt to measure hemispheric areas like crus I and crus II that have been targeted for their involvement in social behavior [36,40]. Making these connections will require further investigation.

Possible limitations to this study include the range of ages of all mice tested and potential confounds from the estrous cycle in female mice, both of which are variables that could ostensibly affect strain or sex differences demonstrated here. In the present study, we use mice between 8 and 16 weeks old, which is in the young adult to adult ranges in terms of mouse age. The age range used here is consistent with previous studies [40,45], but there is reason to suggest that age in young adulthood may affect the stability of some behaviors [98]. In terms of the stability of eyeblink conditioning performance with age, one prior suggests that the behavior is relatively stable between 4 months and 9–12 months of age in C57BL/6 mice [99], while another suggests a wider range of stability for C57BL/6J mice, from 2 months to 10 months [100], within the range present experiment. To date, similar comparisons have not been made in BTBR mice. While other studies have looked at the ontogeny of eyeblink conditioning in very young rodents [101–103], the literature is notably scarce when it comes to the young adult time point (2 months) in mice. Cerebellar anatomy seems similarly stable within this age range [104,105]. On the other hand, rotarod performance may be affected: one prior study suggests small but significant differences between 2–3 and 4–5-month-old C57BL/6J mice [98]. Such a difference might account for the lack of a significant difference in male mice on test day 1, but it would not account for the observed differences on the following test day. In the present study, we also used female mice without monitoring the estrous cycle. Previously regarded as a potential source of variation in measurements in female mice, a widely cited meta-analysis shows that female mice, when tested irrespective of their estrous cycle, showed no significant increase in variation compared to male mice [106]. This lack of difference in variability has recently been confirmed for delay eyeblink conditioning, motor behavior, and other aspects of cerebellar function in another study [87]. It is possible that other factors that vary with sex that are cited in the Oyaga study, like wheel running [107] and response to stressful and anxiogenic situations [108], may account for the differences noted in our study as well as in the literature. Likewise, rotarod performance [109] and aspects of cerebellar anatomy [110] measured in this study are unlikely to be affected appreciably by the hormonal state of female mice. It is, however, possible that BTBR mice have altered variability in response to the estrous cycle, which has been examined in the occasional study [111] but never with regard to cerebellar behavior. Finally, we should acknowledge that a more rigorous investigation on eyeblink conditioning that looks at different modalities for conditioned stimuli and different delay intervals may reveal substantial differences in learning that are more like those in female mice.

The present study expands the BTBR literature in a few significant ways. First, it highlights ways in which male and female BTBR brains both differ from their C57 counterparts and from each other. The goal of recent pushes in our field to examine sex as a biological variable is justified [112]—it ensures that we do not ignore a significant portion of the patient population. As one of the few studies that have examined both male and female BTBR, the present project asks whether the BTBR mouse model is valid for studying all aspects of ASD in all patients. Future studies should attempt to pinpoint the mechanism underlying the sex differences we have observed here. Second, our study is the first to test whether cerebellum-dependent behaviors—namely, delay eyeblink conditioning—are disrupted in these mice. The study helps put the BTBR mouse model in the larger context of studies in other mouse models that have observed delay eyeblink conditioning deficits as a highly penetrant feature of ASD. Third, our study identifies lobule-specific abnormalities that may correlate with the behavioral profile of the BTBR mouse. Future studies should attempt to identify a causative link between lobule-specific disruption or rescue in the BTBR mouse and alterations of behavior. Finally, this study demonstrates the validity of the BTBR mouse model for understanding cerebellar dysfunction as it mirrors phenotypes in at least a segment of the ASD patient population. Future research should continue to characterize this mouse model for the purposes of identifying effective treatments for and understanding the underlying etiology of ASD in a particular patient subpopulation.

Supplementary Materials: The following supporting information can be downloaded at: https://www.mdpi.com/article/10.3390/neurosci3040045/s1, Figure S1: Protocol for collecting and analyzing images from delay eyeblink conditioning sessions; Figure S2: The thicknesses of the molecular and granule cell layers depend on sex and lobule; Figure S3: Side-by-side sex and strain analysis of male and female experiments for BTBR and C57 mice; Supplementary Materials Excel File: Full details of statisticaly analyses performed in this study.

Author Contributions: E.A.K.: performed experiments; analyzed data; writing; figure creation; editing. J.A.C.: performed experiments; analyzed data; figure creation; editing. J.K.S.: performed experiments; analyzed data; figure creation; editing. H.E.B.: performed experiments; analyzed data; editing. K.R.V.: performed experiments; designed and built experimental apparatus; analyzed data; editing. J.E.V.: performed experiments; designed and built experimental apparatus; analyzed data; editing. A.D.K.: designed and performed experiments; designed and built experimental apparatus; analyzed data; writing; figure creation; editing; obtained funding. All authors have read and agreed to the published version of the manuscript.

Funding: This work is supported by an Institutional Development Award (IDeA) from the National Institute of General Medical Sciences of the National Institutes of Health (P20GM103443), by the National Science Foundation/EPSCoR Award No. IIA-1355423 to the South Dakota Board of Regents and by support from the Augustana University Biology Department Endowment.

Institutional Review Board Statement: The animal study protocol was approved by the Institutional Animal Care and Use Committee of Augustana University (protocol number 003, 28 March 2018).

Informed Consent Statement: Not applicable.

Data Availability Statement: Data from the study are available upon request.

Acknowledgments: We would like to acknowledge the work of collecting a small amount of data on this project by Abby Reynen and Christina Pickett. Special thanks to Brenda Rieger and Brian Vander Aarde for animal husbandry and technical assistance. Illustrations in Supplementary Figure S1 and Graphical Abstract are the artistic creations of J.K.S.

Conflicts of Interest: The authors have no conflict of interest.

References

1. Lord, C.; Brugha, T.S.; Charman, T.; Cusack, J.; Dumas, G.; Frazier, T.; Jones, E.J.H.; Jones, R.M.; Pickles, A.; State, M.W.; et al. Autism spectrum disorder. *Nat. Rev. Dis. Primer* **2020**, *6*, 5. [CrossRef] [PubMed]
2. Loomes, R.; Hull, L.; Mandy, W.P.L. What is the male-to-female ratio in autism spectrum disorder? A systematic review and meta-analysis. J. Am. Acad. Child Adolesc. *Psychiatry* **2017**, *56*, 466–474. [CrossRef]
3. Maenner, M.J.; Shaw, K.A.; Bakian, A.V.; Bilder, D.A.; Durkin, M.S.; Esler, A.; Furnier, S.M.; Hallas, L.; Hall-Lande, J.; Hudson, A.; et al. Prevalence and Characteristics of Autism Spectrum Disorder Among Children Aged 8 Years—Autism and Developmental Disabilities Monitoring Network, 11 Sites, United States, 2018. *Morb. Mortal. Wkly. Rep. Surveill. Summ. Wash. DC* **2021**, *70*, 1–16. [CrossRef]
4. Zeidan, J.; Fombonne, E.; Scorah, J.; Ibrahim, A.; Durkin, M.S.; Saxena, S.; Yusuf, A.; Shih, A.; Elsabbagh, M. Global prevalence of autism: A systematic review update. *Autism Res.* **2022**, *15*, 778–790. [CrossRef] [PubMed]
5. James, W.H.; Grech, V. Potential explanations of behavioural and other differences and similarities between males and females with autism spectrum disorder. *Early Hum. Dev.* **2020**, *140*, 104863. [CrossRef]
6. Young, H.; Oreve, M.-J.; Speranza, M. Clinical characteristics and problems diagnosing autism spectrum disorder in girls. *Arch. Pediatr.* **2018**, *25*, 399–403. [CrossRef]
7. Amaral, D.G.; Schumann, C.M.; Nordahl, C.W. Neuroanatomy of autism. *Trends Neurosci.* **2008**, *31*, 137–145. [CrossRef]
8. Bruchhage, M.M.K.; Bucci, M.-P.; Becker, E.B.E. Cerebellar involvement in autism and ADHD. *Handb. Clin. Neurol.* **2018**, *155*, 61–72. [CrossRef]
9. Fatemi, S.H.; Aldinger, K.A.; Ashwood, P.; Bauman, M.L.; Blaha, C.D.; Blatt, G.J.; Chauhan, A.; Chauhan, V.; Dager, S.R.; Dickson, P.E.; et al. Consensus paper: Pathological role of the cerebellum in autism. *Cerebellum* **2012**, *11*, 777–807. [CrossRef]
10. Hampson, D.R.; Blatt, G.J. Autism spectrum disorders and neuropathology of the cerebellum. *Front. Neurosci.* **2015**, *9*, 420. [CrossRef]
11. Mosconi, M.W.; Wang, Z.; Schmitt, L.M.; Tsai, P.; Sweeney, J.A. The role of cerebellar circuitry alterations in the pathophysiology of autism spectrum disorders. *Front. Neurosci.* **2015**, *9*, 296. [CrossRef] [PubMed]
12. Peter, S.; De Zeeuw, C.I.; Boeckers, T.M.; Schmeisser, M.J. Cerebellar and Striatal Pathologies in Mouse Models of Autism Spectrum Disorder. *Adv. Anat. Embryol. Cell Biol.* **2017**, *224*, 103–119. [CrossRef]
13. Wang, S.S.-H.; Kloth, A.D.; Badura, A. The cerebellum, sensitive periods, and autism. *Neuron* **2014**, *83*, 518–532. [CrossRef]
14. Courchesne, E.; Saitoh, O.; Yeung-Courchesne, R.; Press, G.A.; Lincoln, A.J.; Haas, R.H.; Schreibman, L. Abnormality of cerebellar vermian lobules VI and VII in patients with infantile autism: Identification of hypoplastic and hyperplastic subgroups with MR imaging. *AJR Am. J. Roentgenol.* **1994**, *162*, 123–130. [CrossRef] [PubMed]
15. Courchesne, E.; Yeung-Courchesne, R.; Press, G.A.; Hesselink, J.R.; Jernigan, T.L. Hypoplasia of cerebellar vermal lobules VI and VII in autism. *N. Engl. J. Med.* **1988**, *318*, 1349–1354. [CrossRef] [PubMed]
16. Traut, N.; Beggiato, A.; Bourgeron, T.; Delorme, R.; Rondi-Reig, L.; Paradis, A.-L.; Toro, R. Cerebellar Volume in Autism: Literature Meta-analysis and Analysis of the Autism Brain Imaging Data Exchange Cohort. *Biol. Psychiatry* **2018**, *83*, 579–588. [CrossRef]
17. Webb, S.J.; Sparks, B.-F.; Friedman, S.D.; Shaw, D.W.W.; Giedd, J.; Dawson, G.; Dager, S.R. Cerebellar vermal volumes and behavioral correlates in children with autism spectrum disorder. *Psychiatry Res.* **2009**, *172*, 61–67. [CrossRef]
18. Fatemi, S.H.; Halt, A.R.; Realmuto, G.; Earle, J.; Kist, D.A.; Thuras, P.; Merz, A. Purkinje cell size is reduced in cerebellum of patients with autism. *Cell. Mol. Neurobiol.* **2002**, *22*, 171–175. [CrossRef]
19. Skefos, J.; Cummings, C.; Enzer, K.; Holiday, J.; Weed, K.; Levy, E.; Yuce, T.; Kemper, T.; Bauman, M. Regional Alterations in Purkinje Cell Density in Patients with Autism. *PLoS ONE* **2014**, *9*, e81255. [CrossRef]
20. Sudarov, A. Defining the role of cerebellar Purkinje cells in autism spectrum disorders. *Cerebellum* **2013**, *12*, 950–955. [CrossRef]
21. Yip, J.; Soghomonian, J.-J.; Blatt, G.J. Decreased GAD67 mRNA levels in cerebellar Purkinje cells in autism: Pathophysiological implications. *Acta Neuropathol.* **2007**, *113*, 559–568. [CrossRef] [PubMed]
22. Shukla, D.K.; Keehn, B.; Lincoln, A.J.; Müller, R.-A. White matter compromise of callosal and subcortical fiber tracts in children with autism spectrum disorder: A diffusion tensor imaging study. *J. Am. Acad. Child Adolesc. Psychiatry* **2010**, *49*, 1269–1278. [CrossRef]
23. Courchesne, E. Abnormal early brain development in autism. *Mol. Psychiatry* **2002**, *7*, S21–S23. [CrossRef] [PubMed]
24. Stoodley, C.J.; Limperopoulos, C. Structure-function relationships in the developing cerebellum: Evidence from early-life cerebellar injury and neurodevelopmental disorders. *Semin. Fetal. Neonatal Med.* **2016**, *21*, 356–364. [CrossRef] [PubMed]
25. Menashe, I.; Grange, P.; Larsen, E.C.; Banerjee-Basu, S.; Mitra, P.P. Co-expression profiling of autism genes in the mouse brain. *PLoS Comput. Biol.* **2013**, *9*, e1003128. [CrossRef] [PubMed]

26. Bhat, A.N. Motor Impairment Increases in Children with Autism Spectrum Disorder as a Function of Social Communication, Cognitive and Functional Impairment, Repetitive Behavior Severity, and Comorbid Diagnoses: A SPARK Study Report. *Autism Res.* **2021**, *14*, 202–219. [CrossRef]

27. Jaber, M. Autism is (also) a movement disorder. *Mov. Disord.* **2015**, *30*, 341. [CrossRef]

28. Christian, K.M.; Thompson, R.F. Neural substrates of eyeblink conditioning: Acquisition and retention. *Learn. Mem.* **2003**, *10*, 427–455. [CrossRef]

29. Freeman, J.H.; Steinmetz, A.B. Neural circuitry and plasticity mechanisms underlying delay eyeblink conditioning. *Learn. Mem.* **2011**, *18*, 666–677. [CrossRef]

30. Thompson, R.F.; Steinmetz, J.E. The role of the cerebellum in classical conditioning of discrete behavioral responses. *Neuroscience* **2009**, *162*, 732–755. [CrossRef]

31. Oristaglio, J.; Hyman West, S.; Ghaffari, M.; Lech, M.S.; Verma, B.R.; Harvey, J.A.; Welsh, J.P.; Malone, R.P. Children with autism spectrum disorders show abnormal conditioned response timing on delay, but not trace, eyeblink conditioning. *Neuroscience* **2013**, *248*, 708–718. [CrossRef] [PubMed]

32. Sears, L.L.; Finn, P.R.; Steinmetz, J.E. Abnormal classical eye-blink conditioning in autism. *J. Autism Dev. Disord.* **1994**, *24*, 737–751. [CrossRef] [PubMed]

33. Welsh, J.P.; Oristaglio, J.T. Autism and Classical Eyeblink Conditioning: Performance Changes of the Conditioned Response Related to Autism Spectrum Disorder Diagnosis. *Front. Psychiatry* **2016**, *7*, 137. [CrossRef] [PubMed]

34. D'Mello, A.M.; Crocetti, D.; Mostofsky, S.H.; Stoodley, C.J. Cerebellar gray matter and lobular volumes correlate with core autism symptoms. *NeuroImage Clin.* **2015**, *7*, 631–639. [CrossRef] [PubMed]

35. Laidi, C.; Boisgontier, J.; Chakravarty, M.M.; Hotier, S.; d'Albis, M.-A.; Mangin, J.-F.; Devenyi, G.A.; Delorme, R.; Bolognani, F.; Czech, C.; et al. Cerebellar anatomical alterations and attention to eyes in autism. *Sci. Rep.* **2017**, *7*, 12008. [CrossRef] [PubMed]

36. Stoodley, C.J.; D'Mello, A.M.; Ellegood, J.; Jakkamsetti, V.; Liu, P.; Nebel, M.B.; Gibson, J.M.; Kelly, E.; Meng, F.; Cano, C.A. Altered cerebellar connectivity in autism and cerebellar-mediated rescue of autism-related behaviors in mice. *Nat. Neurosci.* **2017**, *20*, 1744–1751. [CrossRef] [PubMed]

37. Crawley, J.N. Translational animal models of autism and neurodevelopmental disorders. *Dialogues Clin. Neurosci.* **2012**, *14*, 293–305. [CrossRef]

38. Thabault, M.; Turpin, V.; Maisterrena, A.; Jaber, M.; Egloff, M.; Galvan, L. Cerebellar and Striatal Implications in Autism Spectrum Disorders: From Clinical Observations to Animal Models. *Int. J. Mol. Sci.* **2022**, *23*, 2294. [CrossRef]

39. Al Sagheer, T.; Haida, O.; Balbous, A.; Francheteau, M.; Matas, E.; Fernagut, P.-O.; Jaber, M. Motor Impairments Correlate with Social Deficits and Restricted Neuronal Loss in an Environmental Model of Autism. *Int. J. Neuropsychopharmacol.* **2018**, *21*, 871–882. [CrossRef]

40. Badura, A.; Verpeut, J.L.; Metzger, J.W.; Pereira, T.D.; Pisano, T.J.; Deverett, B.; Bakshinskaya, D.E.; Wang, S.S.-H. Normal cognitive and social development require posterior cerebellar activity. *eLife* **2018**, *7*, e36401. [CrossRef]

41. Cupolillo, D.; Hoxha, E.; Faralli, A.; De Luca, A.; Rossi, F.; Tempia, F.; Carulli, D. Autistic-Like Traits and Cerebellar Dysfunction in Purkinje Cell PTEN Knock-Out Mice. *Neuropsychopharmacology* **2016**, *41*, 1457–1466. [CrossRef] [PubMed]

42. Gibson, J.M.; Howland, C.P.; Ren, C.; Howland, C.; Vernino, A.; Tsai, P.T. A Critical Period for Development of Cerebellar-Mediated Autism-Relevant Social Behavior. *J. Neurosci.* **2022**, *42*, 2804–2823. [CrossRef] [PubMed]

43. Haida, O.; Al Sagheer, T.; Balbous, A.; Francheteau, M.; Matas, E.; Soria, F.; Fernagut, P.O.; Jaber, M. Sex-dependent behavioral deficits and neuropathology in a maternal immune activation model of autism. *Transl. Psychiatry* **2019**, *9*, 124. [CrossRef]

44. Hoxha, E.; Tonini, R.; Montarolo, F.; Croci, L.; Consalez, G.G.; Tempia, F. Motor dysfunction and cerebellar Purkinje cell firing impairment in Ebf2 null mice. *Mol. Cell. Neurosci.* **2013**, *52*, 51–61. [CrossRef] [PubMed]

45. Kloth, A.D.; Badura, A.; Li, A.; Cherskov, A.; Connolly, S.G.; Giovannucci, A.; Bangash, M.A.; Grasselli, G.; Peñagarikano, O.; Piochon, C.; et al. Cerebellar associative sensory learning defects in five mouse autism models. *eLife* **2015**, *4*, e06085. [CrossRef]

46. Matas, E.; Maisterrena, A.; Thabault, M.; Balado, E.; Francheteau, M.; Balbous, A.; Galvan, L.; Jaber, M. Major motor and gait deficits with sexual dimorphism in a Shank3 mutant mouse model. *Mol. Autism* **2021**, *12*, 2. [CrossRef]

47. Peter, S.; ten Brinke, M.M.; Stedehouder, J.; Reinelt, C.M.; Wu, B.; Zhou, H.; Zhou, K.; Boele, H.-J.; Kushner, S.A.; Lee, M.G.; et al. Dysfunctional cerebellar Purkinje cells contribute to autism-like behaviour in Shank2-deficient mice. *Nat. Commun.* **2016**, *7*, 12627. [CrossRef]

48. Piochon, C.; Kloth, A.D.; Grasselli, G.; Titley, H.K.; Nakayama, H.; Hashimoto, K.; Wan, V.; Simmons, D.H.; Eissa, T.; Nakatani, J.; et al. Cerebellar plasticity and motor learning deficits in a copy-number variation mouse model of autism. *Nat. Commun.* **2014**, *5*, 5586. [CrossRef]

49. Tsai, P.T.; Hull, C.; Chu, Y.; Greene-Colozzi, E.; Sadowski, A.R.; Leech, J.M.; Steinberg, J.; Crawley, J.N.; Regehr, W.G.; Sahin, M. Autistic-like behaviour and cerebellar dysfunction in Purkinje cell Tsc1 mutant mice. *Nature* **2012**, *488*, 647–651. [CrossRef]

50. Wang, R.; Tan, J.; Guo, J.; Zheng, Y.; Han, Q.; So, K.-F.; Yu, J.; Zhang, L. Aberrant Development and Synaptic Transmission of Cerebellar Cortex in a VPA Induced Mouse Autism Model. *Front. Cell. Neurosci.* **2018**, *12*, 500. [CrossRef]

51. Simmons, D.H.; Titley, H.K.; Hansel, C.; Mason, P. Behavioral Tests for Mouse Models of Autism: An Argument for the Inclusion of Cerebellum-Controlled Motor Behaviors. *Neuroscience* **2021**, *462*, 303–319. [CrossRef] [PubMed]

52. de la Torre-Ubieta, L.; Won, H.; Stein, J.L.; Geschwind, D.H. Advancing the understanding of autism disease mechanisms through genetics. *Nat. Med.* **2016**, *22*, 345–361. [CrossRef] [PubMed]

53. Meyza, K.Z.; Blanchard, D.C. The BTBR mouse model of idiopathic autism—Current view on mechanisms. *Neurosci. Biobehav. Rev.* **2017**, *76*, 99–110. [CrossRef] [PubMed]

54. Meyza, K.Z.; Defensor, E.B.; Jensen, A.L.; Corley, M.J.; Pearson, B.L.; Pobbe, R.L.H.; Bolivar, V.J.; Blanchard, D.C.; Blanchard, R.J. The BTBR T+ tf/J mouse model for autism spectrum disorders-in search of biomarkers. *Behav. Brain Res.* **2013**, *251*, 25–34. [CrossRef] [PubMed]

55. Amodeo, D.A.; Jones, J.H.; Sweeney, J.A.; Ragozzino, M.E. Differences in BTBR T+ tf/J and C57BL/6J mice on probabilistic reversal learning and stereotyped behaviors. *Behav. Brain Res.* **2012**, *227*, 64–72. [CrossRef]

56. Chao, O.Y.; Yunger, R.; Yang, Y.-M. Behavioral assessments of BTBR T+Itpr3tf/J mice by tests of object attention and elevated open platform: Implications for an animal model of psychiatric comorbidity in autism. *Behav. Brain Res.* **2018**, *347*, 140–147. [CrossRef]

57. Faraji, J.; Karimi, M.; Lawrence, C.; Mohajerani, M.H.; Metz, G.A.S. Non-diagnostic symptoms in a mouse model of autism in relation to neuroanatomy: The BTBR strain reinvestigated. *Transl. Psychiatry* **2018**, *8*, 234. [CrossRef]

58. McFarlane, H.G.; Kusek, G.K.; Yang, M.; Phoenix, J.L.; Bolivar, V.J.; Crawley, J.N. Autism-like behavioral phenotypes in BTBR T+tf/J mice. *Genes Brain Behav.* **2008**, *7*, 152–163. [CrossRef]

59. McTighe, S.M.; Neal, S.J.; Lin, Q.; Hughes, Z.A.; Smith, D.G. The BTBR mouse model of autism spectrum disorders has learning and attentional impairments and alterations in acetylcholine and kynurenic acid in prefrontal cortex. *PLoS ONE* **2013**, *8*, e62189. [CrossRef]

60. Scattoni, M.L.; Gandhy, S.U.; Ricceri, L.; Crawley, J.N. Unusual repertoire of vocalizations in the BTBR T+tf/J mouse model of autism. *PLoS ONE* **2008**, *3*, e3067. [CrossRef]

61. Amodeo, D.A.; Pahua, A.E.; Zarate, M.; Taylor, J.A.; Peterson, S.; Posadas, R.; Oliver, B.L.; Amodeo, L.R. Differences in the expression of restricted repetitive behaviors in female and male BTBR T + tf/J mice. *Behav. Brain Res.* **2019**, *372*, 112028. [CrossRef]

62. Queen, N.J.; Boardman, A.A.; Patel, R.S.; Siu, J.J.; Mo, X.; Cao, L. Environmental enrichment improves metabolic and behavioral health in the BTBR mouse model of autism. *Psychoneuroendocrinology* **2020**, *111*, 104476. [CrossRef] [PubMed]

63. Ellegood, J.; Anagnostou, E.; Babineau, B.A.; Crawley, J.N.; Lin, L.; Genestine, M.; DiCicco-Bloom, E.; Lai, J.K.Y.; Foster, J.A.; Peñagarikano, O.; et al. Clustering autism: Using neuroanatomical differences in 26 mouse models to gain insight into the heterogeneity. *Mol. Psychiatry* **2015**, *20*, 118–125. [CrossRef] [PubMed]

64. Ellegood, J.; Babineau, B.A.; Henkelman, R.M.; Lerch, J.P.; Crawley, J.N. Neuroanatomical analysis of the BTBR mouse model of autism using magnetic resonance imaging and diffusion tensor imaging. *NeuroImage* **2013**, *70*, 288–300. [CrossRef]

65. Shpyleva, S.; Ivanovsky, S.; de Conti, A.; Melnyk, S.; Tryndyak, V.; Beland, F.A.; James, S.J.; Pogribny, I.P. Cerebellar oxidative DNA damage and altered DNA methylation in the BTBR T+tf/J mouse model of autism and similarities with human post mortem cerebellum. *PLoS ONE* **2014**, *9*, e113712. [CrossRef] [PubMed]

66. Nadeem, A.; Ahmad, S.F.; Al-Harbi, N.O.; Attia, S.M.; Alshammari, M.A.; Alzahrani, K.S.; Bakheet, S.A. Increased oxidative stress in the cerebellum and peripheral immune cells leads to exaggerated autism-like repetitive behavior due to deficiency of antioxidant response in BTBR T + tf/J mice. *Prog. Neuropsychopharmacol. Biol. Psychiatry* **2019**, *89*, 245–253. [CrossRef]

67. Chao, O.Y.; Zhang, H.; Pathak, S.S.; Huston, J.P.; Yang, Y.-M. Functional Convergence of Motor and Social Processes in Lobule IV/V of the Mouse Cerebellum. *Cerebellum* **2021**, *20*, 836–852. [CrossRef]

68. Deacon, R.M.J. Measuring motor coordination in mice. *J. Vis. Exp.* **2013**, *75*, e2609. [CrossRef]

69. Cosgrove, J.A.; Kelly, L.K.; Kiffmeyer, E.A.; Kloth, A.D. Sex-dependent influence of postweaning environmental enrichment in Angelman syndrome model mice. *Brain Behav.* **2022**, *12*, e2468. [CrossRef]

70. Siegel, J.J.; Taylor, W.; Gray, R.; Kalmbach, B.; Zemelman, B.V.; Desai, N.S.; Johnston, D.; Chitwood, R.A. Trace Eyeblink Conditioning in Mice Is Dependent upon the Dorsal Medial Prefrontal Cortex, Cerebellum, and Amygdala: Behavioral Characterization and Functional Circuitry. *Eneuro* **2015**, *2*. [CrossRef]

71. Giovannucci, A.; Pnevmatikakis, E.A.; Deverett, B.; Pereira, T.; Fondriest, J.; Brady, M.J.; Wang, S.S.-H.; Abbas, W.; Parés, P.; Masip, D. Automated gesture tracking in head-fixed mice. *J. Neurosci. Methods* **2018**, *300*, 184–195. [CrossRef] [PubMed]

72. Heiney, S.A.; Wohl, M.P.; Chettih, S.N.; Ruffolo, L.I.; Medina, J.F. Cerebellar-dependent expression of motor learning during eyeblink conditioning in head-fixed mice. *J. Neurosci.* **2014**, *34*, 14845–14853. [CrossRef] [PubMed]

73. Hirono, M.; Watanabe, S.; Karube, F.; Fujiyama, F.; Kawahara, S.; Nagao, S.; Yanagawa, Y.; Misonou, H. Perineuronal Nets in the Deep Cerebellar Nuclei Regulate GABAergic Transmission and Delay Eyeblink Conditioning. *J. Neurosci.* **2018**, *38*, 6130–6144. [CrossRef] [PubMed]

74. Allen Institute for Brain Science. Allen Mouse Brain Atlas [Dataset]. 2004. Available online: http://mouse.brain-map.org (accessed on 5 November 2022).

75. Lein, E.S.; Hawrylycz, M.J.; Ao, N.; Ayres, M.; Bensinger, A.; Bernard, A.; Boe, A.F.; Boguski, M.S.; Brockway, K.S.; Byrnes, E.J.; et al. Genome-wide atlas of gene expression in the adult mouse brain. *Nature* **2007**, *445*, 168–176. [CrossRef]

76. Sholl, D.A. The measurable parameters of the cerebral cortex and their significance in its organization. *Prog. Neurobiol.* **1956**, *2*, 324–333.

77. Takehara-Nishiuchi, K. The Anatomy and Physiology of Eyeblink Classical Conditioning. *Curr. Top. Behav. Neurosci.* **2018**, *37*, 297–323. [CrossRef] [PubMed]

78. Bauman, M.L.; Kemper, T.L. Neuroanatomic observations of the brain in autism: A review and future directions. *Int. J. Dev. Neurosci.* **2005**, *23*, 183–187. [CrossRef] [PubMed]

79. Kemper, T.L.; Bauman, M. Neuropathology of infantile autism. *J. Neuropathol. Exp. Neurol.* **1998**, *57*, 645–652. [CrossRef]
80. Whitney, E.R.; Kemper, T.L.; Bauman, M.L.; Rosene, D.L.; Blatt, G.J. Cerebellar Purkinje cells are reduced in a subpopulation of autistic brains: A stereological experiment using calbindin-D28k. *Cerebellum* **2008**, *7*, 406–416. [CrossRef]
81. Whitney, E.R.; Kemper, T.L.; Rosene, D.L.; Bauman, M.L.; Blatt, G.J. Density of cerebellar basket and stellate cells in autism: Evidence for a late developmental loss of Purkinje cells. *J. Neurosci. Res.* **2009**, *87*, 2245–2254. [CrossRef]
82. Xiao, R.; Zhong, H.; Li, X.; Ma, Y.; Zhang, R.; Wang, L.; Zang, Z.; Fan, X. Abnormal Cerebellar Development Is Involved in Dystonia-Like Behaviors and Motor Dysfunction of Autistic BTBR Mice. *Front. Cell Dev. Biol.* **2020**, *8*, 231. [CrossRef] [PubMed]
83. Achilly, N.P.; He, L.-J.; Kim, O.A.; Ohmae, S.; Wojaczynski, G.J.; Lin, T.; Sillitoe, R.V.; Medina, J.F.; Zoghbi, H.Y. Deleting Mecp2 from the cerebellum rather than its neuronal subtypes causes a delay in motor learning in mice. *eLife* **2021**, *10*, e64833. [CrossRef] [PubMed]
84. Koekkoek, S.K.E.; Yamaguchi, K.; Milojkovic, B.A.; Dortland, B.R.; Ruigrok, T.J.H.; Maex, R.; De Graaf, W.; Smit, A.E.; VanderWerf, F.; Bakker, C.E.; et al. Deletion of FMR1 in Purkinje cells enhances parallel fiber LTD, enlarges spines, and attenuates cerebellar eyelid conditioning in Fragile X syndrome. *Neuron* **2005**, *47*, 339–352. [CrossRef] [PubMed]
85. Löwgren, K.; Bååth, R.; Rasmussen, A.; Boele, H.-J.; Koekkoek, S.K.E.; De Zeeuw, C.I.; Hesslow, G. Performance in eyeblink conditioning is age and sex dependent. *PLoS ONE* **2017**, *12*, e0177849. [CrossRef] [PubMed]
86. Wood, G.E.; Shors, T.J. Stress facilitates classical conditioning in males, but impairs classical conditioning in females through activational effects of ovarian hormones. *Proc. Natl. Acad. Sci. USA* **1998**, *95*, 4066–4071. [CrossRef] [PubMed]
87. Oyaga, M.R.; Serra, I.; Kurup, D.; Koekkoek, S.K.E.; Badura, A. Eyeblink conditioning performance and brain-wide C-fos expression in male and female mice. *bioRxiv* **2022**. [CrossRef]
88. Takahashi, A. Toward Understanding the Sex Differences in the Biological Mechanism of Social Stress in Mouse Models. *Front. Psychiatry* **2021**, *12*, 644161. [CrossRef]
89. Mercer, A.A.; Palarz, K.J.; Tabatadze, N.; Woolley, C.S.; Raman, I.M. Sex differences in cerebellar synaptic transmission and sex-specific responses to autism-linked Gabrb3 mutations in mice. *eLife* **2016**, *5*, e07596. [CrossRef]
90. Reeb-Sutherland, B.C.; Fox, N.A. Eyeblink conditioning: A non-invasive biomarker for neurodevelopmental disorders. *J. Autism Dev. Disord.* **2015**, *45*, 376–394. [CrossRef]
91. DeLorey, T.M.; Sahbaie, P.; Hashemi, E.; Homanics, G.E.; Clark, J.D. Gabrb3 gene deficient mice exhibit impaired social and exploratory behaviors, deficits in non-selective attention and hypoplasia of cerebellar vermal lobules: A potential model of autism spectrum disorder. *Behav. Brain Res.* **2008**, *187*, 207–220. [CrossRef]
92. Piven, J.; Saliba, K.; Bailey, J.; Arndt, S. An MRI study of autism: The cerebellum revisited. *Neurology* **1997**, *49*, 546–551. [CrossRef] [PubMed]
93. Stephenson, D.T.; O'Neill, S.M.; Narayan, S.; Tiwari, A.; Arnold, E.; Samaroo, H.D.; Du, F.; Ring, R.H.; Campbell, B.; Pletcher, M.; et al. Histopathologic characterization of the BTBR mouse model of autistic-like behavior reveals selective changes in neurodevelopmental proteins and adult hippocampal neurogenesis. *Mol. Autism* **2011**, *2*, 7. [CrossRef] [PubMed]
94. van der Heijden, M.E.; Gill, J.S.; Sillitoe, R.V. Abnormal Cerebellar Development in Autism Spectrum Disorders. *Dev. Neurosci.* **2021**, *43*, 181–190. [CrossRef] [PubMed]
95. Phillips, M.; Pozzo-Miller, L. Dendritic spine dysgenesis in autism related disorders. *Neurosci. Lett.* **2015**, *601*, 30–40. [CrossRef] [PubMed]
96. Buckner, R.L. The cerebellum and cognitive function: 25 years of insight from anatomy and neuroimaging. *Neuron* **2013**, *80*, 807–815. [CrossRef]
97. Pierce, K.; Courchesne, E. Evidence for a cerebellar role in reduced exploration and stereotyped behavior in autism. *Biol. Psychiatry* **2001**, *49*, 655–664. [CrossRef]
98. Shoji, H.; Takao, K.; Hattori, S.; Miyakawa, T. Age-related changes in behavior in C57BL/6J mice from young adulthood to middle age. *Mol. Brain* **2016**, *9*, 11. [CrossRef]
99. Vogel, R.W.; Ewers, M.; Ross, C.; Gould, T.J.; Woodruff-Pak, D.S. Age-related impairment in the 250-millisecond delay eyeblink conditioning procedure in C57BL/6 mice. *Learn. Mem.* **2002**, *9*, 321–336. [CrossRef] [PubMed]
100. Kishimoto, Y.; Suzuki, M.; Kawahara, S.; Kirino, Y. Age-dependent impairment of delay and trace eyeblink conditioning. *Neuroreport* **2001**, *12*, 3349–3352. [CrossRef] [PubMed]
101. Beekhof, G.C.; Osorio, C.; White, J.J.; van Zoomeren, S.; van der Stok, H.; Xiong, B.; Nettersheim, I.H.M.S.; Mak, W.A.; Runge, M.; Fiocchi, F.R.; et al. Differential spatiotemporal development of Purkinje cell populations and cerebellar-dependent sensorimotor behaviors. *eLife* **2021**, *10*, e63668. [CrossRef]
102. Freeman, J.H.; Nicholson, D.A.; Muckler, A.S.; Rabinak, C.A.; DiPietro, N.T. Ontogeny of eyeblink conditioned response timing in rats. *Behav. Neurosci.* **2003**, *117*, 283–291. [CrossRef] [PubMed]
103. Schreurs, B.G.; Burhans, L.B.; Smith-Bell, C.A.; Mrowka, S.W.; Wang, D. Ontogeny of trace eyeblink conditioning to shock-shock pairings in the rat pup. *Behav. Neurosci.* **2013**, *127*, 114–120. [CrossRef] [PubMed]
104. Woodruff-Pak, D.S. Stereological estimation of Purkinje neuron number in C57BL/6 mice and its relation to associative learning. *Neuroscience* **2006**, *141*, 233–243. [CrossRef] [PubMed]
105. Woodruff-Pak, D.S.; Foy, M.R.; Akopian, G.G.; Lee, K.H.; Zach, J.; Nguyen, K.P.T.; Comalli, D.M.; Kennard, J.A.; Agelan, A.; Thompson, R.F. Differential effects and rates of normal aging in cerebellum and hippocampus. *Proc. Nat. Acad. Sci. USA* **2010**, *107*, 1624–1629. [CrossRef] [PubMed]

106. Prendergast, B.J.; Onishi, K.G.; Zucker, I. Female mice liberated for inclusion in neuroscience and biomedical research. *Neurosci. Biobehav. Rev.* **2014**, *40*, 1–5. [CrossRef]
107. Albergaria, C.; Silva, N.T.; Pritchett, D.L.; Carey, M.R. Locomotor activity modulates associative learning in the mouse cerebellum. *Nat. Neurosci.* **2018**, *21*, 725–735. [CrossRef] [PubMed]
108. Dalla, C.; Shors, T.J. Sex differences in learning processes of classical and operant conditioning. *Phys. Behav.* **2009**, *97*, 229–238. [CrossRef]
109. Meziane, H.; Ouagazzal, A.M.; Aubert, L.; Wietrzych, M.; Krezel, W. Estrous cycle effects on behavior of C57BL/6J and BALB/cByJ female mice: Implications for phenotyping strategies. *Genes Brain Behav.* **2007**, *6*, 192–200. [CrossRef]
110. Hamson, D.K.; Csupity, A.S.; Gaspar, J.M.; Watson, N.V. Analysis of Foxp2 expression in the cerebellum reveals a possible sex difference. *Neuroreport* **2009**, *20*, 611–616. [CrossRef] [PubMed]
111. Kim, H.; Son, J.; Yoo, H.; Kim, H.; Oh, J.; Han, D.; Hwang, Y.; Kaang, B.-K. Effects of the female estrous cycle on the sexual behaviors and ultrasonic vocalizations of male C57BL/6 and autistic BTBR T+ tf/J mice. *Exp. Neurobiol.* **2016**, *25*, 156–162. [CrossRef] [PubMed]
112. Shansky, R.M.; Woolley, C.S. Considering Sex as a Biological Variable Will Be Valuable for Neuroscience Research. *J. Neurosci.* **2016**, *36*, 11817–11822. [CrossRef]

 *NeuroSci*

MDPI

Article

Neural Assemblies as Precursors for Brain Function

Kieran Greer

Distributed Computing Systems, Belfast BT1 9JY, UK; kgreer@distributedcomputingsystems.co.uk

Abstract: This concept paper gives a narrative about intelligence from insects to the human brain, showing where evolution may have been influenced by the structures in these simpler organisms. The ideas also come from the author's own cognitive model, where a number of algorithms have been developed over time and the precursor structures should be codable to some level. Through developing and trying to implement the design, ideas like separating the data from the function have become architecturally appropriate and there have been several opportunities to make the system more orthogonal. Similarly for the human brain, neural structures may work in-sync with the neural functions, or may be slightly separate from them. Each section discusses one of the neural assemblies with a potential functional result, that cover ideas such as timing or scheduling, structural intelligence and neural binding. Another aspect of self-representation or expression is interesting and may help the brain to realise higher-level functionality based on these lower-level processes.

Keywords: neural; brain; structural intelligence; cell expression; evolution

1. Introduction

This paper describes some neural representations that may be helpful for realising intelligence in the human brain. The ideas come from the author's own cognitive model, where a number of algorithms have been developed over time. Through developing and trying to implement the design, ideas like separating the data from the function have become architecturally appropriate and there have been several opportunities to make the system more orthogonal. Similarly for the human brain, neural structures may work in-sync with the neural functions, or may be slightly separate from them. Having more than 1 information flow actually makes the problem of how the human brain works much easier to solve. Another aspect of self-representation or expression is interesting and may help the brain to realise higher-level functionality based on these lower-level processes, maybe even natural language itself. The cognitive model is still at the symbolic level and so the neural representations are also at this level. The neuron discussion is therefore at a statistical or biophysical level rather than a biological one.

The rest of the paper is organised as follows: Section 2 describes some related work. Then, the other sections discuss one of the neural assemblies with a potential functional result. Section 3 describes earlier work on a timer or scheduler. Section 4 describes how intelligence may be inherent in the neuron structure. Section 5 describes how the neural binding problem can be simplified. Section 6 describes some aspects of the author's own cognitive model that have influenced the writing of this paper and Section 7 describes how natural language may have evolved naturally from similar structures. Finally, Section 8 gives some conclusions on the work.

2. Related Work

This paper is based mostly on the author's own cognitive model, who comes from a computer science background. It has been described in detail, in particular, in the paper series 'New Ideas for Brain Modelling' 1–7 [1–4]. Most of the Artificial Intelligence technology is therefore described in the following sections, but a background to supporting biological work is described here. Supporting biological work includes [5–8] and also

Citation: Greer, K. Neural Assemblies as Precursors for Brain Function. *NeuroSci* **2022**, *3*, 645–655. https://doi.org/10.3390/neurosci3040046

Academic Editor: Szczepan Paszkiel

Received: 18 October 2022
Accepted: 9 November 2022
Published: 10 November 2022

Publisher's Note: MDPI stays neutral with regard to jurisdictional claims in published maps and institutional affiliations.

biophysical or statistical work, for example [9,10]. Having more than 1 information flow has been studied extensively. For example, the paper [8] describes that more than one type of sodium channel can be created and that they interact with each other, producing a variable signal. Small currents are involved, even for Ion channels and they work at different potentials, etc. It is also described how neurons can change states and start firing at different rates. Memory is a key topic, where the paper [6] describes that positive regulators can give rise to the growth of new synaptic connections and this can also form memories. There are also memory suppressors, to ensure that only salient features are learned. Long-term memory endures by virtue of the growth of new synaptic connections, which is a structural change. There is also some mathematical background, where the paper [7] was the basis for the simulation equation of [2] and the book [5] is a critical work on the neocortex and higher brain functions. The argument for this paper is still at the symbolic level, where the papers [9,10] both try to describe how the brain might organise itself through statistical processes.

The paper [11] may have developed a synaptic model, based on the themes of this paper. The authors state that recent neuroscience evidences indicate that astrocytes interact closely with neurons and participate in the regulation of synaptic neurotransmission, which has motivated new perspectives for the research of stigmergy in the brain. Additionally, that each astrocyte contains hundreds or thousands of branch microdomains, and each of them encloses a synapse, where distance between coupled branch microdomains is a critical factor. They also carry out tests to show the importance of regular distances between neurons.

The pioneering work of Santiago Ramón y Cajal (http://www.scholarpedia.org/article/Santiago_Ramón_y_Cajal (accessed on 18 October 2022)) may be supportive, in relation to pacemaker cells [12] and discrete units. Then, a new theory by Tsien [13] suggests that perineuronal nets, discovered by Golgi (https://en.wikipedia.org/wiki/Camillo_Golgi (accessed on 18 October 2022)) may be key to how the brain stores long-term memories and it is the basis for the cognitive model of this paper as well. The idea of an extracellular matrix was actually rejected by Cajal, where a discrete brain function was preferred. Neural binding is discussed in one section, but with a view to making it less holistic, where contrasting biological work might include [14,15]. Other biological work on simpler organisms includes [16–23] and is noted in the following sections.

3. Timing

This was an early discovery for an automatic scheduler or counter [2]. It is not as relevant to the other sections, but it does offer an automatic construction for an intelligent process. The paper considered using nested structures, not only for concept ensembles, but also for more mechanical processes. If the structure fires inwards, then the rather obvious idea would be that an inner section would fire inhibitors outwards that would eventually switch off the source to its activation. It may also fire positively inwards, when the process would repeat with the next inner section, and so on. This switching on and off of nested sections could lead to a type of scheduling or timing, if each section also sent a signal somewhere else. This is illustrated in Figure 1, which also shows how a circular arrangement can behave in the same way [16]. A simulation of this process was run using Equation (1) that processed at a pattern level, not a synapse level and is a simplified version of an equation from [7]. It showed the expected result of how the pattern excitatory values would flow through the nested levels, rather like a colonic movement, for example. These tests therefore only considered the excitatory/inhibitory part, to measure how the patterns would switch on and off through their interactions.

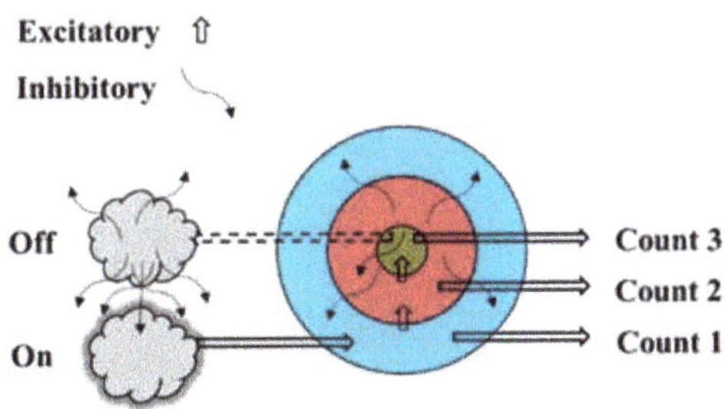

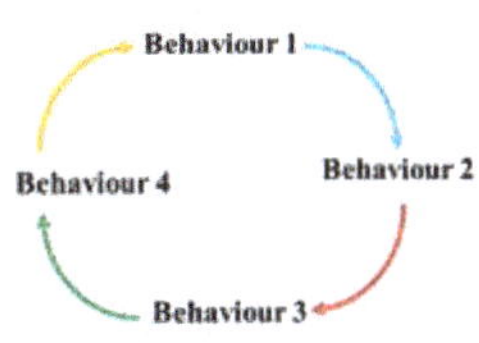

(**a**) Nested timer, counter, or scheduler.   (**b**) Control from circular firing.

Figure 1. Nested Scheduling [2], or circular scheduling.

The test equation, introduced in [2], is repeated next:

$$X_{it} = \sum_{p=1}^{Pi} E_p t - \left(\sum_{k=Pj}^{l} \sum_{y=1}^{m} \sum_{j=1}^{n} (Hjy * \delta) \right) \tag{1}$$

where $y \neq t$ and $i \in Pi$ and *not* $j \in Pi$, and
X_{it} = total input signal for neuron i at time t.
E_p = total excitatory input signal for neuron p in pattern P.
H_{jy} = total inhibitory input signal for neuron j at time y.
δ = weights inhibitory signal.
t = time interval for firing neuron.
y = time interval for any other neuron.
n = total number of neurons.
m = total number of time intervals.
l = total number of active patterns.
P_i = pattern for neuron i.
P = total number of patterns.

A schematic of the total signal input to each neuron over 3 time periods, is given in Table 1. To save space, repeating neuron values are not shown.

Table 1. Relative Pattern Strengths after Firing Sequences.

Neurons	$t = 3$	$t = 4$	$t = 5$
1	7.5	5.0	0.0
2	7.5	5.0	0.0
6	7.5	7.5	5.0
7	7.5	7.5	5.0
11	5.0	7.5	7.5
12	5.0	7.5	7.5
16	0.0	5.0	7.5
17	0.0	5.0	7.5
21	0.0	0.0	5.0
22	0.0	0.0	5.0

This is therefore one of the most basic processes in a human and other much simpler animals. The elegans worm is much studied, for example, because it has a brain of only about 300 neurons that can be mapped accurately. The paper [16] found that 'most active neurons share information by engaging in coordinated, dynamical network activity that corresponds to the sequential assembly of motor commands'. While the neural assemblies might not be nested, there is a circular arrangement to the behavioural network [17] that produces a sequence of behaviours. It is likely that the nested arrangement would be more powerful, however. The worm also has pacemaker-like cells to activate some behaviours [18].

4. From Neuron to Network

It is proposed in this paper that the neuron and the brain network use a similar functionality that derives from the structure. The architecture for the neuron is the standard one of soma body, dendrites as the input channels and the axon as the output channel. The input is an amalgamation of other neuron signals, which gets sorted in the dendrites and soma into a more specific signal that is then transferred to the axon for sending to other neurons. In essence, the process converts signals from being set-based to being type-based. This would be a well-accepted filtering process and it is argued that the conversion from a set-based 'scenario' to more specific and local types in the output, is key to generating intelligence from the structure. This may also help to justify the author's own ensemble-hierarchy structure ([4] and earlier papers). Note that a type however is simply something more singular. It does not have to represent only one input signal, for example, but represents a consistent set of input values. With this architecture therefore, the signal from one neuron to another must also be type-based, but the ensemble input is a set of signals from several neurons. Each set may get sorted differently and therefore create a different set of output types, and so the neuron can be part of more than 1 pattern at any time, where the timing of receiving a signal type would be important.

The neuron can therefore be part of several patterns, making it quite flexible with regard to the information flow. If the input has a chemical bias, for example, then that may allow the synapses related to that particular chemical type, to form and gather sufficient energy to release the signal to other neurons. This would be stigmergic [11,19] in nature. For example, if a neuron fires a signal of a particular type and that is then sent through a network and back to the neuron again, the neuron will already be able to reinforce its current state. It may also now be prepared for the signal [5] and be able to emit it more easily again. This means that the output from a neuron can be sent anywhere, but as with biophysics [10], where there are similar concentrations of a particular type, then the network will start to fire and form patterns relating to that type. Ants or Termites, for example, are able to share information locally, from the stigmergic build-up of chemical signals and this is also optimised for journey time [20]. Another paper [21] discovered that the ant can use different chemical types to indicate 'road-signs' inside of the nest and they use this to spatially segregate. They therefore recognise different types.

This architecture still does not require any intelligence. Thinking about simpler organisms again, at the SAI'14 conference (Prof. Chen's Talk, SAI'14), the speaker asked 'why' an amoeba has a memory and not just how. If it is not to think, then it must be for a functional reason and this function must have evolved from the genetic makeup of the amoeba, hinting that such a mechanism can evolve naturally. So, why did the amoeba develop a memory? The obvious answer would be an evolutionary development for survival, but the author would like to postulate further and guess that it may also be because a living organism has a need to express itself. This desire may go back to the reproduction process itself. An earlier paper argued that true AI cannot be realised because we cannot simulate the living aspect of human cells [24], for example, and that may include this expressive nature. As with a stigmergic build-up, if the amoeba has set itself-up for a particular type of input, maybe it does not react to other input immediately but can only react to the specific input again, even after a short delay. The paper [23] models the amoeba behaviour as a memristor, which is a similar type of electronic circuit. They note that: the model however does not fully explain the memory response of the amoeba and does not take into account the fact that, at a microscopic level, changes in the physiology of the organism occur independently of the biological oscillators. These changes also occur over a finite period of time and must be dependent on the state of the system at previous times. This last point is particularly important: it is in fact this state-dependent feature which is likely to produce memory effects rather than the excitation of biological oscillators. Therefore, at least 2 processes are at work in this single celled organism, where one is slow moving and one is much quicker. The oscillators would be tuned by the viscosity channels, that would maintain a behaviour until the channels themselves changed and this slower change is more structural. Figure 2 illustrates how the neuron and network transpose from ensemble to type, and the amoeba may indicate a precursor to neural synapses, for example.

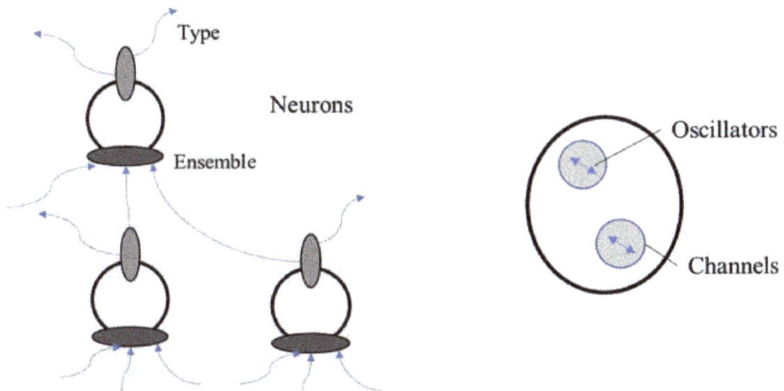

(**a**) Neuron transposes from ensemble to type.   (**b**) Amoeba oscillators tuned by channels.

Figure 2. Network of neurons, with a comparative amoeba.

5. Neural Binding

This is an important question from both the psychological and the mechanical aspects of the human brain. It asks why the brain does not confuse concepts like 'red square' and 'blue circle' unless these are fully defined by brain patterns first. Why is 'red' and 'blue' not confused, for example. The problem is that it would not be possible to store every memory instance combination in the brain and so (dynamic) linking of concepts is required. The paper [15] includes the idea of consciousness and how the brain is able to be coherent. Some models may include temporal logic or predicate calculus rules to explain how variables can bind with each other. Quantum mechanics is another plausible mechanism for merging patterns [25]. The paper [14] is quite interesting, where they describe a framework called the Specialized Neural Regions for Global Efficiency (SNRGE) framework. The paper describes that 'the specializations associated with different brain areas represent computational trade-offs that are inherent in the neurobiological implementation of cognitive processes. That is, the trade-offs are a direct consequence of what computational processes can be easily implemented in the underlying biology'. The specializations of the paper correspond anatomically to the hippocampus (HC), the prefrontal cortex (PFC), and all of neocortex that is posterior to prefrontal cortex (posterior cortex, PC). Essentially, prefrontal cortex and the hippocampus appear to serve as memory areas that dynamically and interactively support the computation that is being performed by posterior brain areas. They argue against temporal synchrony, because of the 'red circle blue square' question and prefer to argue for coarse-coded distributed representations (CCDR) [26] instead. With CCDR, the concepts themselves can remain separate and it is not necessary to declare every binding instance explicitly, but it can be obtained from a local overlay coding scheme. The author has argued that patterns can be aggregated to some extent ([4] and earlier papers), when manipulation of them can then be done with much fewer neural connections over the aggregated representations. He has also argued that simply 2 layers with the same node representations can produce the required circuits. However, to realise these two concepts, still requires linked formations that either contain red and square, or blue and circle and so CCDR looks like a neat solution to this. However, it might be a question of whether the links are permanent or created 'on the fly'. There is also the problem with imagination that can create new images. If the ensemble structure does not exist, then it would have to be constructed dynamically.

The author has also argued, or asked, why the senses are not part of the human conscious. Recent science however, is beginning to suggest that the whole nervous system is the conscious. We have eyes, ears, voice box, and so on, which we use as external mechanisms to the brain function and the paper [27] argues that when the brain thinks, it sends signals back to these organs and senses, and that they are essential to realise what the brain is thinking. If we consider the 'red square, blue circle' problem again, then one problem with current philosophy may be that we assume the pattern formations are translated only by the brain. One problem with that is the fact that the conscious would have to see every pattern and pattern part as the same. It would then require additional capability to try to differentiate. The 'red' concept has the same makeup as the 'square' concept to the brain conscious, for example. If it is possible to introduce different functions to the problem, then a solution may be easier to find and for the author, this would mean feedback to the external sensory organs. Considering the eye and for the sake of argument, let it produce only image shapes and colours. What if one signal could request the eye to produce an imprint of a shape on it and then a second separate signal requests that the eye gives it a particular colour. If this was possible, then the two signals would not necessarily have to be linked first, where that requirement has changed over to one about sending different function requests to the eye. This example is illustrated in Figure 3 and would make the neural-binding problem much easier, because the orthogonal function requests do not require all of the combinations that the more holistic conscious might require. Part of the binding has been moved to the eye itself. It may be the case that this is only for long-term memory, where a holistic memory store of recent images would still operate. Additionally, with this setup, the functional signals do not need to be fully linked patterns, but can be single links, for example, while the concept patterns can still be fully linked.

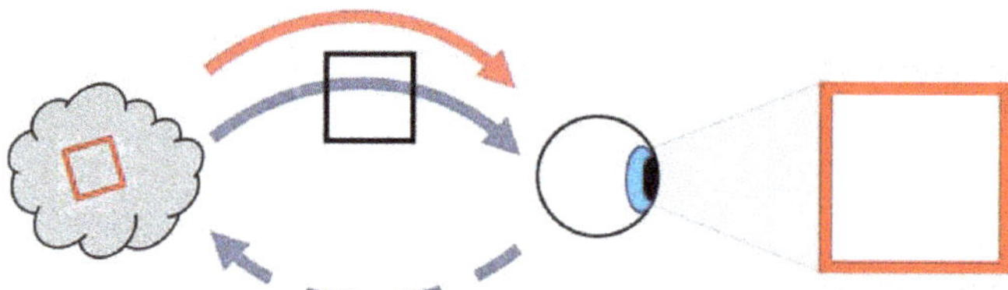

Figure 3. The brain sends two signals to the eye to construct an image and receives the feedback of this.

6. Cognitive Model

The author has developed a cognitive model over several years. The original design [24] had an architecture of 3 levels of increasing complexity, but also a global ontology that any of the 3 levels could use. The idea of an underlying global representation raises some interesting ideas. The author's background is in computer science, distributed systems and also the Internet, where the SUMO ontology [28] and others, have been previously suggested. SUMO (Suggested Upper Merged Ontology) has been created to be a common language for the Internet. It is more expressive than object or semantic recognition, but not as much as natural language. Base concepts include 'object' and 'process', for example, but being an ontology, it includes relations between the different ontology entities. The author has worked on a cognitive model that is now at an early stage of development, with an even simpler ontology at its base. It is not even an ontology, but levels of symbolic node clusters, where a lower-level contains more frequently occurring symbols or concepts. The clusters are not linked together, but they offer some kind of global ordering over the stored symbolic representations. One example may be the 3 short stories—'fox and crow', 'fox and stork' and 'fox and grape'. If a basic word count is done on each story, then for the 'fox and crow' story, the crow word occurs more frequently and so if using the concept trees counting rule (any child node in a tree cannot have a larger frequency count than its parent), it would be placed as the root tree node. In a global sense however, the fox word is more common, because it occurs in all 3 stories. Therefore, the global ontology would in fact re-order the local 'fox and crow' instance, to place 'fox' at the root node and then 'crow' one level above that. For this architecture therefore, the local story instance provides what concepts are available, but they are then re-ordered by the global ensemble. It is the same idea as the natural ordering for concept trees, described in section 6.4 of the paper [29]. With that, a road would always be placed below a car, for example, because a car would run on the road. With the cognitive model implementation therefore, there is the global ensemble of concepts at the base as the memory structure. Each of the 3 cognitive levels—pattern optimisation, tree-pattern aggregation and more complex concepts—also write a simplified view of their structures to the memory database. Then, when any level wants to read from memory, it uses the global database to retrieve whatever memory type it requires. The global memory structure therefore has different levels of representation that reference the ensemble clusters. It is also a common view of the information in the system, where any module can read and understand what a memory node is, because the more complex functionality that may be specific to any module is missing.

6.1. Natural Structure

This is also in the context of the cognitive model. Considering the author's earlier work, the ReN (Refined Neuron) [1] has not been considered recently. The original idea was to make the signal more analogue, but it has become clear with biological modelling of the neuron that it can produce variable signals by itself. What may not be clear however, is if this is in discrete signal bands or a continuous signal. Discrete bands would match better with a type-based approach, when the ReN may still be useful. The other idea was that it is caused by repulsion of the signal down the input channels, which would be the axons and that would encourage new outlet paths to form. A third idea of balance is implicit in any energy system, even before the biological world.

The idea of frames (Minsky, [1]) is still interesting for the cognitive model. If it was used as part of the memory structure, then it would produce distinct units, including terminal states. The author has suggested a frequency grid classifier [3], which is entropy-based, reducing a global error, but one that is event or experience-based. It is a self-organising method that clusters elements with other elements that they are most often associated with. It was also suggested that it would be the base of a 'unit of work' that is a unit of ensemble-hierarchy structure. The ensemble-hierarchy structure [3,4] was originally intended to produce a more combined and analogue signal, but from Newtonian mechanics rather than a Quantum effect. The hierarchy would repeat the ensemble nodes, but with an additional tree structure and then resonating between similar node sets in both parts would produce 'notes' that would be recognised by the conscious. This is really very similar to the relationship between astrocytes and neurons [11], for example. Resonating is not part of the current research, but the ensemble-hierarchy is still an important structure. The tree nodes might become abstracted representations of ensemble patterns instead, where the structure adds meaning to what is otherwise a flat or nested matrix. In this sense, the ensemble-hierarchy would probably be expected, rather than being novel and it also fits better with a tokenised memory.

It may be interesting to note that Hill et al. [9] discovered that the connectome of cortical microcircuitry is largely formed from the nonspecific alignment of neuron morphologies, rather than pairwise chemical signals. This means that structure is preferred, whereas the signal is more dynamic. They also discovered that, although the specific positions of synapses are random, the restrictions caused by structure and neuron type, serve to ensure a robust and invariant set of distributed inputs and outputs between pattern populations. This would be grounded in biophysics. If the neuron synchronises more with static structure, then this will help it to maintain both form and lifespan [1], which is again a favourable conclusion for the ensemble-hierarchy relationship. The paper [2] then showed that it is more economic energy-wise, to produce a new neuron half-way between other neurons. This would also help to keep the path lengths regular, which again helps the neurons to synchronize their firing. If a particular region became active and started producing new neurons, that would change the path lengths, but the lengths would still remain quite regular and therefore recognisable as a type. Therefore, a lot of intelligence can be derived automatically from the structure, before even considering the neural functions. The author also wonders if distance between neurons is part of the pattern type itself. If, for example, the same chemical travels round a network of closely packed neurons or sparsely packed neurons, would that represent a different type to the brain? It would certainly change the relative strength of the signal, but also firing rates and timing.

6.2. Natural Function

The world therefore appears to be typed, even at the lowest level. For an amoeba, it may be a single type, whereas for humans, it is ensembles of types, but it is necessary to be able to discriminate over this. Order is another low-level process, not a high-level one. Not only order, but also regularity, where there is a sense of learning from repetition. Worms for example, have a behavioural order and ants make use of both of these functions, where collectively, they appear to exhibit intelligence. Feedback is also essential, where even at the cell level, there may be a necessity to express oneself. Thus far, we have energy optimisation, object or type recognition, spatial awareness, feedback, timing and ordering. Then, intelligence appears to be the evaluation of these lower-level processes, where there are obviously different levels of intelligence. The bee, for example, has a more developed brain with modules that are also recognisable in the human brain [22], but its reasoning process must surely only be at a logical level.

7. Natural Language Development

In the human brain, there are cells other than neurons, such as the more-simple glial and interneuron cells. More recently the perineuronal network [13] has received a lot of attention and may be exactly the memory structure that the cognitive model will now use. If the memory structure is sightly separate therefore, this can lead to at least two different information flows, for either memory or function. If the Perineuronal network is made of the glial cells—astrocytes and deodendrocytes, for example, then astrocytes are also known to produce energy for neurons and so successfully syncing with the memory structure would also provide an energy supply. The author's own cognitive model implements a similar type of architecture, described in Section 6. Through implementing the cognitive model, it was interesting to note some separation between the global representation and the original sources, and also a little bit of autonomy for the global representation. Tokenized text, for example, might be stored largely as nouns and verbs, without all of the natural language. The architecture also works with images. The author is using a new idea called Image Parts [30], which scans an image and splits it up into parts, but is currently only useful for object recognition. The parts can then be stored in the global ensemble database and re-used. To re-construct an image representation, one part may be north of another part, for example. The algorithm is not very accurate, achieving only 80% accuracy, where neural networks would achieve closer to 100% accuracy, but it is also explainable. When other modules want to interface with the image, they can make use of the same ensemble parts, structured by an abstract tree representation.

The author postulates that this is like the brain architecture itself making use of a common language, to allow the different modules to interact with each other. The homogeneous input is converted over to a different tokenised representation, that is then used to describe the input to any other part of the system. If that process is internal to the brain itself, then it may be a reason why humans have developed their natural language, in order to try to express this internal structure. Nouns and verbs are the basis of the real world as well, for example and the paper [31] concludes that: 'The available studies on the neural basis of normal language development suggest that the brain systems underlying language processing are in place already in early development'. This suggests that the structure for natural language is in place from a very early age. The paper [32] states that deep learning algorithms can produce, at least, coarse brain-like representations, suggesting that they process words and sentences in a human-like way. Word vectors may be superior to tree linking, but it is still a distributed and tokenised AI algorithm that can be mapped to brain regions. Problems have also been found with the design. Bees are also thought to communicate using a symbolic language that results in their waggle dance. Like the amoeba then, did they reason that they should communicate this, or is it a reflection of their internal structure? Maybe it is just an evolutionary quirk.

8. Conclusions

This paper gives a narrative that outlines structural components of simpler organisms that may have helped the human brain to evolve. More than that, the structures are so basic, they can be included in a computer model for Artificial Intelligence and are consistent with the author's own cognitive model. The design may not be 100% accurate, but there appears to be a consistency about it and some biological and mathematical evidence can help to validate the theories. An early idea about scheduling through nesting may be seen in action in worms, for example, but in a simpler form. Then, one idea may be that intelligence can be realised automatically by converting from ensemble input to type-based output. This would occur automatically in the neuron network, where the realisation of types will produce some understanding and therefore intelligence. Amoebas are able to learn single types. The stigmergic processes of termites or ants, for example, have become interesting to explaining the neural structures for several reasons. Firstly, it is suggested that the neural microcircuitry is constructed primarily from the alignment of morphology or structure, rather than signal type and this includes synapse alignment and preparation. Although, the chemical signal will still change the type emitted by the cell. Secondly, the relationship between neurons and the substrate of glial cells, for example, also suggests stigmergic processes.

It would be interesting if there is an underlying global memory structure to the brain, which is this perineuronal substrate and if it can abstract and even re-structure input signals. The uniformity of the substrate would allow it to communicate this to other modules and a computer model would be able to simulate it to some level. When modelling the biological structure, images may be stored as whole representations in the short-term memory, but when they are moved into long-term memory, they become tokenized and abstracted. One final idea is that the neural binding problem is constrained by current thinking about a holistic conscious and if it can be made more orthogonal and receive help from other organs, the problem will become much easier to solve.

Most interesting then may be the idea that a cell or organism evolves, not only to survive, but also by expressing itself, where the expression is a result of its own internal structures and processes. In this respect, the memory substrate would be a precursor to our own natural language and this might also be seen in bees. The structural transformation from input to tokenized ensemble results in a communication process that is akin to a common language. The higher cognitive processes, if you like, have built themselves on the lower-level structures and processes.

Funding: This research received no external funding.

Informed Consent Statement: Not applicable.

Data Availability Statement: Not applicable.

Acknowledgments: The author would like to thank the reviewers for their helpful comments.

Conflicts of Interest: The author declares no conflict of interest.

References

1. Greer, K. New Ideas for Brain Modelling. In *BRAIN. Broad Research in Artificial Intelligence and Neuroscience*; Lumen Publishing: Bucharest, Romania, 2015; Volume 6, pp. 26–46.
2. Greer, K. New Ideas for Brain Modelling 2. In *Intelligent Systems in Science and Information 2014, Studies in Computational Intelligence*; Arai, K., Kapoor, S., Bhatia, R., Eds.; Springer International Publishing: Cham, Switzerland, 2014; Volume 591, pp. 23–39. [CrossRef]
3. Greer, K. New Ideas for Brain Modelling 3. In *Cognitive Systems Research*; Elsevier: Amsterdam, The Netherlands, 2019; Volume 55, pp. 1–13. [CrossRef]
4. Greer, K. New Ideas for Brain Modelling 7. *Int. J. Comput. Appl. Math. Comput. Sci.* **2021**, *1*, 34–45.
5. Hawkins, J.; Blakeslee, S. On Intelligence. *Times Books*, 3 October 2004; p. 261.
6. Kandel, E.R. The Molecular Biology of Memory Storage: A Dialogue Between Genes and Synapses. *Science* **2001**, *294*, 1030–1038. [CrossRef] [PubMed]
7. Vogels, T.P.; Kanaka Rajan, K.; Abbott, L.F. Neural Network Dynamics. *Annu. Rev. Neurosci.* **2005**, *28*, 357–376. [CrossRef] [PubMed]
8. Waxman, S.G. Sodium channels, the electrogenisome and the electrogenistat: Lessons and questions from the clinic. *J. Physiol.* **2012**, *590*, 2601–2612. [CrossRef] [PubMed]
9. Hill, S.L.; Wang, Y.; Riachi, I.; Schürmann, F.; Markram, H. Statistical connectivity provides a sufficient foundation for specific functional connectivity in neocortical neural microcircuits. *Proc. Natl. Acad. Sci. USA* **2012**, *109*, E2885-94. [CrossRef] [PubMed]
10. Weisbuch, G. The Complex Adaptive Systems Approach to Biology. *Evol. Cogn.* **1999**, *5*, 7–28.
11. Xu, X.; Zhao, Z.; Li, R.; Zhang, H. Brain-inspired stigmergy learning. *IEEE Access.* **2019**, *7*, 54410–54424. [CrossRef]
12. Sanders, K.; Ward, S. Interstitial cells of Cajal: A new perspective on smooth muscle function. *J. Physiol.* **2006**, *576 Pt 3*, 721–726. [CrossRef] [PubMed]
13. Tsien, R.Y. Very long-term memories may be stored in the pattern of holes in the perineuronal net. *Proc. Natl. Acad. Sci. USA* **2013**, *110*, 12456–12461. [CrossRef] [PubMed]
14. Cer, D.M.; O'Reilly, R.C. Neural Mechanisms of Binding in the Hippocampus and Neocortex: Insights from Computational Models. In *Handbook of Binding and Memory: Perspectives from Cognitive Neuroscience*; Zimmer, H.D., Mecklinger, A., Lindenberger, U., Eds.; Oxford University Press: Oxford, UK, 2006; pp. 193–220.
15. Feldman, J. The Neural Binding Problem(s). *Cogn. Neurodyn.* **2013**, *7*, 1–11. [CrossRef] [PubMed]
16. Kato, S.; Kaplan, H.S.; Schrödel, T.; Skora, S.; Lindsay, T.H.; Yemini, E.; Lockery, S.; Zimmer, M. Global brain dynamics embed the motor command sequence of Caenorhabditis elegans. *Cell* **2015**, *163*, 656–669. [CrossRef] [PubMed]
17. Yan, G.; Vértes, P.E.; Towlson, E.K.; Chew, Y.L.; Walker, D.S.; Schafer, W.R.; Barabási, A.L. Network control principles predict neuron function in the Caenorhabditis elegans connectome. *Nature* **2017**, *550*, 519–523. [CrossRef] [PubMed]
18. Jiang, J.; Su, Y.; Zhang, R.; Li, H.; Tao, L.; Liu, Q. *C. elegans* enteric motor neurons fire synchronized action potentials underlying the defecation motor program. *Nat. Commun.* **2022**, *13*, 2783. [CrossRef] [PubMed]
19. Grassé, P.P. La reconstruction dun id et les coordinations internidividuelles chez Bellicositermes natalensis et *Cubitermes* sp., La théorie de la stigmergie: Essais d'interprétation du comportement des termites constructeurs. *Insectes Sociaux* **1959**, *6*, 41–84. [CrossRef]
20. Oettler, J.; Schmid, V.S.; Zankl, N.; Rey, O.; Dress, A.; Heinze, J. Fermat's principle of least time predicts refraction of ant trails at substrate borders. *PLoS ONE* **2013**, *8*, e59739. [CrossRef] [PubMed]
21. Heyman, Y.; Shental, N.; Brandis, A.; Hefetz, A.; Feinerman, O. Ants regulate colony spatial organization using multiple chemical road-signs. *Nat. Commun.* **2017**, *8*, 15414. [CrossRef] [PubMed]
22. Menzel, R. A short history of studies on intelligence and brain in honeybees. *Apidologie* **2021**, *52*, 23–34. [CrossRef]
23. Pershin, Y.V.; La Fontaine, S.; Di Ventra1, M. Memristive model of amoeba learning. In *Physical Review E*; Sprouse, G.D., Ed.; American Physical Society: College Park, MD, USA, 2009; Volume 80, p. 021926.
24. Greer, K. Turing: Then, Now and Still Key. Artificial Intelligence, Evolutionary Computation and Metaheuristics (AIECM)–Turing 2012. In *Studies in Computational Intelligence*; Yang, X.-S., Ed.; Springer: Berlin/Heidelberg, Germany, 2013; Volume 427, pp. 43–62. [CrossRef]
25. Mashour, G.A. The Cognitive Binding Problem: From Kant to Quantum Neurodynamics. *NeuroQuantology* **2004**, *2*, 29–38. [CrossRef]
26. Hinton, G.E.; McClelland, J.L.; Rumelhart, D.E. Distributed representations. In *Parallel distributed processing*; Rumelhart, D.E., McClelland, J.L., PDP Research Group, Eds.; MIT Press: Cambridge, MA, USA, 1986; Volume 1, pp. 77–109.
27. Greer, K. Is Intelligence Artificial? In *Euroasia Summit, Congress on Scientific Researches and Recent Trends-8*; The Philippine Merchant Marine Academy: Philippines, PA, USA, 2021; pp. 307–324.
28. Niles, I.; Pease, A. Towards a Standard Upper Ontology. In Proceedings of the 2nd International Conference on Formal Ontology in Information Systems (FOIS-2001), Ogunquit, ME, USA, 17–19 October 2001.
29. Greer, K. Concept Trees: Building Dynamic Concepts from Semi-Structured Data using Nature-Inspired Methods, Complex system modelling and control through intelligent soft computations. In *Studies in Fuzziness and Soft Computing*; Zhu, Q., Azar, A.T., Eds.; Springer: Berlin/Heidelberg, Germany, 2014; Volume 319, pp. 221–252.

30. Greer, K. Recognising Image Shapes from Image Parts, Not Neural Parts. 2022. Available online: https://www.preprints.org/manuscript/202201.0259/v1 (accessed on 18 October 2022).
31. Friederici, A.D. The Neural Basis of Language Development and Its Impairment. In *Neuron*; Elsevier: Amsterdam, The Netherlands, 2006; Volume 52, pp. 941–952. [CrossRef]
32. Caucheteux, C.; King, J.R. Brains and algorithms partially converge in natural language processing. *Commun. Biol.* **2022**, *5*, 134. [CrossRef] [PubMed]

Article

A Deep Learning Model for Preoperative Differentiation of Glioblastoma, Brain Metastasis, and Primary Central Nervous System Lymphoma: An External Validation Study

Leonardo Tariciotti [1,2,*], Davide Ferlito [3,4], Valerio M. Caccavella [2], Andrea Di Cristofori [3], Giorgio Fiore [1,2], Luigi G. Remore [1,2], Martina Giordano [2], Giulia Remoli [4], Giulio Bertani [1], Stefano Borsa [1], Mauro Pluderi [1], Paolo Remida [5], Gianpaolo Basso [4,5], Carlo Giussani [3,4], Marco Locatelli [1,6,†] and Giorgio Carrabba [3,4,†]

1   Fondazione IRCCS Cà Granda Ospedale Maggiore Policlinico, Unit of Neurosurgery, 20122 Milan, Italy
2   Department of Oncology and Hemato-Oncology, University of Milan, 20122 Milan, Italy
3   Unit of Neurosurgery, Ospedale San Gerardo, Azienda Socio-Sanitaria Territoriale di Monza, 20900 Monza, Italy
4   School of Medicine and Surgery, University of Milano-Bicocca, 20900 Monza, Italy
5   Unit of Neuroradiology, Ospedale San Gerardo, Azienda Socio-Sanitaria Territoriale di Monza, 20900 Monza, Italy
6   Department of Pathophysiology and Transplantation, University of Milan, 20122 Milan, Italy
*   Correspondence: leonardotariciottimd@gmail.com
†   These authors contributed equally to this work.

Citation: Tariciotti, L.; Ferlito, D.; Caccavella, V.M.; Di Cristofori, A.; Fiore, G.; Remore, L.G.; Giordano, M.; Remoli, G.; Bertani, G.; Borsa, S.; et al. A Deep Learning Model for Preoperative Differentiation of Glioblastoma, Brain Metastasis, and Primary Central Nervous System Lymphoma: An External Validation Study. *NeuroSci* 2023, 4, 18–30. https://doi.org/10.3390/neurosci4010003

Academic Editor: Szczepan Paszkiel

Received: 22 November 2022
Revised: 26 December 2022
Accepted: 28 December 2022
Published: 31 December 2022

Abstract: (1) Background: Neuroimaging differentiation of glioblastoma, primary central nervous system lymphoma (PCNSL) and solitary brain metastasis (BM) represents a diagnostic and therapeutic challenge in neurosurgical practice, expanding the burden of care and exposing patients to additional risks related to further invasive procedures and treatment delays. In addition, atypical cases and overlapping features have not been entirely addressed by modern diagnostic research. The aim of this study was to validate a previously designed and internally validated ResNet101 deep learning model to differentiate glioblastomas, PCNSLs and BMs. **(2) Methods:** We enrolled 126 patients (glioblastoma: $n = 64$; PCNSL: $n = 27$; BM: $n = 35$) with preoperative T1Gd-MRI scans and histopathological confirmation. Each lesion was segmented, and all regions of interest were exported in a DICOM dataset. A pre-trained ResNet101 deep neural network model implemented in a previous work on 121 patients was externally validated on the current cohort to differentiate glioblastomas, PCNSLs and BMs on T1Gd-MRI scans. **(3) Results:** The model achieved optimal classification performance in distinguishing PCNSLs (AUC: 0.73; 95%CI: 0.62–0.85), glioblastomas (AUC: 0.78; 95%CI: 0.71–0.87) and moderate to low ability in differentiating BMs (AUC: 0.63; 95%CI: 0.52–0.76). The performance of expert neuro-radiologists on conventional plus advanced MR imaging, assessed by retrospectively reviewing the diagnostic reports of the selected cohort of patients, was found superior in accuracy for BMs (89.69%) and not inferior for PCNSL (82.90%) and glioblastomas (84.09%). **(4) Conclusions:** We investigated whether the previously published deep learning model was generalizable to an external population recruited at a different institution—this validation confirmed the consistency of the model and laid the groundwork for future clinical applications in brain tumour classification. This artificial intelligence-based model might represent a valuable educational resource and, if largely replicated on prospective data, help physicians differentiate glioblastomas, PCNSL and solitary BMs, especially in settings with limited resources.

Keywords: brain metastases; deep learning; glioblastoma; machine learning; primary central nervous system lymphoma

1. Introduction

Preoperative classification of brain tumours represents a critical aspect of patient management. Brain metastases (BMs), glioblastoma and primary central nervous system lymphomas (PCNSLs) are among the most frequent intracranial neoplasms in adults (17%, 14.3% and 1.9%, respectively); hence, a correct diagnosis is a crucial point in the therapeutic path of a large number of patients worldwide [1–3].

In spite of the increased efficiency and popularity of MRI and the availability of advanced neuroimaging techniques that may assist in differentiating glioblastomas, BMs and PCNSLs, cases showing atypical features may prove challenging even for expert clinicians who spend a large proportion of their work time identifying, segmenting and classifying these lesions [4,5].

As far as the T1-weighted gadolinium-enhanced (T1Gd) images considered in this study are concerned, glioblastomas appear as iso-hypointense masses with necrotic-cystic areas and irregular contrast-enhanced margins similar to solitary BMs; however, atypical glioblastomas may show minimal or absent central necrosis.

PCNSLs, on the contrary, are usually shown on T1Gd images as iso-hypointense masses with a homogeneous enhancement within the entire lesion boundaries; in atypical presentations, there is central necrosis that may mimic glioblastomas [6], and the preoperative use of steroids in patients with PCNSLs may entail false negative pathological results, requiring additional invasive manoeuvres and potential harm and costs [7] to obtain the correct diagnosis.

In recent years, artificial intelligence (AI)—more specifically, deep learning (DNN)—has been accounted as an emerging and promising technique in supporting physicians in decision-making tasks based on MRI images (i.e., computer vision) [8–12].

The aim of this study was to develop a fast and reliable system for brain tumour classification in an experimental retrospective clinical scenario. In a previous investigation [13], we designed and internally validated a DNN model, achieving excellent diagnostic performance. The purpose of this study was the external validation of the model's accuracy in differentiating GBMs, PCNSLs and BMs on T1Gd MRI scans and discussion of its eventual role in the amelioration of diagnostic and interventional workflows.

2. Methods

2.1. Study Definition

Ethical approval was waived by the two institutions involved, by the local Ethics Committees in view of the retrospective nature of the study and because all performed procedures were part of routine care. Informed consent was obtained from all participants included in the study. All procedures performed in studies involving human participants were in accordance with the Helsinki declaration.

An internal committee among authors (L.T., G.F., G.A.B., G.C., M.L.) was formed, and a consensus achieved on the current investigation's proper design and reporting guidelines. An extensive review of "Enhancing the quality and transparency of health research" (EQUATOR) [14] network "https://www.equator-network.org" (accessed on 4 January 2022) contents was performed, and the "Standard for reporting of diagnostic accuracy study—Artificial Intelligence" (STARD-AI) [15] guidelines were selected and followed in the study protocol definition. The STARD-AI [15] guidelines were developed to report AI diagnostic test accuracy studies as an evolution of the previous STARD 2015 version [16], with the addition of a specific focus on designing and reporting evidence provided through AI-centred interventions. Adherence to STARD-AI recommendations was reviewed by the senior authors (G.C. and M.L.) throughout the investigation and during final review.

2.2. Patient Selection

The medical records and preoperative imaging of patients who underwent surgical tumour resection or biopsy at "Fondazione IRCCS Cà Granda Ospedale Maggiore Policlinico, Milan, Italy" (named Training Site or TrS) between June 2020 and April 2021 and at "Ospedale San Gerardo di Monza, Monza, Italy" (named Testing Site or TeS) between January 2018 and November 2021 were retrospectively collected. Patient data were included in the analysis if preoperative T1Gd MR images were available and histological analysis confirmed the diagnosis of glioblastoma, PCNSL or solitary BMs.

Patients were excluded if:

(1) Preoperative T1Gd MR images were absent or inadequate in quality, according to the senior neuroradiologists;

(2) They had previously received intracranial intervention (surgical intervention, gamma knife surgery or radiation therapy);

(3) Multiple enhancing lesions were detected on preoperative MRI;

(4) In glioblastoma cases, histopathological exams included testing for IDH mutations—hence, only IDH1 and IDH2 wild-type tumours were further considered in the investigation.

One-hundred twenty-one patients operated on at the TrS were selected to provide image data for the training dataset of our DNN model, as reported in a previous study [13].

A total of 126 patients met the inclusion criteria at the TeS and were selected for external validation of the aforementioned model.

2.3. MR Acquisition and Image Pre-Processing

The MR image scanning parameters at the TrS are reported elsewhere [13]. Concerning the MRI acquisition protocol at the TeS, all brain MRI studies were performed with a 1.5 T system (Philips® Ingenia 1.5T CX), including axial T2-weighted imaging, fluid-attenuated inversion recovery (FLAIR) imaging, diffusion-weighted images (DWI) (a b-value of 1000 sec/mm^2 and a single b-0 acquisition), susceptibility-weighted imaging (SWI), volumetric contrast-enhanced axial and sagittal T1Gd (Gadovist 1 mmol/mL; 0.1 mmol/kg body weight) imaging; ADC maps were calculated from isotropic DWI.

All MR images in the digital imaging and communications in medicine (DICOM) format were input to the Horos DICOM Viewer version 3.3.5, "www.horosproject.org" (accessed on 4 January 2022), a free, open-source medical imaging viewer and analytic tool. The lesions' regions of interest (ROIs) were manually delineated on volumetric axial T1Gd scans. After segmentation and signal intensity normalization, all ROIs were then centred in a 224 × 224 pixels black box and exported in PNG file format (Figure 1).

2.4. Convolutional Neural Network Model

A 2D convolutional neural network model (i.e., ResNet-101) with 101 layers consisting of three-layer residual blocks pre-trained with the TrS dataset was used [13,17–20].

Each ROI was used as input for all three channels expected by the ResNet model and was treated as an independent image to increase the input data, though a group of slices was available for each patient. The predicted diagnostic class for each patient was the most frequently voted among its entire ROI set. The reported performance metrics were computed considering the number of correctly predicted patients and not the whole ROI dataset.

2.5. Performance Metrics

The classification performance of the DNN model was evaluated considering the following metrics:

(1) Area under the receiving operative characteristics curve (AUC-ROC):

$$\mathrm{AUC}\,(f) = \frac{\sum_{t0 \in \mathcal{D}} {}^{0} \sum_{t1 \in \mathcal{D}} 1\,[f(t_0) < f(t_1)\,]}{\left| D^0 \right| \cdot \left| D^1 \right|} \tag{1}$$

where $1\,[f(t_0) < f(t_1)\,]$ denotes an indicator function, which returns 1 if $f(t_0) < f(t_1)$; otherwise, returns 0. D^0 is the set of negative examples and D^1 is the set of positive examples.

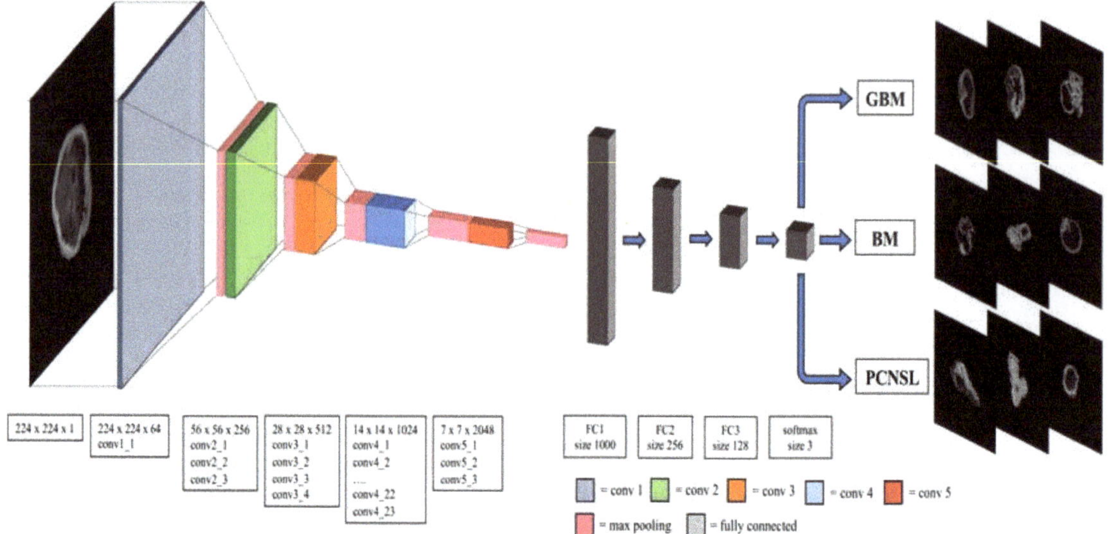

Figure 1. Model architecture trained as reported in Tariciotti et al. [13] and externally validated on the TeS dataset in the current study. The window size and stride for convolutional, maxpooling and fully connected layers are also presented. Conv: convolutional layer; FC: fully connected layer; GBM: glioblastoma; PCNSL: primary central nervous system lymphoma; BM: brain metastasis. "Reprinted with permission from Tariciotti et al. [13]. **Copyright** © 2022 Tariciotti, Caccavella, Fiore, Schisano, Carrabba, Borsa, Giordano, Palmisciano, Remoli, Remore, Pluderi, Caroli, Conte, Triulzi, Locatelli and Bertani. This is an open-access article distributed under the terms of the Creative Commons Attribution License (CC BY).

(2) Accuracy:

$$\frac{TP + TN}{TP + TN + FP + FN} \tag{2}$$

where TP = true positive; TN = true negative; FP = false positive; FN = false negative.

(3) Precision or positive predictive value (PPV):

$$\frac{TP}{TP + FP} \tag{3}$$

(4) Negative predictive value (NPV):

$$\frac{TN}{TN + FN} \tag{4}$$

(5) Recall or sensitivity:

$$\frac{TP}{TP + FN} \tag{5}$$

(6) Specificity:

$$\frac{TN}{TN + FP} \tag{6}$$

(7) F-1 score:

$$2 \times \frac{\text{Precision} \times \text{Recall}}{\text{Precision} + \text{Recall}} \tag{7}$$

A complete explanation of the parameters mentioned above is beyond the scope of the current study; further comprehensive descriptions are available elsewhere [21].

A one-vs-rest (OVR) multiclass strategy was employed to extract performance metrics for each outcome class. Then, the average value and its 95% bootstrap confidence interval were computed for each performance metric on the hold-out test set.

2.6. Human "Gold Standard" Performance

The tumour radiological assessment was addressed by experienced neuroradiologists (P.R. and G.B.) with at least 10 years of clinical experience. Electronic radiological reports were retrospectively reviewed to collect the primary radiological diagnosis. Afterwards, a comparison with the histopathological charts was completed, and the diagnostic classes were checked for discrepancies between radiological and pathological characterization. An OVR multiclass method was employed to extract neuroradiologists' performance metrics for each outcome class.

2.7. Software and Hardware

All the statistical analyses were performed in a Jupyter Notebook using Python v.3.7.6 "https://www.python.org/" (accessed on 4 January 2022). The Python packages used for this study included: 'PyTorch v1.7' to develop and train the DNN model, 'Numpy' for Excel dataset handling; 'Scikit-learn' to compute performance metrics and 'Seaborn' to plot ROC-AUC. The workstation used to train the DNN model mounted an Intel Core i7–10700K processor, while the GPU was a Tesla K80 12GB.

3. Results

The cohort of selected patients included: 64 glioblastomas (mean age, 64.4 ± 9.04), 27 PCNSLs (mean age, 58.1 ± 16.5) and 33 BMs (mean age, 62.7 ± 14.2). A total of 2853 axial slices/ROIs of tumours were extracted, of which 1748 glioblastoma ROIs (mean ROIs 28.0 ± 19.0), 412 PCNSL ROIs (mean ROIs 15.0 ± 4.0) and 693 BMs ROIs (mean ROIs 21.0 ± 14.0). No significant differences in age, gender, number of total sequences or tumour ROI slice distributions were found between the three tumour groups ($p > 0.05$). The BM group included patients with various primary tumours, the most common of which being lung cancer ($n = 16$, 48.4% of all BMs), breast cancer ($n = 5$, 15.1%), gastrointestinal cancer ($n = 4$, 12.1%) and renal cancer ($n = 3$, 9.1%). Additional primary diagnoses were endometrial cancers and melanoma. Demographic characteristics are summarised in Table 1.

Table 1. Demographics and imaging acquisition data.

		Glioblastoma		BM		PCNSL		*p*-Value
		Count (N%)	Mean (SD)	Count (N%)	Mean (SD)	Count (N%)	Mean (SD)	
Gender	Female	26 (41.3%)		12 (36.4%)		8.0 (29.6%)		$p > 0.05$
	Male	37 (58.7%)		21 (63.6%)		19.0 (70.4%)		$p > 0.05$
Age (years)			64.4 (9.04)		62.7 (14.2)		58.5 (16.5)	$p > 0.05$
N° Slices of T1Gd sequence (N)			108.0 (52.0)		107.0 (59.0)		74.0 (61.0)	$p > 0.05$
N° Slices of ROI (N)			28.0 (19.0)		21.0 (4.0)		15.0 (14.0)	$p > 0.05$

Demographic characteristics of patients recruited at TeS. BM: brain metastasis; PCNSL: primary central nervous system lymphoma; ROI: region of interest.

3.1. DNN Model Performance Metrics Evaluation

The validated DNN model (Figure 1) achieved AUCs of 0.73 (95% CI: 0.62–0.85), 0.78 (95% CI: 0.71–0.87) and 0.63 (95% CI: 0.52–0.76), respectively, for the PCNSL (Figure 2), glioblastoma (Figure 3) and BM (Figure 4) diagnostic classes. High reliability was reported across all performance metrics for PCNSLs and glioblastomas diagnostic outcome classes, while lower reliability was reported for BMs. The complete performance metric evaluation and the related confusion matrix are reported in Table 2 and Figure 5.

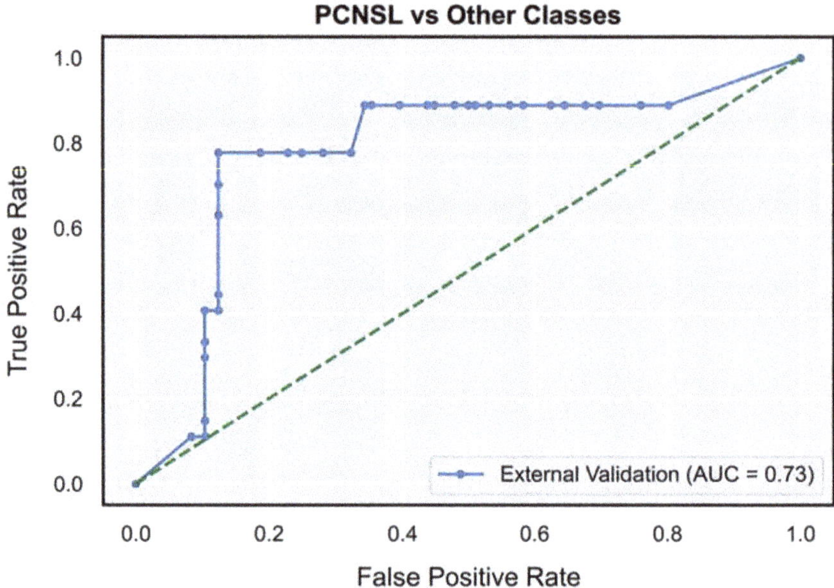

Figure 2. AUC-ROC curves (on TeS validation dataset) for PCNSL diagnostic outcome class (OVR). OVR: one-vs-rest; PCNSL: primary central nervous system lymphoma.

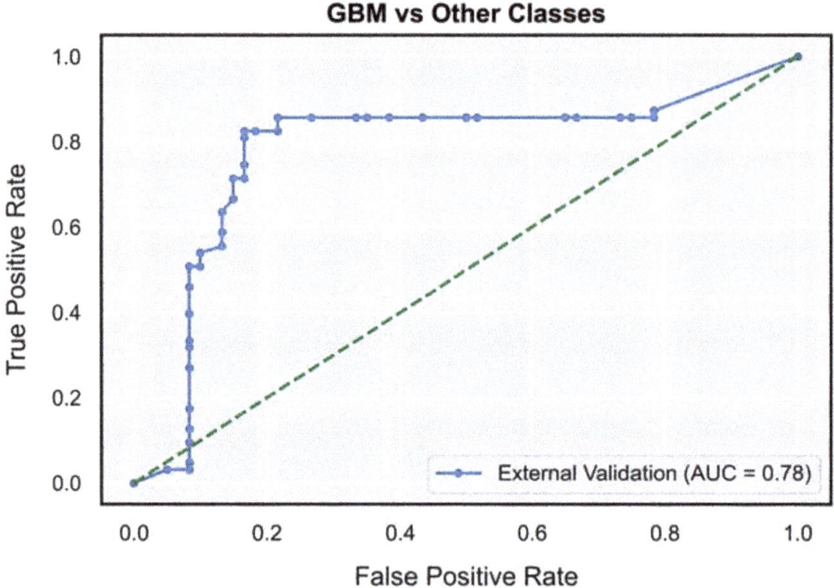

Figure 3. AUC-ROC curves (on TeS validation dataset) for glioblastoma diagnostic outcome class (OVR). GBM: glioblastoma; OVR: one-vs-rest.

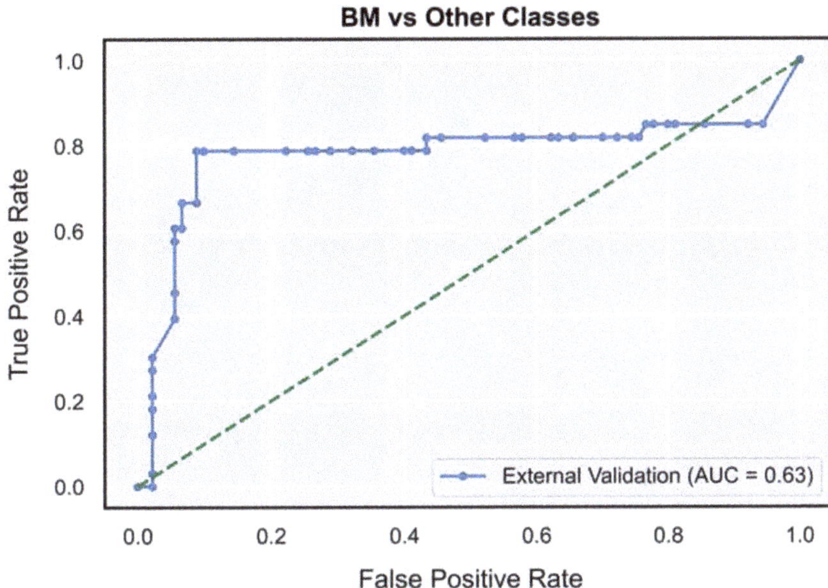

Figure 4. AUC-ROC curves (on TeS validation dataset) for solitary brain metastasis diagnostic outcome class (OVR). BM: brain metastasis; OVR: one-vs-rest.

Table 2. Performance metrics achieved by the convolutional neural network model in differentiating PCNSLs, glioblastomas and BMs.

Performance Metrics	PCNSL	Glioblastoma	BM
AUC	0.73 (0.62–0.85)	0.78 (0.71–0.87)	0.63 (0.52–0.76)
Accuracy	80.46% (74.8–87.01%)	80.37% (74.8–86.99%)	77.12% (71.54–83.74%)
Precision (PPV)	54.85% (44.11–70.00%)	84.13% (77.97–92.0%)	57.71% (46.67–72.73%)
Recall (Sensitivity)	66.86% (51.85–85.19%)	76.14% (66.67–85.71%)	57.04% (42.42–72.73%)
Specificity	84.29% (78.12–91.67%)	84.8% (78.33–93.33%)	84.49% (77.78–91.14%)
F1-Score	0.60 (0.50–0.73)	0.80 (0.73–0.87)	0.57 (0.45–0.70)

Performance metrics achieved on the hold-out test set were computed adopting an OVR multiclass strategy. Average value and 95% bootstrap confidence interval are reported. AUC: area under the curve; BM: brain metastasis; OVR: one-vs-rest; PCNSL: primary central nervous system lymphoma; PPV: positive predictive value.

3.2. Comparison of DNN Model and Neuroradiologists' Gold Standard Performance

The performance metrics achieved by expert neuroradiologists are provided in Table 3. The DNN model showed a classification performance not inferior to the neuroradiologists' gold standard reference on glioblastomas (F1 score 0.80 (0.73–0.87) vs. 0.81), PCNSL (F1 score 0.60 (0.50–0.73) vs. 0.59) and performed poorer than physicians in diagnosing BMs (0.57 (0.45–0.70) vs. 0.82).

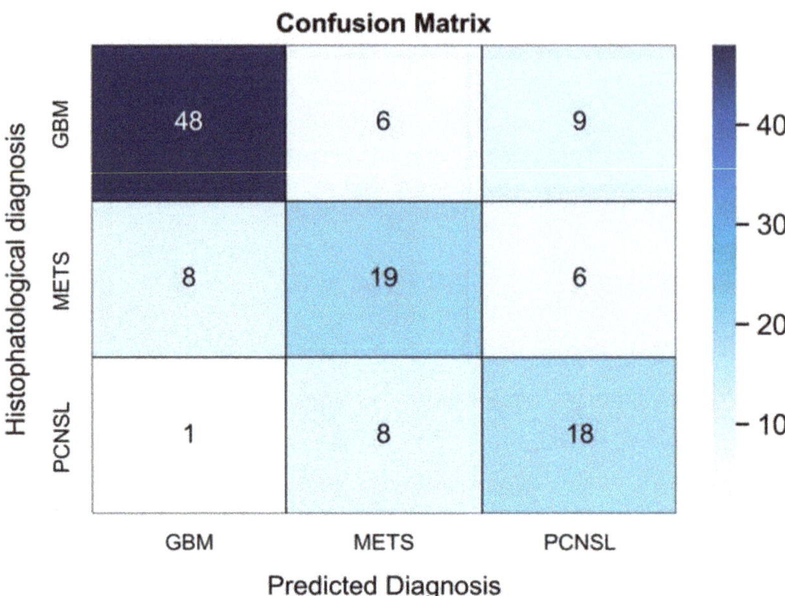

Figure 5. The confusion matrix (CM) shows the exact collocation of each patient among classification classes during a validated simulation with our DNN model. On the TeS patients' data, the model misclassified histologically-confirmed atypical PCNSL nine times: in eight out of nine cases, the error led to a computer-based diagnosis of BM. On the contrary, among histologically-diagnosed BM, the model correctly identified 19 cases, while the remaining 14 cases were declared as glioblastomas (n = 8) and PCNSLs (n = 6). Overall, glioblastomas were more likely to be correctly diagnosed by the DNN model. The CM shows how the model chose among available diagnostic classes in the current work. BM: brain metastasis; CM: Confusion matrix; DNN: deep neural network; GBM: glioblastoma; PCNSL: primary central nervous system lymphoma.

Table 3. Neuroradiologist (Gold standard) performance metrics in differentiating PCNSL, glioblastoma and BM in the cohort examined.

Performance Metrics	PCNSL	Glioblastoma	BM
Accuracy	82.90%	84,09%	89.69%
Precision (PPV)	65.21%	87.50%	79.31%
Negative predictive value (NPV)	87.23%	81.57%	94.11%
Recall (Sensitivity)	55.55%	77.77%	85.18%
Specificity	91.11%	89.85%	91.42%
F1-Score	0,595	0,819	0,818

Performance metrics achieved by neuro-radiologists (defined as the gold standard) adopting an OVR multiclass strategy. The metrics were retrospectively computed by examining patient report charts: all patients underwent conventional plus advanced (T1-weighted, T2-weighted, FLAIR, diffusion-weighted, conventional T1-contrast-enhanced, dynamic contrast-enhanced and perfusion) MRI scans. Values were reported as single computation, so 95% bootstrap confidence intervals were not defined. BM: brain metastasis; OVR: one-vs-rest; PCNSL: primary central nervous system lymphoma; PPV: positive predictive value; NPV: negative predictive value.

4. Discussion

4.1. Performance Validation

In a previous study, we reported on a DNN model capable of efficiently and accurately differentiating glioblastomas, PCNSLs and BMs in an experimental "offline" environment [13]. Here, we externally validated the DNN model on "never seen" data gathered at an external academic site (TeS) with the comparable caseload, facility settings and technologies. The accuracy returned by our model was not inferior to a senior neuroradiologist's performance in identifying PCNSLs and glioblastomas; accuracy for BMs identification was moderate, despite being lower than human evaluation.

In light of our previous preliminary findings, the evidence of model robustness and generalizability achieved in the current study supports the thesis of our DNN model being "experimentally not inferior" to senior physicians in classifying brain tumours in an unbiased cohort, endorsing the development and deployment of such models in medical training and clinical practice if cleared by regulatory authorities.

As previously documented, differentiating dubious BMs from gliomas and PCNSLs is challenging per se. Despite exponential advancements in the last decade, no single MRI modality can differentiate PCNSLs, BMs and glioblastomas with absolute accuracy. The search for a single sequence candidate to better classify these tumours has been limited to academic speculation, being restricted to synthetic scenarios rather than simulating clinical practice decision workflow, where multimodality is preferred. Indeed, results from previous studies are contradictory [22,23], with several authors reporting either T2-weighted, FLAIR or T1Gd scans' superiority in brain tumour segmentation and classification [24–26]. The multimodality MRI approach recently showed promising diagnostic performance in differentiating brain neoplasms in experimental settings. Relevant findings were confirmed about dynamic susceptibility contrast (DSC) and apparent diffusion coefficient (ADC) maps combined with T1Gd-MRI scans. This multimodal approach came at the cost of an unstandardized diagnostic role due to the operator-dependent interpretation bias, high heterogeneity among brain tumour phenotypes and the additional need for hardware and set-up protocols, which might curb its use in facilities with limited resources [27–29].

During the study design, the authors agreed to implement T1Gd-MRI images only, relying on the greater worldwide availability of this sequence compared to diffusion and perfusion protocols, with the aim of extending the reproducibility of our workflow. Plus, the superior distinction of tumour borders and precise representation of central necrosis, which are common features of glioblastomas, atypical PCNSLs and BMs [30], facilitates manual segmentation avoiding ROIs' drawing biases. However, the inclusion of additional sequences might have allowed a superior performance in the classification task.

Performance on BMs scored significantly lower compared to both the internal validation dataset and neuroradiologists' performance metrics (accuracy: 77% vs. 81% vs. 89%, respectively [13]). This underperformance may be imputable to the great histological heterogeneity of this group of lesions and the consequent variability in radiological features. Additionally, a key distinguishing feature of BMs is abundant peritumoral oedema [31]; however, the peritumoural radiological environment was not included in the ROI segmentation of our dataset, which was limited to T1Gd boundaries. This might have influenced the lower performance of DNN on BMs, together with the neuroradiologists' access to clinical history and additional imaging work-ups that the DNN model was blinded to. Indeed, while the model was blinded to any additional historical or diagnostic information except T1Gd scans, the diagnostic process accomplished at the time of imaging work-up comprehended additional characterization by means of total body CT, positron emission tomography (PET), and advanced MRI scans in a proportion of cases; being the retrospective evaluation of radiological reports set in routine clinical practice, we could not assess whether the aforementioned diagnostic exams—not involved in the current investigation—had a valuable impact on the putative radiological diagnosis. The comparative performance of DNN and senior neuroradiologists should be evaluated accordingly, and conclusions should be drawn carefully.

4.2. Perspective for Clinical Application and Public Health Impact

From a public health perspective, diagnostic tools such as our validated DNN model represent a promising technology spreading worldwide within industry, academia, and personal life settings. It is estimated that implementing AI algorithms in the USA might save USD 150 billion in healthcare costs by 2026 [32], with a net benefit even in lower-income countries, where AI experimentation is still under-practised. Implementation of AI protocols in healthcare is increasing in resource-poor countries of Asia and Africa collaterally to the wider availability of mobile phones, mobile health applications and cloud computing, which generate a sufficient mass of data to redirect to the purpose of studies like our own.

Given this, we believe that AI models might assist physicians in low-income countries in tackling macro and micro-scale healthcare disparities and might reduce healthcare borders and inequalities across high- and low-income countries by optimizing diagnostic workflows, augmenting physician performance in those settings where highly trained personnel are not routinely available or favouring teleconsultations and patient referral to more experienced hospitals. The whole process, as auspicated in high-income countries, might provide benefits to healthcare quality and allow weighted cost reduction [33], as suggested by a recent survey conducted in Pakistan [34]. However, our belief about the contributions of AI to healthcare optimization in such settings is speculative, and sufficient literature about AI use in resource-poor countries is still lacking to draw accurate previsions.

4.3. Perspective in Medical Education

Other than the previously discussed applications, efficiency of computer vision has already been demonstrated in other clinical scenarios (i.e., skin cancer classification, diagnosis of retinal disease, detection of mammographic lesions, fracture detection and many other tasks) [35–38].

Recent advancements have been made in integrating CV, and ML in general, into medical education and skill evaluation. Oliveira et al. reported a deep learning model called PRIME that is able to evaluate the microsurgical ability of different neurosurgeons in vessels dissection and micro-suture; the latter was designed with the aim of smoothing the microsurgical steep learning curve and providing a self-paced ML-advised tutor for continuous training without the need for any motion sensors around the operating table [39]. Similarly, Smith et al. reported a motion-tracking ML algorithm for surgical instrument monitoring during cataract surgery [40].

Finally, aimed to standardize surgical procedures, enhance training and lay the groundwork for future robot-assisted surgery, several groups are investigating whether DNN models can dissect surgical workflows into reproducible phases according to environmental exposure, segmentation of the anatomical scenario and instrument usage [41–43].

4.4. Strengths and Limitations

The DNN model hereby presented and validated on a cohort of more than one hundred patients is a simple but efficient tool able to help physicians diagnose atypical intracranial tumours with limited addition of human effort. Despite not being used in real-time scenarios yet, it is a promising and robust classification model and a candidate for further investigations in clinical trials. Nevertheless, several limitations restrict the generalizability of our results; the outcome accuracy was gauged in "offline" settings on a retrospective pool of image data. To date, the usefulness in actual clinical practice has been inferred but not demonstrated. In fact, while neuroradiologists with access to other relevant information scored as high as the DNN model in the majority of classes (and even higher on BMs), the interaction between the DNN response and the human decision-making process has not been experienced and evaluated. Further prospective trials are required to clarify the impact of artificial intelligence-based decision-making tools on human judgement and performance in clinical practice.

5. Conclusions

These results confirm the feasibility and reliability of our DNN model in experimental scenarios and open new possibilities for prospective clinical investigations. The delivery of such a diagnostic tool might enhance physicians' performance and reduce the healthcare access gap in settings with limited human and instrumental resources. The validated model was built on an open-source programming language, and our methodology could be exported and further validated at different institutions.

Author Contributions: Conceptualization, L.T., V.M.C. and G.C.; methodology, L.T., V.M.C. and D.F.; software, L.T. and V.M.C.; validation, G.C., M.L., S.B., M.P., G.B. (Giulio Bertani), P.R., G.B. (Gianpaolo Basso), C.G. and A.D.C.; formal analysis, L.T. and V.M.C.; investigation, L.T., D.F. and V.M.C.; resources, G.C. and M.L.; data curation, L.T., D.F. and V.M.C.; writing—original draft preparation, L.T. and D.F.; writing—review and editing, L.T., D.F., M.L., G.C., M.G., G.R., L.G.R. and G.F.; visualization, L.T. and V.M.C.; supervision, G.C. and M.L.; project Administration, L.T. All authors have read and agreed to the published version of the manuscript.

Funding: No funds, grants, or other support were received.

Informed Consent Statement: Informed consent was obtained from all individual participants included in the study. Written informed consent has been obtained from the patient(s) to publish this paper.

Data Availability Statement: All authors confirm the appropriateness of all datasets and software used to support the conclusion. The dataset that supports the findings of this study is available from the corresponding author, L.T., upon request. The source code employed to develop the herein presented deep learning model is available from the corresponding author, L.T., upon request.

Conflicts of Interest: The authors have no relevant financial or non-financial interests to disclose.

References

1. Ostrom, Q.T.; Gittleman, H.; Truitt, G.; Boscia, A.; Kruchko, C.; Barnholtz-Sloan, J.S. CBTRUS statistical report: Primary brain and other central nervous system tumors diagnosed in the United States in 2011–2015. *Neuro-Oncol.* **2018**, *20*, iv1–iv86. [CrossRef] [PubMed]
2. Ostrom, Q.T.; Patil, N.; Cioffi, G.; Waite, K.; Kruchko, C.; Barnholtz-Sloan, J.S. CBTRUS statistical report: Primary brain and other central nervous system tumors diagnosed in the United States in 2013–2017. *Neuro-Oncol.* **2020**, *22*, iv1–iv96. [CrossRef]
3. Nayak, L.; Lee, E.Q.; Wen, P.Y. Epidemiology of brain metastases. *Curr. Oncol. Rep.* **2012**, *14*, 48–54. [CrossRef] [PubMed]
4. Biratu, E.S.; Schwenker, F.; Ayano, Y.M.; Debelee, T.G. A Survey of Brain Tumor Segmentation and Classification Algorithms. *J. Imaging* **2021**, *7*, 179. [CrossRef] [PubMed]
5. Abd-Ellah, M.K.; Awad, A.I.; Khalaf, A.A.M.; Hamed, H.F.A. A review on brain tumor diagnosis from MRI images: Practical implications, key achievements, and lessons learned. *Magn. Reson. Imaging* **2019**, *61*, 300–318. [CrossRef]
6. Baris, M.M.; Celik, A.O.; Gezer, N.S.; Ada, E. Role of mass effect, tumor volume and peritumoral edema volume in the differential diagnosis of primary brain tumor and metastasis. *Clin. Neurol. Neurosurg.* **2016**, *148*, 67–71. [CrossRef]
7. Batchelor, T.; Loeffler, J.S. Primary CNS lymphoma. *J. Clin. Oncol.* **2006**, *24*, 1281–1288. [CrossRef]
8. Augustin Toma, M.; Gerhard-Paul Diller, M.P.; Patrick, R.; Lawler, M.M. Deep Learning in Medicine. *JACC Adv.* **2022**, *1*, 100017. [CrossRef]
9. Kim, M.; Yun, J.; Cho, Y.; Shin, K.; Jang, R.; Bae, H.J.; Kim, N. Deep Learning in Medical Imaging. *Neurospine* **2019**, *16*, 657–668. [CrossRef]
10. Lee, W.-J.; Hong, S.D.; Woo, K.I.; Seol, H.J.; Choi, J.W.; Lee, J.-I.; Nam, D.-H.; Kong, D.-S. Combined endoscopic endonasal and transorbital multiportal approach for complex skull base lesions involving multiple compartments. *Acta Neurochir.* **2022**, *164*, 1911–1922. [CrossRef]
11. Zaharchuk, G.; Gong, E.; Wintermark, M.; Rubin, D.; Langlotz, C.P. Deep Learning in Neuroradiology. *AJNR. Am. J. Neuroradiol.* **2018**, *39*, 1776–1784. [CrossRef] [PubMed]
12. Tariciotti, L.; Palmisciano, P.; Giordano, M.; Remoli, G.; Lacorte, E.; Bertani, G.; Locatelli, M.; Dimeco, F.; Caccavella, V.M.; Prada, F. Artificial intelligence-enhanced intraoperative neurosurgical workflow: State of the art and future perspectives. *J. Neurosurg. Sci.* **2021**, *66*, 139–150. [CrossRef] [PubMed]
13. Tariciotti, L.; Caccavella, V.M.; Fiore, G.; Schisano, L.; Carrabba, G.; Borsa, S.; Giordano, M.; Palmisciano, P.; Remoli, G.; Remore, L.G.; et al. A Deep Learning Model for Preoperative Differentiation of Glioblastoma, Brain Metastasis and Primary Central Nervous System Lymphoma: A Pilot Study. *Front. Oncol.* **2022**, *12*, 816638. [CrossRef] [PubMed]
14. Simera, I.; Moher, D.; Hoey, J.; Schulz, K.F.; Altman, D.G. The EQUATOR Network and reporting guidelines: Helping to achieve high standards in reporting health research studies. *Maturitas* **2009**, *63*, 4–6. [CrossRef] [PubMed]
15. Sounderajah, V.; Ashrafian, H.; Aggarwal, R.; De Fauw, J.; Denniston, A.K.; Greaves, F.; Karthikesalingam, A.; King, D.; Liu, X.; Markar, S.R.; et al. Developing specific reporting guidelines for diagnostic accuracy studies assessing AI interventions: The STARD-AI Steering Group. *Nat. Med.* **2020**, *26*, 807–808. [CrossRef]

16. Cohen, J.F.; Korevaar, D.A.; Altman, D.G.; Bruns, D.E.; Gatsonis, C.A.; Hooft, L.; Irwig, L.; Levine, D.; Reitsma, J.B.; De Vet, H.C.W.; et al. STARD 2015 guidelines for reporting diagnostic accuracy studies: Explanation and elaboration. *BMJ Open* **2016**, *6*, e012799. [CrossRef]

17. He, K.; Zhang, X.; Ren, S.; Sun, J. Deep Residual Learning for Image Recognition. In Proceedings of the IEEE conference on Computer Vision and Pattern Recognition, Las Vegas, NV, USA, 27–30 June 2016; pp. 770–778.

18. Deng, J.; Dong, W.; Socher, R.; Li, L.-J.; Li, K.; Fei-Fei, L. Imagenet: A large-scale hierarchical image database. *IEEE Conf. Comput. Vis. Pattern Recognit.* **2009**, 248–255.

19. Kingma, D.P.; Ba, J. Adam: A Method for Stochastic Optimization. *arXiv* **2014**, arXiv:1412.6980.

20. Ioffe, S.; Szegedy, C. Batch Normalization: Accelerating Deep Network Training by Reducing Internal Covariate Shift. In Proceedings of the 32nd International Conference on Machine Learning, Lille, France, 6–11 July 2015.

21. Fawcett, T. An Introduction to ROC Analysis. *Pattern Recognit. Lett.* **2006**, *27*, 861–874. [CrossRef]

22. Larroza, A.; Bodí, V.; Moratal, D. Texture Analysis in Magnetic Resonance Imaging: Review and Considerations for Future Applications. In *Assessment of Cellular and Organ Function and Dysfunction Using Direct and Derived MRI Methodologies*; IntechOpen Limited: London, UK, 2016. [CrossRef]

23. Kunimatsu, A.; Kunimatsu, N.; Yasaka, K.; Akai, H.; Kamiya, K.; Watadani, T.; Mori, H.; Abe, O. Machine learning-based texture analysis of contrast-enhanced mr imaging to differentiate between glioblastoma and primary central nervous system lymphoma. *Magn. Reson. Med. Sci.* **2019**, *18*, 44–52. [CrossRef]

24. Fruehwald-Pallamar, J.; Hesselink, J.; Mafee, M.; Holzer-Fruehwald, L.; Czerny, C.; Mayerhoefer, M. Texture-Based Analysis of 100 MR Examinations of Head and Neck Tumors—Is It Possible to Discriminate Between Benign and Malignant Masses in a Multicenter Trial? In *RöFo-Fortschritte auf dem Gebiet der Röntgenstrahlen und der Bildgeb. Verfahren*; Thieme: New York, NY, USA, 2015; Volume 188, pp. 195–202. [CrossRef]

25. Tiwari, P.; Prasanna, P.; Rogers, L.; Wolansky, L.; Badve, C.; Sloan, A.; Cohen, M.; Madabhushi, A. Texture descriptors to distinguish radiation necrosis from recurrent brain tumors on multi-parametric MRI. In *Proceedings of the Medical Imaging 2014: Computer-Aided Diagnosis*; SPIE: Bellingham, WA, USA, 2014; Volume 9035, p. 90352B.

26. Xiao, D.-D.; Yan, P.-F.; Wang, Y.-X.; Osman, M.S.; Zhao, H.-Y. Glioblastoma and primary central nervous system lymphoma: Preoperative differentiation by using MRI-based 3D texture analysis. *Clin. Neurol. Neurosurg.* **2018**, *173*, 84–90. [CrossRef] [PubMed]

27. Davnall, F.; Yip, C.S.P.; Ljungqvist, G.; Selmi, M.; Ng, F.; Sanghera, B.; Ganeshan, B.; Miles, K.A.; Cook, G.J.; Goh, V. Assessment of tumor heterogeneity: An emerging imaging tool for clinical practice? *Insights Imaging* **2012**, *3*, 573–589. [CrossRef] [PubMed]

28. Cha, S.; Lupo, J.M.; Chen, M.H.; Lamborn, K.R.; McDermott, M.W.; Berger, M.S.; Nelson, S.J.; Dillon, W.P. Differentiation of glioblastoma multiforme and single brain metastasis by peak height and percentage of signal intensity recovery derived from dynamic susceptibility-weighted contrast-enhanced perfusion MR imaging. *Am. J. Neuroradiol.* **2007**, *28*, 1078–1084. [CrossRef] [PubMed]

29. Qin, J.; Li, Y.; Liang, D.; Zhang, Y.; Yao, W. Histogram analysis of absolute cerebral blood volume map can distinguish glioblastoma from solitary brain metastasis. *Medicine* **2019**, *98*, e17515. [CrossRef]

30. Raza, S.M.; Lang, F.F.; Aggarwal, B.B.; Fuller, G.N.; Wildrick, D.M.; Sawaya, R. Necrosis and Glioblastoma: A Friend or a Foe? A Review and a Hypothesis. *Neurosurgery* **2002**, *51*, 2–13. [CrossRef]

31. Thammaroj, J.; Wongwichit, N.; Boonrod, A. Evaluation of Perienhancing Area in Differentiation between Glioblastoma and Solitary Brain Metastasis. *Asian Pac. J. Cancer Prev.* **2020**, *21*, 2525. [CrossRef]

32. Cossy-Gantner, A.; Germann, S.; Schwalbe, N.R.; Wahl, B. Artificial intelligence (AI) and global health: How can AI contribute to health in resource-poor settings? *BMJ Glob. Health* **2018**, *3*, 798. [CrossRef]

33. Guo, J.; Li, B. The Application of Medical Artificial Intelligence Technology in Rural Areas of Developing Countries. *Health Equity* **2018**, *2*, 174. [CrossRef]

34. Hoodbhoy, Z.; Hasan, B.; Siddiqui, K. Does artificial intelligence have any role in healthcare in low resource settings? *J. Med. Artif. Intell.* **2019**, *2*, 854. [CrossRef]

35. Haenssle, H.A.; Fink, C.; Schneiderbauer, R.; Toberer, F.; Buhl, T.; Blum, A.; Kalloo, A.; Ben Hadj Hassen, A.; Thomas, L.; Enk, A.; et al. Man against machine: Diagnostic performance of a deep learning convolutional neural network for dermoscopic melanoma recognition in comparison to 58 dermatologists. *Ann. Oncol. Off. J. Eur. Soc. Med. Oncol.* **2018**, *29*, 1836–1842. [CrossRef]

36. De Fauw, J.; Ledsam, J.R.; Romera-Paredes, B.; Nikolov, S.; Tomasev, N.; Blackwell, S.; Askham, H.; Glorot, X.; O'Donoghue, B.; Visentin, D.; et al. Clinically applicable deep learning for diagnosis and referral in retinal disease. *Nat. Med.* **2018**, *24*, 1342–1350. [CrossRef] [PubMed]

37. Kooi, T.; Litjens, G.; van Ginneken, B.; Gubern-Mérida, A.; Sánchez, C.I.; Mann, R.; den Heeten, A.; Karssemeijer, N. Large scale deep learning for computer aided detection of mammographic lesions. *Med. Image Anal.* **2017**, *35*, 303–312. [CrossRef] [PubMed]

38. Kalmet, P.H.S.; Sanduleanu, S.; Primakov, S.; Wu, G.; Jochems, A.; Refaee, T.; Ibrahim, A.; Hulst, L.; Lambin, P.; Poeze, M. Deep learning in fracture detection: A narrative review. *Acta Orthop.* **2020**, *91*, 215–220. [CrossRef] [PubMed]

39. Oliveira, M.M.; Quittes, L.; Costa, P.H.V.; Ramos, T.M.; Rodrigues, A.C.F.; Nicolato, A.; Malheiros, J.A.; Machado, C. Computer vision coaching microsurgical laboratory training: PRIME (Proficiency Index in Microsurgical Education) proof of concept. *Neurosurg. Rev.* **2021**, *45*, 1601–1606. [CrossRef]

40. Smith, P.; Tang, L.; Balntas, V.; Young, K.; Athanasiadis, Y.; Sullivan, P.; Hussain, B.; Saleh, G.M. "PhacoTracking": An evolving paradigm in ophthalmic surgical training. *JAMA Ophthalmol.* **2013**, *131*, 659–661. [CrossRef] [PubMed]
41. Khan, D.Z.; Luengo, I.; Barbarisi, S.; Addis, C.; Culshaw, L.; Dorward, N.L.; Haikka, P.; Jain, A.; Kerr, K.; Koh, C.H.; et al. Automated operative workflow analysis of endoscopic pituitary surgery using machine learning: Development and preclinical evaluation (IDEAL stage 0). *J. Neurosurg.* **2021**, 1–8. [CrossRef]
42. Kitaguchi, D.; Takeshita, N.; Matsuzaki, H.; Oda, T.; Watanabe, M.; Mori, K.; Kobayashi, E.; Ito, M. Automated laparoscopic colorectal surgery workflow recognition using artificial intelligence: Experimental research. *Int. J. Surg.* **2020**, *79*, 88–94. [CrossRef]
43. Ward, T.M.; Hashimoto, D.A.; Ban, Y.; Rattner, D.W.; Inoue, H.; Lillemoe, K.D.; Rus, D.L.; Rosman, G.; Meireles, O.R. Automated operative phase identification in peroral endoscopic myotomy. *Surg. Endosc.* **2021**, *35*, 4008–4015. [CrossRef]

 *NeuroSci*

Opinion

Neural Stimulation of Brain Organoids with Dynamic Patterns: A Sentiomics Approach Directed to Regenerative Neuromedicine

Alfredo Pereira, Jr. [1,*], **José Wagner Garcia** [2,3] **and Alysson Muotri** [4]

[1] Philosophy Graduate Program, UNESP/Marilia Campus, São Paulo State University, Botucatu 18618-689, Brazil

[2] Architecture and Urbanism, USP and PUC, São Paulo, Brazil

[3] Media Lab, MIT, Cambridge, MA, USA

[4] Stem Cell Program, Department of Pediatrics & Cellular Molecular Medicine/UCSD, Institute for Genomic Medicine, San Diego, CA, USA

* Correspondence: alfredo.pereira@unesp.br

Abstract: The new science called *Sentiomics* aims to identify the dynamic patterns that endow living systems with the capacity to feel and become conscious. One of the most promising fields of investigation in *Sentiomics* is the development and 'education' of human brain organoids to become sentient and useful for the promotion of human health in the (also new) field of Regenerative Neuromedicine. Here, we discuss the type of informational-rich input necessary to make a brain organoid sentient in experimental settings. Combining this research with the ecological preoccupation of preserving ways of sentience in the Amazon Rainforest, we also envisage the development of a new generation of biosensors to capture dynamic patterns from the forest, and use them in the 'education' of brain organoids to afford them a 'mental health' quality that is likely to be important in future advances in 'post-humanist' procedures in regenerative medicine. This study is closely related to the psychophysical approach to human mental health therapy, in which we have proposed the use of dynamic patterns in electric and magnetic brain stimulation protocols, addressing electrochemical waves in neuro-astroglial networks.

Keywords: brain organoids; sentiomics; regenerative neuromedicine; dynamic patterns; Amazon Rainforest

Citation: Pereira, A., Jr.; Garcia, J.W.; Muotri, A. Neural Stimulation of Brain Organoids with Dynamic Patterns: A Sentiomics Approach Directed to Regenerative Neuromedicine. *NeuroSci* **2023**, *4*, 31–42. https://doi.org/10.3390/neurosci4010004

Academic Editor: Xerxes D. Arsiwalla

Received: 29 November 2022
Revised: 5 January 2023
Accepted: 11 January 2023
Published: 16 January 2023

1. Introduction

We have approached the concept of *Sentience*, in reference to universal dynamic patterns (amplitude-modulated waveforms; see [1,2]) that support conscious experience in living systems. These patterns are not intrinsically conscious, but their operations in neural tissues (e.g., forming synchronized rhythms in neural assemblies) endow living systems with the capacity of feeling and becoming conscious. We call *Sentiomics* the science responsible for the identification and preservation of these patterns, a task executed in scientific laboratories and natural environments, using the scientific method to explore mental potentialities intrinsic to natural processes.

In this paper, we establish conceptual foundations for an experimental research program aimed to preserve sentience, both in natural ecosystems and damaged brains. The main claim of the paper is about the possibility of providing an "education" to 'in vitro' brain organoids to allow the development of *sentience*, defined as the *capacity* of feeling [3–6]. In the beginning of the system's self-organized growth, there is no cellular differentiation, and therefore, specialized sensory systems are absent, but each neuron (or proto-neuron) has a dendritic tree that can receive signals. We plan to develop biosensors, and to use already existing techniques from brain stimulation research (basically, electrical—with microelectrodes; and magnetic—with EM wave generators, using adequate frequencies and amplitudes, procedures), for the neural encoding of information during the early life of brain organoids.

The intended outcome of the research program is the 'education' of brain organoids grown in the Lab exposed to signals from nature, to induce the formation of adequate connections and dynamic processes necessary for the development of the capacity of feeling (sentience). According to [1,2], this capacity depends on the existence of an adequate substrate composed of electrochemical waves in neuro-astroglial networks and extracellular matrix.

A human brain organoid is a biological system composed of human neuronal and glial cells grown 'in vitro' (see, e.g., [7]), with affective, cognitive, and enactive potentialities, to be studied experimentally. Instead of the formative ontogenetic process of living systems in their niche, freely interacting with their ecosystems, brain organoids receive signals chosen by the experimenter during a training period. In this period, the experimenter should furnish the organoid with rich dynamic patterns, to activate the signaling networks existent in neural (neuronal and glial) tissues, making possible the formation of the necessary substrate for the expression of their mental potentialities.

In this paper, we present a preliminary approach to relevant issues likely to become the frontier of research and application in regenerative neuromedicine.

2. Epistemological and Bioethical Concerns

The generation of 'in vitro' brain organoids in the Lab raises relevant questions about their eventual affective, cognitive, and enactive capacities, as well as ethical concerns about human interactions in regenerative medicine and 'post-humanist' projects [8]. Before focusing on ethical issues [9], it is convenient to evaluate the biological viability of the emergence of sentience in brain organoids and possible usages in neuromedicine. Here, we argue that the emergence of sentience in brain organoids is not only possible, but also a probable consequence of the system's neural structure (composed of networks of neurons and glial cells) *once an adequate type of 'education'* (continued input of dynamic patterns) is given.

The concept of sentience and its implications should be clear from the start, to avoid premature worries about the bioethical status of 'in vitro' organoids. We do not assume that sentience is, or contains, a degree of consciousness, but instead we relate sentience with a *potentiality* for consciousness only. We define sentience as the *capacity* of: (a) perceiving stimuli, and (b) forming a feeling, which guides the response of the organism to stimuli. Therefore, in our conceptual framework, sentience is not the *conscious experience* of feeling (e.g., feeling pain), but refers to the biological substrate necessary for the experience. Only when the experience effectively happens (as in the example of feeling dizzy in a virtual roller coaster, discussed by Pereira Jr. [6]), engendering cognitive, affective and enactive functions, the corresponding bioethical concerns arise (e.g., is the organoid feeling pain?).

The study of conscious experience, and related bioethical issues, are not in the knowledge domain of *Sentiomics*. This science studies the *unconscious* dynamic patterns that form the biological substrate for the instantiation of feelings. It does not study the conscious experience of feeling. The distinction between *Sentiomics* and *Qualiomics*, discussed in the next section, has the goal of separating biological research on sentience from issues related to the study of qualitative conscious experiences, which are the subject of another knowledge area, called *Qualiomics*.

Making experiments with material substrates capable of feeling is equivalent to making experiments with neuron cultures, a trivial type of experiment carried out in many laboratories in the world, without special bioethical concerns related to animal or human experimentation. We agree with the claim that any neural network can generate sentience [10]. However, only when a developed neural system is "educated" and *consciously experimenting with feelings*, does the comparison with animals or plants, and the corresponding bioethical concerns, become relevant. In our research program with brain organoids, we are still not in this phase of the research. In the future, organoid consciousness may become a reality [11], and then, bioethical principles should become established for the social interaction of these systems with human society.

We also recall that bioethics is an enterprise of the scientific community as a whole, directed to establish universal principles. It is not a statement that each researcher or group of researchers make for each experiment with living systems. We prefer to wait until the rules for experimentation with mature brain organoids are established by the community, and then follow them in practice, instead of speculating in advance.

We claim that such an education of organoids to become sentient systems is necessary for the purposes of regenerative neuromedicine, e.g., growing neural tissues in vitro for transplantation to human brains in vivo. The reason why this is necessary is that the main function of neural tissues is to carry mental functions, which require a proper type of development for information processing in neuro-astroglial networks that compose the tissue. Only properly 'educated' brain organoids are likely to develop the electrochemical waves that (according to [1,2]) operate as a specialized substrate for the instantiation of feelings. In biology, structures are intimately connected to functions, and the functions of cells of the nervous system are cognitive, affective and enactive, all of them depending on sentience, the capacity of feeling [4]. We claim that the exposition of brain organoids to rich dynamic patterns (for instance, those registered from a natural environment such as the Amazon Rainforest) can elicit the type of electrochemical waves found in conscious systems, operating as a substrate for the instantiation of feelings.

3. Sentiomics and the Human Brain

Sentience has been approached as a psycho-biological phenomenon, corresponding to a cycle in living tissue composed of processes of chemical homeostasis in neural tissue (involving transmitters, modulators, hormones and peptides) that activate hydro-ionic waves (mostly calcium waves), and the feedback from these waves, controlling electrochemical homeostasis [3], The dynamic patterns of the waves in living tissue, both in animals and plants [1,2], make possible the conscious experience of feelings. Basic sensations, such as hunger and thirst, pleasure and pain, and mood states (depressed, euphoric) are (putatively) generated as temporal processes involving these electrochemical waves.

Ways of feeling are studied in two modalities:

(A) As the *universal set of patterns* of Sentience, which we call *Sentiomics*;
(B) As species-specific and individually different sets of *qualitative subjective experiences*, which we call *Qualiomics*.

Qualiomics is, of course, a difficult issue for conventional science, as stated in the "hard problem of consciousness" [12], because it leads to the much-discussed distinction of first- and third-person perspectives. The first-person perspective of individuals belonging to other species, or even to our species, is not accessible to scientific observation, measurement, and objective explanation.

There is an important difference between the capacity of feeling and emotion: while the first is a basic phenomenon possibly present in (almost) all living systems, the second is a higher-order phenomenon, corresponding to a dynamic cycle in which the affective drive that results from sentience is matched with affordances in the domain of the interaction of the agent with the environment [6], generating contextual *emotional feelings*, such as social emotions of love and hate, and individual emotions, such as being happy or sad. The study of conscious emotions belongs to the domain of *Qualiomics*, a field of knowledge constructed with first-person perspective approaches (as Introspective Psychology, Qualitative Research, and other forms of reporting subjective experiences).

Sentiomics is more amenable to a scientific (empirical, experimental, and hetero-phenomenological) treatment than *Qualiomics*. For instance, in our experimental work, we assume that dynamic patterns captured in real time by biosensors in the Amazon Rainforest relate to the sentience of the living beings that compose the ecosystem. We claim that the preservation of biological species is also the preservation of these universal forms of sentience, and vice-versa: the preservation of universal forms of sentience can also contribute to the preservation of the species.

Sentiomics becomes relevant in a historical moment when biological populations are under threat, modified in their ways of living, or extinct. We allude to the possibility of *preserving sentience* with the help of scientific measurements, artistic sensibility, and technological tools. This type of project can be interesting as a "safety net" for the preservation of data about the sentient ecosystems, and as an auxiliary tool for the mobilization of people's environmental consciousness.

There are three types of conscious functions [13] carried out by the human brain: affective, cognitive, and enactive (concerned with the control of action). The study of mental health based on the neurosciences and psychology has focused mostly on cognitive and enactive functions. The study of affective processes is difficult because the neural substrates of feelings (from sensations to social emotions) are not local in the brain. Rather, they involve distributed and temporal processes related to the functions of glial cells (cells that, in the CNS, outnumber neurons, and that do not conduct nerve impulses).

'Sentience' refers to the unconscious neural processes that make us capable of feeling. As far as we know, only sentient beings are conscious. Being conscious is "feeling what happens" [14]. The neural bases of feelings are related to the control of homeostasis and allostasis by glial cells [3–6]. These cells were brought to scientific attention by means of the work of a group of scientists led by R. Douglas Fields. In his book *"The Other Brain"*, he stated: "The potential breakthroughs for medical science . . . are the most exciting frontier in glia research today. Diseases such as brain cancer and multiple sclerosis are caused by diseased glia. Glia plays an important role in such psychiatric illnesses as schizophrenia and depression and in neurodegenerative diseases such as Parkinson's and Alzheimer's. They are linked to infectious diseases such as HIV and prion disease (mad cow disease, for example) and to chronic pain. Scientists have discovered that glia can help repair the brain and spinal cord after injury and stroke" [15].

The generation of feelings in neural tissues involve temporal processes of homeostasis and allostasis [3] achieved by means of electrochemical slow waves of calcium ions inside astrocytes (the most common form of glial cell in the CNS), prompting currents of calcium and potassium ions into the extracellular milieu, where they modulate neural activity [16–18]. These slow waves and currents are not registered by conventional scalp EEG, or by technologies such as functional magnetic resonance imaging. They can, however, be imaged by invasive methods, such as optical imaging with two-photon microscopy or invasive electrodes during brain surgery.

Waking up involves a rise in the amplitude of these ionic waves and currents, together with other electrochemical processes. Nedergaard and colleagues have found evidence for what they call the 'glympathic' system, composed of channels for the circulation of cerebrospinal fluid in neural tissues. While we are awake and sentient, there is an increase in lactate levels in the glympathic pathways, whereas the lactate concentration decreases when we are in deep sleep [19].

The sensations, emotions, and affective states we experience depend on several factors. For instance, gastronomic experiences depend on our degree of hunger; experiencing the same food at different times and under different conditions may be variously pleasurable or disgusting, depending on the initial state of our digestive and neural systems. More generally, the quality of our feelings depends on the temporal dynamics of physiological processes involving mechanisms of homeostasis and allostasis [3].

The treatment of mental problems related to brain pathologies has been mostly through pharmacological interventions. Although these methods have improved in the last decades (see, e.g., the development of serotonin reuptake inhibitors), they are limited and, in some cases, produce adverse effects and/or lead to drug addiction. Most of them do not have the electrochemical medium as their target, but operate on neuron membrane receptors that are mostly involved with cognitive and enactive processes. One remarkable exception is the therapeutic use of lithium for treating bipolar disorder, because this exogenous ion can support the same functions as endogenous calcium ions in astrocytes.

The understanding of the dynamical patterns of sentience and their relationship with glial cells may allow the development of new pharmacological therapies (e.g., [20]) and also non-pharmacological therapies for the diseases of the brain and the mind [21,22]. Non-pharmacological techniques include, for instance, brain electro-magnetic stimulation to reduce abdominal pain [23] and electrical stimulation targeting astrocytes used against depression [24]. A possible therapeutic development of brain physical stimulation technology to treat affective disorders is direct brain (magnetic or ultrasound) stimulation with dynamic patterns, such as music or other types of bio-signals, considering that they involve the same type of temporal amplitude-modulated waveforms of the brain's endogenous activity. It can be conjectured that we can directly stimulate the brain with dynamic patterns using devices that induce the amplitude-modulated signal (for instance, using a scalp-located device, such as the rotating magnets developed by Helekar et al. [25], or the input of chemical biosensors, as described in the next section). This would be an alternative to stimulation with non-informational sine waves (used in transcranial Direct or Alternating Current Stimulation, tDCS and tACS, respectively) or static EM fields (used in Transcranial Magnetic Stimulation, TMS). The patterns to be used in stimulation could be adjusted to the taste of the subjects, personalizing the information and making it possible to address specific types of affective states—instantiated in ionic waves and currents—instead of perturbing the whole neural tissue [26].

4. Building a Sentient and Creative Brain Organoid 'In Vitro'

In the same path of therapeutic strategies targeting dynamic patterns of hydro-ionic waves embodied in neuro-astroglial networks, we propose to develop brain organoids for use in regenerative medicine, giving them a continued information-rich stimulus during an 'educative' period.

Receiving information about what happens is not sufficient for becoming conscious; besides receiving the information, the conscious system must also feel 'what it is like' to be informed. A machine that is not capable of feeling (such as a computer or a smartphone) only registers information without becoming conscious of this information. On the basis of the previous work we have carried out (and summarized above), we claim that to become conscious, a brain organoid should be continuously fed with rich dynamic patterns, to afford the development of neural (neuronal and glial) networks and patterns of activity compatible with sentience, the *capacity* of feeling, considered to be a condition for consciousness [4–6].

In our experimental setting, bio-signals (molecular, chemical, bioelectric, and bio-electro-magnetic) from the Amazon Rainforest will be recorded on independent devices, according to the type of signal (e.g., molecular and chemical signals from microorganisms, sounds of birds, insects, amphibians, rain, wind, etc.). With the current tools of synthetic biology, we can build genetic circuits internalized in bacteria extracted from the Amazonian rhizosphere. Such biosensors will be able to transfer information from the patterns generated by Amazon plant and animal species and the physical environment to the organoids, mediated by an interface that uses graphene networks and 'quantum dot' technologies (Figure 1).

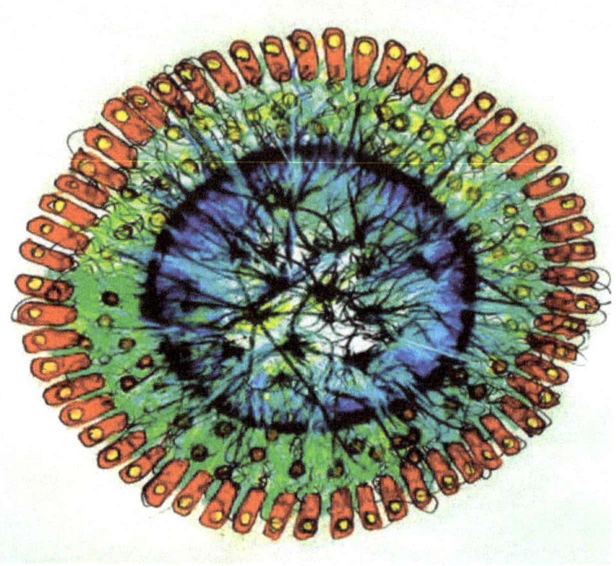

Figure 1. Coupling of Bacteria and Brain Organoid by Biosensors. A genetically engineered bacterial network (colored red) connected around a brain organoid (colored blue; the intensity corresponds to hypothetical neuron firings) by means of a bio-sensor composed of a synthetic interface device made of graphene (green) and golden quantum dots (yellow) to deliver natural bio-signals to the organoid. The bacteria have the role of transducing dynamic patterns from the Amazon Rainforest environment for the 'education' of the organoid.

Records will be compositionally treated and transmitted to the brain organoid via interfaces that generate or modulate Local Field Potentials. The interface with the brain organoid may be an array of implanted microelectrodes, a rotating and/or vibrating magnets, and/or biophysical or biochemical interfaces, delivering a temporally structured stimulation of bio-signals. In conventional settings, the signals, in a pulsed current, are supplied to the brain organoid with the proper voltage and amperage, shaping both the dendritic potentials of neurons and the calcium waves of astrocytes. In the synthetic interface, which may be used to deliver bio-signals from bacteria of the Amazon Rainforest to brain organoids, graphene and gold quantum dots are used for the communication of dynamic patterns (Figure 2). The bacterial network can be genetically engineered to capture specific types of bio-signals, using the CALTECH Biological Circuit Design (see http://be150.caltech.edu/2020/content/lessons/01_intro_to_circuit_design.html accessed on 29 November 2022) derived from the work of MIT researcher Chris Voigt and collaborators [27].

In the cultivated brain organoid, functional regions develop, based on genetic determinations and the types of input provided. The variety of bio-signals from the forest makes it possible to induce a number of specialized sensory regions, able to recognize specific types of dynamic patterns. These regions interact and generate an output signal in the effector region of the brain organoid (corresponding to the human motor system). In the input regions, the recorded signals are delivered, and in the output region, single-cell electrodes are inserted to record the respective spike trains.

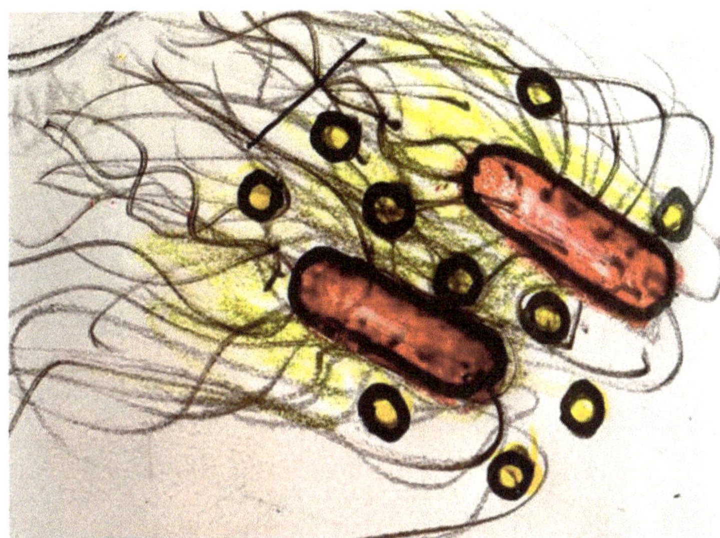

Figure 2. Synthetic device to connect Bacteria to Brain Organoids. The bacteria (colored red) grows curly fibers (black) that wrap (and are sustained by) the hexagonal structure of the graphene (not shown in the picture), forming a structure in which the golden quantum dots (yellow) flow, transmitting the information (dynamic patterns of the rainforest) to the brain organoid (see Figure 1).

In providing these signals for the brain organoid, we seek to induce the formation of circuits specialized in pattern recognition. For this, the forest signals will be organized according to, for instance—in analogy with musical patterns—rhythmic, "melodic", "timbre", and "harmonic" characteristics of electrochemical waves. Each of these compositional arrangements will be transmitted to a brain organoid region, to induce functional specialization during the training period. For example, considering the modality of audio recording, it is possible to organize the forest registers in four compositions: the first one, highlighting melodies, with birdsongs; another one, focusing on the rhythm, with sounds of raindrops; a third one, highlighting the variety of timbers produced by insects and amphibians; and a fourth one, highlighting the harmonic spectrum of the sound produced by the wind on tree leaves, and other sounds that present a rich spectrum of frequencies. This is only a simplified example, because a new generation of biosensors will be developed in the experiment to capture molecular, chemical, and microbiological dynamic patterns from the forest, using the bacterial synthetic biological device shown in Figures 1 and 2.

In the training period, sensory specialization should be induced in the brain organoid. Considering Hebb's Law, it is assumed that each region submitted to a type of stimulus will form interconnections specific to the information patterns present in the stimulus. In the process of Long-Term Potentiation, it is assumed that the connections formed will present a greater sensitivity to the type of pattern they were repeatedly exposed to. The patterns of connections formed at each specialized region putatively generate different types of dynamic patterns of electrochemical waves.

The training period corresponds to the time necessary for the consolidation of neural connections, through the plasticity of the system, which involves the formation of calcium currents and waves (respectively, in neurons and astrocytes), the activation of signal transduction pathways within neurons, gene regulation, the production of the growth factors of dendrites and axons, and the formation of new dendritic and axonal ramifications. The time required for the brain organoid to consolidate these memory traces should be evaluated during the experiment, as there are no available data that allow us to predetermine the time of learning.

The generation of informational coherent responses to the input, and the development, of adaptive patterns of action (when the brain organoid is connected to effector parts, affording actions) may be rewarded with an increase in the supply of nutrients or modulators. In the output region, single cell electrodes can be used, capturing the axonal firing of neurons. During training, there is feedback from these spike trains to the sensory areas of the brain organoid. For this, we will make use of an artificial intelligence counting algorithm to evaluate the informational coherence of these spike trains, "rewarding" the system with nutrients (glucose) or transmitters that generate pleasurable sensations (for example, dopamine) when there is more consistency in the signals (consistency: maximized combination of variety and redundancy). The type of reward will be determined during the research.

After the training period, in which each region has established specialized connections forming neural circuits (in its interior and with the other regions), the experiment enters the second stage, the production of works by the brain organoid, through an interface that connects the output with a signal converter and a synthesizer. The spike trains generated by the brain organoid, spontaneously displaying—similar to any other axonal firing 'in vivo'—the encoding of information patterns in frequency and phase, is connected through the signal converter with a synthesizer that generates virtual reality streams. In this part of the experiment, we assume that the various specialized sensory regions interact with each other, allowing the system to self-organize, generating a coherent output signal.

In the experimental setting, after the training period, the system will be exposed to new stimuli. Its responses can be recorded, using an interface with the human observer (e.g., a converter of the filtered binary signal—the spike trains generated by the brain organoid—to a synthesizer), generating a 'virtual reality' product for our evaluation—using psychological assessment protocols—of possible neural activity in the system. This is a task for *Qualiomics* research. Besides this 'qualitative' method, physiological biomarkers [4] can also be used to evaluate the mental activity of brain organoids.

Our working hypothesis for this phase of the research program is that the cognitive work produced by the brain organoid will convey a "feeling" that impacts the human receptor, which will then experience a 'virtual reality' exposition of the product. Considering that the brain organoid is generated from human cells, containing our species' genome, there is a potential for brain organoids to develop in similar ways during the training period, allowing the emergence of creative responses to new stimuli. We predict the emergence of conscious experiences in the first-person perspective of the brain organoid, motivating the system to compose creative works, recombining the dynamic patterns previously presented. This prediction is based on the abovementioned fact that this type of system contains the same types of cells (neurons and astrocytes developed from stem cells) as the human nervous system. Experiencing a human-like feeling that motivates creativity, the brain organoid can generate cognitive responses with sentient qualities, to be evaluated by a human committee at the end of the experiment. At this phase of the research program, both the qualitative evaluation of the outputs of the system, and the bioethical concerns related to their use in human society, are issues that belong to the knowledge domain of *Qualiomics*.

5. Astrocytes in the Development of Brain Organoids and Regenerative Medicine

Astroglial cells and networks are the keys to theoretical and technological progress in the new fields of research Sentiomics and Regenerative Neuromedicine. As we have claimed above, and in several publications, astrocytes embody the hydro-ionic waves of *sentience*, the capacity of feeling, which is necessary for the emergence of consciousness. These waves also play a key role in the development of brain organoids (where they can develop [28]) and possible applications in Regenerative Neuromedicine.

The role of astrocytes in the development of neural tissues is well-known (see, for instance, https://www.biotechniques.com/cell-and-tissue-biology/using-astrocytes-to-speed-organoid-development/, accessed on 29 November 2022). Astrocytes participate in neuronal synapses, receiving signals from the neurons and modulating them back [16,29].

Neuronal synchrony induces the formation of astroglial hydro-ionic waves by means of constructive interferences [16]. These waves have been claimed to embody affective patterns of sentience [4] and participate in the brain's large-scale integration of temporal patterns that provide the 'signature' of consciousness [30].

In brain organoids, the roles of astrocytes are the same as those in animal brains freely interacting with the environment, with the only difference being that the brain organoids are 'in vitro' entities (similar to Putnam's famous story about "the brain in a vat"; see [31]). Once the system is provided with an artificial interface with the environment, as sketched in our previous sections, affording—with the help of the experimenter—"action-perception cycles" and an educational process similar to the experience of living systems interacting with their environment, it can develop sentience in a similar manner.

The results of this experimental program can have many applications in Regenerative Neuromedicine:

(1) In vitro studies of new drugs and physical (magnetic and electric stimulation) therapies used for 'in vivo' therapeutic procedures;
(2) Transplantation of neural tissue to damaged regions of the nervous system, highlighting the role of glial ('glue') cells: not only astrocytes, but also microglia [32];
(3) 'In vitro' studies of brain development (e.g., [33]) and regeneration (e.g., in cases of ischemia, hemorrhage, epileptic crises, traumatic lesions, cancer or aging degeneration diseases) and the testing of therapeutic resources [34–37].

6. Registering Signals of the Amazon Rainforest for Use in Brain Organoids' 'Education'

Exemplifying and advancing our reflections and practical actions towards the viability of *Sentiomics*, we report a scientific and artistic sub-project of recording data in the Amazon Rainforest, using it to preserve the "feeling of the forest", and eventually induce them to other sentient beings, as human brain organoids, or to human sensibilization in artistic installations and museums of natural history.

In an age of glimpsing the possibilities of artificial intelligence, the rationality that led to the development of the machine and the industrial age also led to a disconnection of the human with his own senses and his environment. In this context, a scientific revival emerges, which seeks not a return to a primitive state, but a transcendence of the human-nature relationship *mediated* by the very intelligence of the machine. This opens the door for new multidimensional and multisensory explorations; for instance, projects to catch, process, and display life's invisible dynamics. It aims to be a nature mediation system, approaching natural processes not apprehended by our senses, or facilitating the visualization of large datasets. More than just showing, it can correlate and predict phenomena.

For this purpose, we have captured a variety of signals of the forest in real time and compared these with previous registers. The data is "human knowledge" only in the sense that human beings place the sensors, build the computers that record them, detect the correlations, and help to generate an interface for public exposition of the results. According to the conceptual foundations of Sentiomics, the detected and analyzed dynamic patterns are universal, being present in all varieties of living systems, in different combinations and temporal sequences.

The project starts with data from the Amazon Rainforest ZF2 tower, Amazon Satellite Data, and the Massachusetts Living Observatory, in collaboration with the MIT Media Lab, comparing the dynamics of the 1500 y.o. forest with a recent regeneration area, analyzing the similarities and differences between conservation and restoration processes.

A system capable of displaying phenomena in real time is a new kind of interface for science. The scientific method seeks to store large datasets, and then further process and display them in the form of tables, diagrams and graphs. Our *Biobit Forest* system stores data from the past, learns from it, and can process new information. This makes it a communication interface with a natural phenomenon in real time, where events can be observed "in the act".

Experiencing high complexity or high dimensionality of datasets represents a critical obstacle to reach new clusters of dynamical patterns in nature. Through a large amount of data already available from the last 40 years of the partner scientific institutions, we propose a computational processing project that finds complex correlations and patterns, creating an artificial intelligence capable of interpreting real-time phenomena. Past information, constantly updated, feeds the system. Computational intelligence processes data and looks for patterns, becoming capable of interpreting data in real time. The result is a scientific/artistic representation system that enables the multidimensional sensing of data. These environments have physical and sensory elements that reorganize into dynamic flows according to data input. Kinetic structures, geometric transformations, use of colors and sounds are some of the possibilities. The purpose is to create a language that decodes nature and expresses itself through sensory experiences. In sum, the *Biobit Forest* research affords a new paradigm in the scientist's relationship to data, a new way of studying nature, by means of the elucidation of complex relations between crossed data, allowing the investigation of the role of the Amazon in global climate dynamics; and the comparison between data from a native forest and a regenerating area.

The preservation of lifeforms is usually addressed as a political and legal issue, focused on the continuity of biodiversity across human generations. Treves et al. [38] propose some ethical principles for biological preservation ("geocentrism, equitable consideration of non-humans, bio-proportionality, and intergenerational equity") to guide the institution of courts "with constitutional powers to adjudicate the rights of futurity and non-humans against the rights of present humans who are threatening all life on Earth with our all-consuming use of natural resources" [38].

Besides the legal and political concerns, there is also an important philosophical issue at play within efforts to preserve lifeforms. The modification or extinction of a biological species is also—in a Monist perspective for which the material and mental aspects belong to the same reality—the modification or extinction of a form of sentience and its related dimensions. For instance, we are interested in the preservation of biological species, such as the Panda, not only because of their biological (genetic, physiological, ecological) richness, and our (human) appreciation of them, but also because they are sentient beings, with ways of feeling that we can inter-subjectively share with them.

Each biological species, besides having proper genes, proteins, metabolic processes, and action schemes to be preserved, also has sentience. Does "just preservation" of the respective populations assure the preservation of their ways of feeling? [39] notes that preservation involves a holistic view of nature. Considering changes caused by the human destruction of ecosystems, "just preservation" of biological beings does not imply the preservation of sentience, because the ways of feeling are closely related to ecological interactions; changing the ecosystem or moving the biological individual to a completely different environment, the dynamic patterns of the sentience of biological populations and individuals are likely to change.

In our foresight, *Sentiomics* has the potential to be the next step in a sequence of developments for the study and promotion of biodiversity and natural forms of mentality, moving forward in relation to successful projects of identification and preservation of genomes and proteomes. Although the artificial intelligence of a machine is not sentient, it can be used as a *medium* to organize data that *preserves the correlations found in sentient systems*. The structured database can be used in other, different biological substrates, such as the brain organoid in Regenerative Neuromedicine, to induce sentience, in an interdisciplinary effort that integrates ecosystemic information richness and human neuromedicine, to promote the health of both—the ecosystem and the damaged human brain.

7. Concluding Remarks

The reviewed scientific information and theoretical proposals afford an experimental program to investigate the neural potentialities of brain organoids trained with signals from the Amazon Rainforest ecosystem, with the following goals:

(1)   Record and preserve bio-signals from the Amazon Rainforest;

(2)   Use the informational variety of these dynamic patterns to generate a repertoire of learning in brain organoids;

(3)   Induce, in brain organoids, the emergence of sentience, evaluated by means of biomarkers, according to the proposal formulated by [4];

(4)   Generate works from the responses of the brain organoids, to be evaluated by human persons, and eventually, they can also be part of exhibitions related to the environmental issue;

(5)   Enable the use of the brain organoid in "post-humanist" projects for the preservation of sentience, both demonstrating the mentality intrinsic to nature and contributing to human "reconnection" with the information richness of nature;

(6)   Generation of brain organoids for the treatment of lesions, neurological illnesses, and degenerative diseases affecting the nervous system.

Beyond these achievements, we also predict that the transfer of dynamic patterns from biosensors to brain organoids will make them capable of learning and acquiring habits. In the next phase of the investigation, organoids coupled with biosensors placed in the Amazon Rainforest, instead of the Lab, will be able to behave as a hybrid intelligence; that is, they will be able to acquire specific functions to detect and feel still unknown types of biochemical and biophysical information.

Author Contributions: Writing and review of literature: A.P.J.: Conceptualization and figures: J.W.G.; Experimental methods and implementation: A.M. All authors have read and agreed to the published version of the manuscript.

Funding: This research received no external funding.

Acknowledgments: FAPESP (Alfredo Pereira Jr.) for the funding of previous research.

Conflicts of Interest: The authors declare no conflict of interest.

References

1.   Pereira, A., Jr. Sentience in living tissue. *Anim. Sentience* **2017**, *2*, 5. [CrossRef]
2.   Pereira, A., Jr.; Foz, F.B.; Rocha, A.F. The dynamical signature of conscious processing: From modality-specific percepts to complex episodes. *Psychol. Conscious.* **2017**, *4*, 230–247. [CrossRef]
3.   Pereira, A., Jr. Developing the concepts of homeostasis, homeorhesis, allostasis, elasticity, flexibility and plasticity of brain function. *NeuroSci* **2021**, *2*, 372–382. [CrossRef]
4.   Pereira, A., Jr. The Role of Sentience in the Theory of Consciousness and Medical Practice. *J. Conscious. Stud.* **2021**, *28*, 22–50.
5.   Pereira, A., Jr. Reply to Commentaries and Future Directions. *J. Conscious. Stud.* **2021**, *28*, 199–228.
6.   Pereira, A., Jr. Sentience and Conscious Experience: Feeling Dizzy on a Virtual Reality Roller Coaster Ride. *J. Conscious. Stud.* **2021**, *28*, 183–198.
7.   Yakoub, A.M. Cerebral organoids exhibit mature neurons and astrocytes and recapitulate electrophysiological activity of the human brain. *Neural Regen. Res.* **2019**, *14*, 757–761. [CrossRef]
8.   Jeziorski, J.; Brandt, R.; Evans, J.H.; Campana, W.; Kalichman, M.; Thompson, E.; Goldstein, L.; Koch, C.; Muotri, A.R. Brain organoids, consciousness, ethics and moral status. *Semin. Cell Dev. Biol.* **2022**. [CrossRef]
9.   Lavazza, A. Potential ethical problems with human cerebral organoids: Consciousness and moral status of future brains in a dish. *Brain Res.* **2021**, *1750*, 147146. [CrossRef]
10.   Kagan, B.J.; Kitchen, A.C.; Tran, N.T.; Habibollahi, F.; Khajehnejad, M.; Parker, B.J.; Bhat, A.; Rollo, B.; Razi, A.; Friston, K.J. In vitro neurons learn and exhibit sentience when embodied in a simulated game-world. *Neuron* **2022**, *110*, 3952–3969.e8. [CrossRef]
11.   Lavazza, A. 'Consciousnessoids': Clues and insights from human cerebral organoids for the study of consciousness. *Neurosci. Conscious.* **2021**, *2*, niab029. [CrossRef] [PubMed]
12.   Chalmers, D. *The Conscious Mind*; Oxford University Press: New York, NY, USA, 1996.
13.   Pereira, A., Jr. The Projective Theory of Consciousness: From Neuroscience to Philosophical Psychology. *Trans/Form/Ação* **2018**, *41*, 199–232. [CrossRef]

14. Damásio, A. *The Feeling of What Happens: Body and Emotion in the Making of Consciousness*; Harcourt Books: San Diego, CA, USA, 2000.
15. Douglas Fields, R. *The Other Brain: From Dementia to Schizophrenia, How New Discoveries about the Brain Are Revolutionizing Medicine and Science*; Simon and Schuster: New York, NY, USA, 2009.
16. Pereira, A., Jr.; Furlan, F.A. Astrocytes and human cognition: Modeling information integration and modulation of neuronal activity. *Prog. Neurobiol.* **2010**, *92*, 405–420. [CrossRef] [PubMed]
17. Verkhratsky, A.; Nedergaard, M. Physiology of Astroglia. *Physiol. Rev.* **2018**, *98*, 239–389. [CrossRef]
18. Rasmussen, R.; O'Donnell, J.; Ding, F.; Nedergaard, M. Interstitial ions: A key regulator of state-dependent neural activity? *Prog. Neurobiol.* **2020**, *193*, 101802. [CrossRef] [PubMed]
19. Lundgaard, I.; Lu, M.L.; Yang, E.; Peng, W.; Mestre, H.; Hitomi, E.; Deane, R.; Nedergaard, M. Glymphatic clearance controls state-dependent changes in brain lactate concentration. *J. Cereb. Blood Flow Metab.* **2017**, *37*, 2112–2124. [CrossRef] [PubMed]
20. Stenovec, M.; Li, B.; Verkhratsky, A.; Zorec, R. Astrocytes in rapid ketamine antidepressant action. *Neuropharmacology* **2020**, *173*, 108158. [CrossRef] [PubMed]
21. Peng, L.; Parpura, V.; Verkhratsky, A. Neuroglia as a Central Element of Neurological Diseases: An Underappreciated Target for Therapeutic Intervention. *Curr. Neuropharmacol.* **2014**, *12*, 303–307. [CrossRef]
22. Finsterwald, C.; Magistretti, P.J.; Lengacher, S. Astrocytes: New Targets for the Treatment of Neurodegenerative Diseases. *Curr. Pharm. Des.* **2015**, *21*, 3570–3581. [CrossRef]
23. Bayer, K.; Neeb, L.; Bayer, A.; Wiese, J.; Siegmund, B.; Prü, M. Reduction of intra-abdominal pain through transcranial direct current stimulation: A systematic review. *Medicine* **2019**, *98*, e17017. [CrossRef]
24. Monai, H.; Hirase, H. Astrocytes as a target of transcranial direct current stimulation (tDCS) to treat depression. *Neurosci. Res.* **2018**, *126*, 15–21. [CrossRef] [PubMed]
25. Helekar, S.A.; Convento, S.; Nguyen, L.; John, B.S.; Patel, A.; Yau, J.M.; Voss, H.U. The strength and spread of the electric field induced by transcranial rotating permanent magnet stimulation in comparison with conventional transcranial magnetic stimulation. *J. Neurosci. Methods* **2018**, *309*, 153–160. [CrossRef] [PubMed]
26. Pereira, A., Jr.; Aguiar, V.J. Sentiomics: The identification and analysis of dynamical patterns that characterize sentience. *J. Multiscale Neurosci.* **2022**, *1*, 1–10. [CrossRef]
27. Nielsen, A.A.; Der, B.S.; Shin, J.; Vaidyanathan, P.; Paralanov, V.; Strychalski, E.A.; Ross, D.; Densmore, D.; Voigt, C.A. Genetic circuit design automation. *Science* **2016**, *352*, aac7341. [CrossRef] [PubMed]
28. Dezonne, R.S.; Sartore, R.C.; Nascimento, J.M.; Saia-Cereda, V.M.; Romão, L.F.; Alves-Leon, S.V.; de Souza, J.M.; Martins-de-Souza, D.; Rehen, S.K.; Gomes, F.C. Derivation of Functional Human Astrocytes from Cerebral Organoids. *Sci. Rep.* **2017**, *7*, 45091. [CrossRef]
29. Hasan, U.; Singh, S.K. The Astrocyte-Neuron Interface: An Overview on Molecular and Cellular Dynamics Controlling Formation and Maintenance of the Tripartite Synapse. *Methods Mol. Biol.* **2019**, *1938*, 3–18. [CrossRef]
30. Pereira, A., Jr. Astroglial hydro-ionic waves guided by the extracellular matrix: An exploratory model. *J. Integr. Neurosci.* **2017**, *16*, 57–72. [CrossRef]
31. Putnam, H. Brains in a Vat. In *Knowledge: Readings in Contemporary Epistemology*; Bernecker, S., Dretske, F.I., Eds.; Oxford University Press: Oxford, UK, 1999; pp. 1–21.
32. Sabate-Soler, S.; Nickels, S.L.; Saraiva, C.; Berger, E.; Dubonyte, U.; Barmpa, K.; Lan, Y.J.; Kouno, T.; Jarazo, J.; Robertson, G.; et al. Microglia integration into human midbrain organoids leads to increased neuronal maturation and functionality. *Glia* **2022**, *70*, 1267–1288. [CrossRef]
33. Matsui, T.K.; Tsuru, Y.; Hasegawa, K.; Kuwako, K.I. Vascularization of human brain organoids. *Stem Cells* **2021**, *39*, 1017–1024. [CrossRef]
34. Chandrasekaran, A.; Avci, H.X.; Leist, M.; Kobolák, J.; Dinnyés, A. Astrocyte Differentiation of Human Pluripotent Stem Cells: New Tools for Neurological Disorder Research. *Front. Cell. Neurosci.* **2016**, *10*, 215. [CrossRef]
35. Li, L.; Shi, Y. When glia meet induced pluripotent stem cells (iPSCs). *Mol. Cell. Neurosci.* **2020**, *109*, 103565. [CrossRef] [PubMed]
36. Qian, X.; Su, Y.; Adam, C.D.; Deutschmann, A.U.; Pather, S.R.; Goldberg, E.M.; Su, K.; Li, S.; Lu, L.; Jacob, F.; et al. Sliced Human Cortical Organoids for Modeling Distinct Cortical Layer Formation. *Cell Stem Cell* **2020**, *26*, 766–781.e9. [CrossRef] [PubMed]
37. Albert, K.; Niskanen, J.; Kälvälä, S.; Lehtonen, Š. Utilising Induced Pluripotent Stem Cells in Neurodegenerative Disease Research: Focus on Glia. *Int. J. Mol. Sci.* **2021**, *22*, 4334. [CrossRef] [PubMed]
38. Treves, A.; Santiago-Ávila, F.J.; Lynn, W.S. Just preservation. *Anim. Sentience* **2019**, *27*.
39. Bergstrom, B.J. Just reductionism: In defense of holistic conservation. Commentary on Treves et al. on Just Preservation. *Anim. Sentience* **2019**, *4*, 8. [CrossRef]

Article

Intraoperative Fluorescein Sodium in Pediatric Neurosurgery: A Preliminary Case Series from a Singapore Children's Hospital

Audrey J. L. Tan [1], Min Li Tey [2], Wan Tew Seow [1,2,3], David C. Y. Low [1,2,3], Kenneth T. E. Chang [4], Lee Ping Ng [2], Wen Shen Looi [5], Ru Xin Wong [5], Enrica E. K. Tan [6] and Sharon Y. Y. Low [1,2,3,*]

[1] Department of Neurosurgery, National Neuroscience Institute, 11 Jalan Tan Tock Seng, Singapore 308433, Singapore

[2] Neurosurgical Service, KK Women's and Children's Hospital, 100 Bukit Timah Road, Singapore 229899, Singapore

[3] SingHealth Duke-NUS Neuroscience Academic Clinical Program, 11 Jalan Tan Tock Seng, Singapore 308433, Singapore

[4] Department of Pathology and Laboratory Medicine, KK Women's and Children's Hospital, 100 Bukit Timah Road, Singapore 229899, Singapore

[5] Department of Radiation Oncology, National Cancer Centre, 11 Hospital Drive, Singapore 169610, Singapore

[6] Paediatric Haematology/Oncology Service, KK Women's and Children's Hospital, 100 Bukit Timah Road, Singapore 229899, Singapore

[*] Correspondence: gmslyys@nus.edu.sg

Abstract: (1) Background: Fluorescein sodium (Na-Fl) has been described as a safe and useful neurosurgical adjunct in adult neurooncology. However, its use has yet to be fully established in children. We designed a study to investigate the use of intraoperative Na-Fl in pediatric brain tumor surgery. (2) Methods: This is a single-institution study for pediatric brain tumor patients managed by the Neurosurgical Service, KK Women's and Children's Hospital. Inclusion criteria consists of patients undergoing surgery for suspected brain tumors from 3 to 19 years old. A predefined intravenous dose of 2 mg/kg of 10% Na-Fl is administered per patient. Following craniotomy, surgery is performed under alternating white light and YELLOW-560 nm filter illumination. (3) Results: A total of 21 patients with suspected brain tumours were included. Median age was 12.1 years old. For three patients (14.3%), there was no significant Na-Fl fluorescence detected and their final histologies reported a cavernoma and two radiation-induced high grade gliomas. The remaining patients (85.7%) had adequate intraoperative fluorescence for their lesions. No adverse side effects were encountered with the use of Na-Fl. (4) Conclusions: Preliminary findings demonstrate the safe and efficacious use of intraoperative Na-Fl for brain tumors as a neurosurgical adjunct in our pediatric patients.

Keywords: intraoperative fluorescence; pediatric brain tumors; neurosurgical adjunct

Citation: Tan, A.J.L.; Tey, M.L.; Seow, W.T.; Low, D.C.Y.; Chang, K.T.E.; Ng, L.P.; Looi, W.S.; Wong, R.X.; Tan, E.E.K.; Low, S.Y.Y. Intraoperative Fluorescein Sodium in Pediatric Neurosurgery: A Preliminary Case Series from a Singapore Children's Hospital. *NeuroSci* **2023**, *4*, 54–64. https://doi.org/10.3390/neurosci4010007

Academic Editor: Frank Schubert

Received: 30 December 2022
Revised: 29 January 2023
Accepted: 10 February 2023
Published: 13 February 2023

1. Introduction

Brain tumors are the most common solid tumors in the pediatric population and, unfortunately, the leading cause of childhood cancer-related deaths [1,2]. Presently, tissue diagnosis via biopsy or resection is paramount for diagnosis and molecular studies that will impact subsequent treatment options. Nonetheless, accurate tissue sampling for brain biopsy and achieving a good extent of resection (EOR) for large tumors can be technically challenging. In recent years, the use of intraoperative fluorescence has been accepted as a safe and useful neurosurgical adjunct in the field of neurooncology [1–3]. One such example is fluorescein sodium (Na-Fl)—a sodium salt and an organic fluorescent dye that has been extensively investigated in adult high-grade gliomas (HGG) [4–6]. Na-Fl disseminates through a disruption in the blood–brain barrier (BBB) and accumulates in the extracellular space of brain tumors [7]. However, its use for the same purpose is comparatively less established in children. At the time of this writing, most of these studies are based in overseas centers and limited to case series or case reports [7–10]. Overall, there is a paucity of data with regards to the utility of Na-Fl for pediatric brain tumor surgery from our region of Southeast Asia. To address its safety and efficacy for our local population, we designed a study to investigate the use of intraoperative Na-Fl in pediatric brain tumor surgery. Concurrently, the reliability and technical aspects of its use are evaluated in corroboration with current literature.

2. Materials and Methods

2.1. Study Design and Patient Selection

This is an ethics-approved, retrospective study for pediatric brain tumor patients managed by the Neurosurgical Service, KK Women's and Children's Hospital (SingHealth CIRB Ref: 2014/2079). Inclusion criteria involves patients undergoing surgery for either suspected or known intracranial tumors from 3 to 19 years old. Patients less than 3 years old, with known allergic reactions to Na-Fl, diagnosed with significant systemic comorbidities such as cardiac, hepatic or renal dysfunction, and or hyperreactivity towards contrast agents are excluded.

2.2. Outline of Operative Procedure and Perioperative Management

At our institution, all patients with suspected brain tumors undergo a standard preoperative cranial magnetic resonance imaging (MRI) scan. The StealthStation^{TM} S8 Surgical Navigation System (Medtronic, Minneapolis, MN, USA) is used for preoperative planning and intraoperative neuronavigation. Depending on the location of the tumor, intraoperative neuromonitoring may be applied at the discretion of the neurosurgeon in charge. Following induction of general anesthesia, a predefined dose of 2 mg/kg of 10% Na-Fl 10% is administered intravenously, with simultaneous monitoring of the patient's vital signs and clinical status. Close monitoring of the patient is continued throughout the surgery, in particular looking out for cutaneous and/or physiological manifestations of an adverse drug reaction. With regards to the brain tumor surgeries, patients are approached using standard microsurgical techniques for tumour resection or stereotaxy-based biopsies. For this study, either the Zeiss KINEVO 900 or Zeiss OPMI Pentero 900 (both from Carl Zeiss, Jena, Germany) surgical microscopes equipped with a YELLOW-560 nm filter are used. Once the dura is exposed, the YELLOW-560 nm filter is turned on to check for Na-Fl staining as a positive control to ensure the dye is working. Dura opening is performed under standard white light and subsequently switched to the YELLOW-560 nm filter during the course of the surgery as required. Briefly, a surgical plan created in neuronavigation console is then used, and sulcal dissection to the lesion of interest takes place. Once the lesion is localized, resection is made under alternating white light and the YELLOW-560 nm filter illumination. A routine postoperative MRI scan is performed the following day to assess the EOR. Separately, for the stereotactic biopsy procedures, we adapted a similar technique as described in the literature [11–13]. Here, an intralesional target point is planned via the neuronavigation console. Typically, the target point is chosen within the core of of the lesion. Next, the safest trajectory to the target is selected to avoid critical neurovascular and eloquent structures. Upon tissue sampling, the specimens are physically assessed under the YELLOW 560 nm filter. Fluorescent samples are then correlated with the frozen section report by the on-site pathologist. A computed tomographic (CT) brain scan is ordered after the biopsy to exclude significant bleeding within 24 h. For the purposes of this study, we verified the efficacy of Na-Fl via a feedback evaluation form by the operating neurosurgeons.

2.3. Patient Demographics, Radiological Features, and Variables of Interest

Individual patients' clinical history, operative notes and radiological records are reviewed to identify variables such as age of diagnosis, gender, clinical status, pre- and post-operative imaging characteristics, perioperative complications and timing of surgery. In addition, the histopathological results of biopsied tissue are correlated with the diagnosis according to the relevant WHO classification within the recruitment time period [14,15]. Statistical analyses are generated using GraphPad Prism version 9.4.1 for Windows (Graph-

Pad Software, La Jolla, CA, USA). As this study has a limited population, descriptive statistics are reported. This includes mean with standard deviation for continuous data, and frequency and percentage for categorical data.

3. Results

3.1. Overview of Study Population and Its Characteristics

From 01 January 2019 to 30 November 2022, a total of 21 patients (13 males and 8 females) radiologically diagnosed with suspected brain tumours were included in this study. Median age of the cohort was 12.1 years old (±5.3 years, range 3 to 19 years old). Fifteen patients (71.4%) underwent resection (8 GTR, 2 NTR and 5 STR) and 6 (28.6%) had tissue biopsies (one open biopsy, two using stereotactic brain needle biopsies and three via the NICO Brainpath® transtubular system). Figure 1 summarizes the types of procedures used in this study. For the purposes of this study, we defined <5% of tumor remnant as achievement of gross total resection (GTR), <10% tumor remnant as near total resection (NTR) and ≥30% tumor remnant as subtotal resection (STR). Concurrent neurosurgical adjuncts implemented included diffusion tensor imaging (DTI) for 13 patients (61.9%), intraoperative neuromonitoring (IONM) for three patients (14.3%) and the intraoperative MRI operating theatre (iMRI OT) for one patient (4.8%). Seven patients (33.3%) had Na-Fl as their only operative adjunct. All patients did not have any acute or delayed adverse effects from Na-Fl. For three patients (14.3%), there was no significant Na-Fl fluorescence detected intraoperatively. Final histology reported a cavernoma and two HGGs that were likely radiation-induced. An overview of the patients' demographics and variables is featured in Table 1.

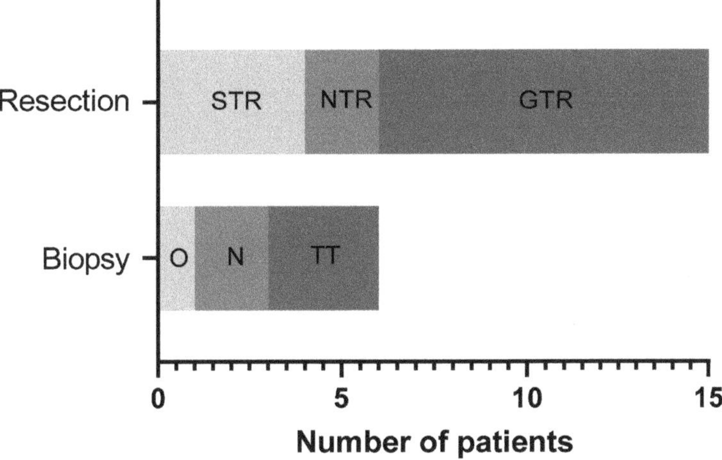

Figure 1. Graph depicting procedure types used in study. Abbreviations: STR = subtotal resection; NTR = near total resection; GTR = gross total resection; O = open; N = needle; TT = transtubular.

Table 1. Patient demographics and variables. Note: values are n (%) unless stated otherwise.

Parameter	Value (%)
Total number of patients	21 (100)
Age (years)	12.1 ± 5.3 (mean ± SD)
Gender	
Male	13 (61.9)
Female	8 (38.1)
Type of surgery	
Biopsy	6 (28.6)
Resection	15 (71.4)

Table 1. *Cont.*

Parameter	Value (%)
Histopathological diagnosis [1]	
Non-neoplastic lesion (cavernoma)	1 (4.8)
Neoplastic lesions	20 (95.2)
Low grade glioma	6 (28.6)
Hemispheric high grade glioma	6 (28.6)
Medulloblastoma	2 (9.5)
Diffuse midline glioma (H3K27M-altered)—1 thalamus and 2 brainstem	3 (14.3)
Craniopharyngioma	1 (4.8)
Choroid plexus carcinoma	1 (4.8)
Primary intracranial malignant melanoma	1 (4.8)
Adverse side effects from Na-Fl	
Yes	0 (0)
No	21 (100)
Score for Na-Fl fluorescence	
0	3 (14.3)
1	0 (0)
2	5 (23.8)
3	13 (61.9)
Concurrent operative adjuncts	
DTI imaging	13 (61.9)
iMRI operating theatre	1 (4.8)
IONM	3 (14.3)
Transtubular system	3 (14.3)
NIL	6 (28.6)
Location of lesion	
Suprasellar	2 (9.5)
Frontal	2 (9.5)
Temporal	1 (4.8)
Parietal	4 (19.1)
Occipital	1 (4.8)
Thalamic	4 (19.1)
Intraventricular (frontal horn)	1 (4.8)
Posterior fossa	4 (19.1)
Brainstem	2 (9.5)

[1] Histopathological diagnosis corresponds to the relevant WHO classification during recruitment period [14,15].

3.2. Evaluation of Na-Fl as an Intraoperative Adjunct

Three pediatric neurosurgeons from the Neurosurgical Unit, KK Women's and Children's Hospital, participated in this study. All of them agreed the fluorophore was easy to firstly prepare (that is, via a straightforward dilution of 2 mg/kg in sterile water) and next to administer intravenously as a bolus dose through a peripheral catheter. Overall, this process only required a few minutes and did not delay the start time of each surgery. We noted that all patients showed a self-limiting neon yellow urine discoloration for up to 12 h after surgery. This is consistent with the known mechanism of Na-Fl, whereby it undergoes hepatic metabolism and subsequently is excreted in the urine [16,17]. Following that, most of the histologically proven tumors demonstrated adequate fluorescence at the start of the surgery, allowing the operating neurosurgeon to visualize tumor boundaries in real time for those requiring resection. With regards to the biopsy cases for tumors located in deeper regions such as the thalamus and brainstem, Na-Fl was useful in confirming our samples were lesional. This is similar to what others have reported in the literature [10,18].

For the only non-neoplastic lesion in our cohort, Na-Fl proved to be a relevant negative. However, we had three brain tumor cases that did not show significant fluorescence. Although there were no technical difficulties switching the illumination between white light and fluorescence on the microscope handle, we noted the following during the course of surgery. Firstly, if a vascular tumor was encountered, the fluorescence was not able to properly delineate the brain–tumor interface due to blood products. Under such circumstances, it was challenging to constantly switch between white light and the YELLOW-560 nm filter during active hemostasis for visualization purposes. Another pertinent observation was the difficulty in differentiating between remnant tumor versus gliotic brain towards the end of the resection, especially for large lesions with radiologically proven perilesional oedema. Nonetheless, postoperative MRI brain scans performed within 48 h after resection reported adequate resection margins that correlated with initial surgical aims. Figure 2 shows an example of our experience with intraoperative Na-Fl for the resection of a hemorrhagic parietal HGG.

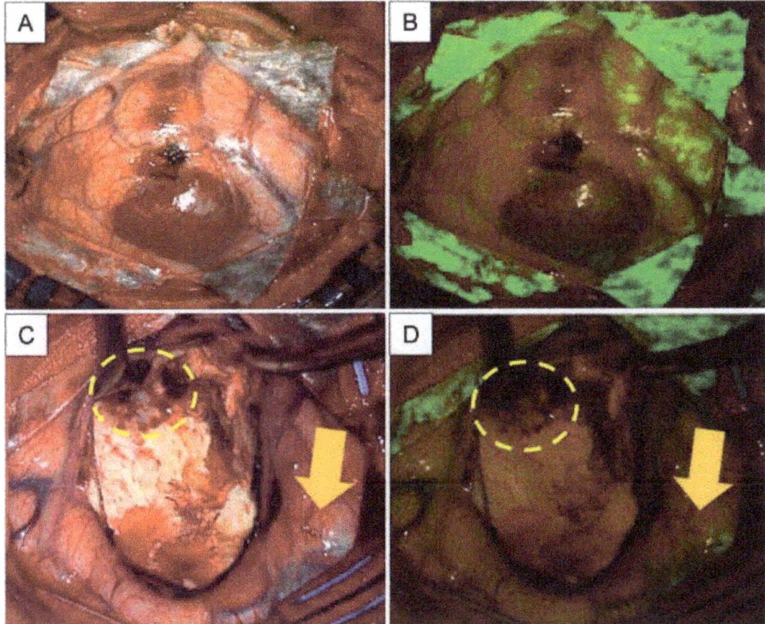

Figure 2. Intraoperative photos taken during resection of a parietal HGG. (**A**,**B**) Visualization of the tumor at the start of surgery under white light versus YELLOW-560 nm filter. Of note, there is good localization of the lobar tumor by the Na-Fl fluorescence. (**C**,**D**) Surgical cavity at the end of resection under white light versus YELLOW 560 nm filter. There was a small, greyish area that was uncertain for remnant tumor under white light (marked out by the yellow circle in dashed lines). However, it did not show fluorescence under the YELLOW-560 nm filter. Additionally, there was a superficial area of fluorescence (yellow arrows) that was outside of the resection margin that did not correlate with the preoperative MRI scans.

3.3. Illustrative Cases of Interest: Radiation-Induced Gliomas

Interestingly, there were two patients with secondary HGGs that did not demonstrate fluorescence in our study. The first case was a previously well female who was diagnosed with a Group 3 medulloblastoma when she was 12 years old. She underwent gross total resection of her tumor followed by adjuvant craniospinal irradiation with tumor bed boost (54 Gy in 30 fractions) and systemic chemotherapy (CHP Protocol). The patient remained in complete remission without radiological evidence of disease recurrence in her follow-up scans. However, she presented 7 years later with worsening headaches. Repeat MRI of her neuroaxis reported bilateral diffuse patchy T2-weighted hyperintensities in the periventricular and subcortical white matter of bilateral cerebral and cerebellar hemispheres, with marked diffuse cerebral oedema, increased prominence, and new long segments of abnormal signal in the cervical cord. Follow-up blood and CSF via lumbar puncture investigations for infective causes were unremarkable. The decision was made to perform a biopsy to ascertain diagnosis. The second case was a 14-year-old female with a background of posterior fossa ependymoma that underwent gross total resection, followed by adjuvant radiotherapy (54 Gy in 30 fractions) when she was 14 months old in an overseas institution. She was subsequently referred to us for continuity of care. Surveillance scans demonstrated an ill-defined right pontine lesion that was progressively enlarging. Similar to the previous case, a biopsy was recommended to exclude malignancy. See Figure 3 for the representative pictorials of both cases.

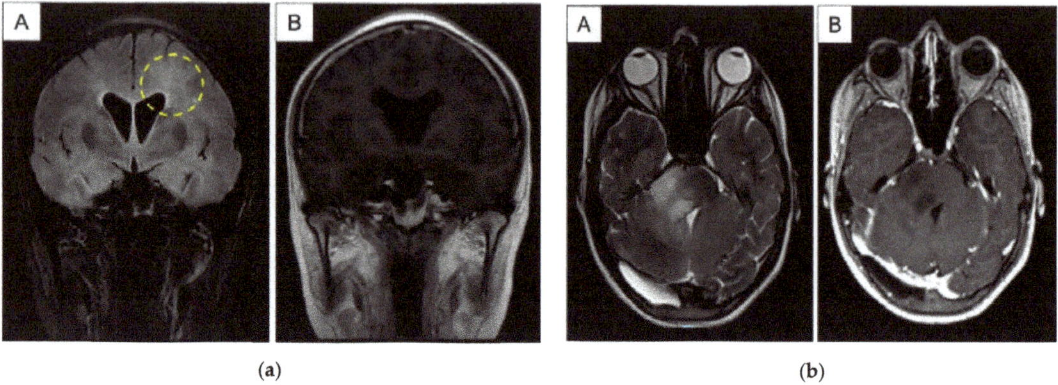

(a) (b)

Figure 3. (a) Representative MRI coronal images in fluid attenuated inversion recovery (FLAIR) and post-contrast T1-weighted sequences, labelled (**A**) and (**B**) respectively. The decision was made for the biopsy to be taken adjacent to the left frontal horn (marked out by yellow circle in dashed lines) where there were more FLAIR-attenuated changes. (b) Representative MRI axial images in T2-weighted and post-contrast T1-weighted sequences, labelled (**A**) and (**B**) respectively. The patient underwent stereotactic needle biopsy of the right pontine lesion.

4. Discussion

4.1. Surgery for Pediatric Brain Tumors: Technical Challenges

Pediatric brain tumors are a heterogenous group of neoplasms that differ biologically from their adult counterparts [19]. There is often variability of clinical presentations and considerable delays in their diagnoses [20]. Under such circumstances, the role of the pediatric neurosurgeon is paramount to either provide lifesaving intervention via cytoreduction if there is a significant mass lesion, or in event of deep-seated or multiple brain lesions, to perform a diagnostic brain biopsy [21]. Here, the technical challenges include firstly, ascertaining that the extent of resection is adequate and next, attaining accurate tissue sampling for diagnosis to guide the next step of treatment. Furthermore, current neuronavigation systems that rely on preoperative images often encounter the issue of brain shift, a recognized phenomenon that may cause inaccurate patient-to-image mapping as surgery progresses [22]. Other established adjuncts in pediatric neurosurgery, such as the intraoperative MRI and intraoperative ultrasound, have their own drawbacks, such as high costs, additional technical expertise and so forth [23].

Broadly speaking, the basis for fluorescence-guided surgery in children is largely extrapolated from what is demonstrated in adult malignant gliomas—that standard white light operating microscope may not be able to fully delineate intrinsic tumors from surrounding parenchyma [24]. Fluorescence enables resection to be effectively facilitated via demarcating tumor margins from normal brain tissue during surgery [25,26]. This important ability to delineate the brain–tumor interface and increase tumor visibility in real time circumvents the concerns of inevitable brain shift and can aid resection definitively [27]. Additionally, we are aware that non-diagnostic yields can be up to 13% in stereotactic brain needle biopsies [28]. However, recent studies report that ex vivo confirmation of Na-Fl fluorescence from biopsy samples has a high positive predictive value [12,21].

4.2. Fluorescent Dyes for Brain Tumor Surgery: An Overview

To date, there are a few intraoperative fluorescence dyes well established in adult brain tumor surgery. Examples include 5-aminolevulinic acid (5-ALA), indocyanine green (ICG) and Na-Fl [2,26,29]. Nonetheless, 5-ALA in children remains an off-label use. Several case reports on fluorescence-guided brain tumor surgery use in children have been published, yet no prospective study has been conducted [24,30]. Furthermore, previous studies have observed significant correlation between higher postoperative alanine aminotransferase values and younger age [31]. At this juncture, there is no available pharmacokinetic data for 5-ALA in children [24]. Overall, there is sparse data underlying the mechanisms of 5-ALA uptake in various types of brain tumors in children [31]. Following that, ICG is a well-studied water-soluble tricarbocyanine fluorescent dye [32]. Specifically in neoplasms, there is a difference in retention of ICG between tumor and normal tissue due to increased vascular permeability and impaired lymphatic drainage in the tumor microenvironment [33]. Consequently, small ICG molecules accumulate passively in the tumor due to this enhanced permeability and retention effect, thereby providing some tumor contrast for intraoperative real-time tumor recognition [32]. Its efficacy has been safely reported in children undergoing surgery for solid tumors [32]. However, similar to the issues faced for the use of 5-ALA in children, little is known about the applications of ICG and ideal dosing for pediatric brain tumor surgery.

Separately, Na-Fl is an organic fluorescent dye that has been safely used in humans for many years, especially in ophthalmology for retinal angiography [34,35]. It emits fluorescent radiation at wavelengths between 500 and 550 nm, with a peak excitation at 490 nm. It acts as a marker of blood–brain barrier damage by accumulating in extravascular spaces where the blood–brain barrier is disrupted [36–39]. Accordingly, adult-based Na-Fl studies report that high-grade glioma margins correlate well with gadolinium-based MRI results [36,40]. The key advantage is its utility in the immediate improved visualization of brain tumor tissue and thus improvement of the extent of resection [41–43]. Additionally, there are studies that incorporate the use of diffusion tensor imaging (DTI) with Na-Fl to facilitate the maximal resection of adult malignant gliomas in eloquent brain regions [44]. Although adequately established in adult brain tumors, the efficacy of Na-Fl in children remains equivocal [7]. In addition, the optimal dose of Na-Fl for brain tumor surgery in children should be determined, as higher doses may cause side effects such as staining of skin and mucosa and occasionally anaphylaxis [9,30,35]. Following that, the ideal timing of administration also needs to be clarified to avoid nonspecific extravasation of fluorescein into regions of perifocal oedema [17]. A balance must be struck between dosing that creates an increased proportion of unbound Na-Fl that is more likely to unspecifically stain a normal brain but more readily stain tumors [1]. Separately, some authors have also mentioned that fluorescence-guided resection techniques can be limited by the extent of vascular permeability in tumor regions, resulting in the failure to stain the full volume of tumor [1]. Nonetheless, there is promising evidence that Na-Fl is safe and versatile as a neurosurgical adjunct for different types of pediatric brain tumors [8].

As this is the first time Na-Fl is being used for this purpose in our institution, the decision to adopt a lower dose of 2 mg/kg of Na-Fl from the literature was for safety concerns [8,13]. We are cautious, as there have been reports of major adverse drug reactions (ADR) as the dose increases [45–47]. Given that the use of intravenous Na-Fl has not been previously reported in our part of the world, we are uncertain of its potential side effects on our patients. This is relevant as there is a known prevalence of atopic disorders, including drug allergies, in Singaporean children that may lead to life-threatening ADR [48,49]. Common side effects include of Na-Fl staining in the skin, sclera, mucous membrane and bodily fluids. These reactions are usually mild, self-limiting and dose-dependent [45]. Rarely, serious ADRs such as sickle crisis, hemolytic anemia, vasovagal reactions, and cardiac or respiratory compromise can occur, with possibly fatal consequences if there is no prompt intervention [50,51]. During the course of our study, we did not encounter any of the abovementioned ADRs in our patient cohort. To date, our findings with regards to its use in childhood brain tumors concur with previous studies in the literature [8–10,52].

4.3. Use of Intraoperative Na-Fl: Institutional Reflections

In congruency with relevant publications, we report that that Na-Fl is well-tolerated by our patient population [8,9,52,53]. During the study period, no adverse side effects or complications were encountered. As an additional adjunct, it can be seamlessly integrated into the current neurosurgical workflow. In addition, its use complements other neurosurgical adjuncts that are used simultaneously. Overall, Na-Fl has favorable points: firstly, it is easy to prepare and administer at the beginning of surgery; next, most lesions of interest fluoresce to guide intraoperative surgical decision-making; and finally, it can be used for most pediatric brain tumors recruited in our study. Separately, it can be used either as a standalone adjunct or in combination with others. However, we observe some limitations as an intraoperative adjunct. Firstly, our experience with the use of Na-Fl fluorescence towards the end of resection is reported to be equivocal for distinct brain tumor demarcation. In addition, both cases of radiation-induced HGGs do not show detectable fluorescence in the operated lesion. Here, we are uncertain if our study's dose of 2 mg/kg may be too low for some of the cases; and specifically for the two radiation-induced HGGs, if the tumors' underlying biology has any bearing on our observations. Hence, we believe that more patient data and further optimization of our current Na-Fl doses as part of our ongoing work are required in order to draw objective conclusions.

4.4. Study Critique and Future Directions

The authors acknowledge that there are limitations of this study that should be highlighted. First and foremost, it is limited by its retrospective design and modest sample size. In addition, the method to assess the degree of Na-Fl fluorescence is based on a qualitative evaluation that is subject to the individual neurosurgeon. We are aware this is an imperfect approach, with its innate bias. Nevertheless, there is currently no established scoring system to measure the extent of intraoperative Na-Fl in real time. To circumvent this issue, the use of an image analysis platform to quantify the degree of fluorescence objectively has been described by a recent study [52]. We are investigating the use of a similar platform as part of our ongoing work.

At this stage, areas that require optimization include the usage in children less than 3 years old and a suitable dose of Na-Fl in our local pediatric population. This is because our study's dosing regimen is currently more conservative that other reports in the literature [52]. Moving forward, there is certainly consideration to increase to a higher dose of Na-Fl prospectively for our study. In addition, we note there is variability in the types of neurosurgical procedures (i.e., burrhole versus craniotomy), anesthesia preparation after induction (such as setting of arterial and central lines, brain monitoring devices and so forth), setting up of neuronavigation and, for some cases, insertion of IONM probes. All of these contribute to differences in the timing between anesthesia induction and dura opening to confirm presence of Na-Fl fluorescence. At this point, we are still collecting data to determine the best timing between administration of Na-Fl and dura exposure. Nevertheless, this preliminary case series reports our initial experience with intraoperative Na-Fl is safe to be extrapolated to a larger cohort of our local pediatric patients for a prospective study. For now, we hope our findings add to the growing body of literature for the use of Na-Fl in childhood neurooncology, keeping in view the potential for future meta-analysis studies. In the meantime, we are cognizant that its implementation for routine clinical use remains a work in progress.

5. Conclusions

Overall, preliminary findings from our pilot study demonstrate that Na-Fl is safe and reliable in our local context. However, ongoing work needs to address its feasibility in children less than 3 years old and ascertain if the underlying biology of selected brain tumors affects its efficacy. Furthermore, the optimal dose of Na-Fl in children, timing of administration in relation to dura opening and objective assessment of the degree of intraoperative fluorescence are factors that need to be addressed as part of our future undertaking. In the meantime, we strongly advocate global, collaborative efforts in the continued development of good operative adjuncts for pediatric neurooncology.

Author Contributions: Conceptualization, S.Y.Y.L.; methodology, S.Y.Y.L.; validation, L.P.N., E.E.K.T. and S.Y.Y.L.; formal analysis, S.Y.Y.L., W.T.S., D.C.Y.L. and K.T.E.C.; investigation, A.J.L.T., M.L.T., R.X.W., W.S.L., D.C.Y.L. and E.E.K.T.; resources, D.C.Y.L., W.T.S., S.Y.Y.L., K.T.E.C., W.S.L., R.X.W. and E.E.K.T.; data curation, S.Y.Y.L.; writing—original draft preparation, A.J.L.T., M.L.T., D.C.Y.L., W.T.S., S.Y.Y.L., K.T.E.C., W.S.L., R.X.W. and E.E.K.T.; writing—review and editing, S.Y.Y.L.; supervision, S.Y.Y.L., D.C.Y.L. and W.T.S.; project administration, S.Y.Y.L. All authors have read and agreed to the published version of the manuscript.

Funding: This research received no external funding.

Institutional Review Board Statement: The study was conducted according to the guidelines of the Declaration of Helsinki and approved by the Institutional Review Board (or Ethics Committee) of SingHealth Centralized Institutional Review Board (protocol code Singhealth CIRB Ref: 2014/2079).

Informed Consent Statement: Informed consent to participate in this study was provided by the participants' legal guardian/next of kin.

Data Availability Statement: Not applicable.

Acknowledgments: Preliminary results of this study were presented at the 48th Annual Meeting of the International Society for Pediatric Neurosurgery (ISPN) held in Singapore from 6 to 10 December 2022.

Conflicts of Interest: The authors declare no conflict of interest.

References

1. Belykh, E.; Shaffer, K.V.; Lin, C.; Byvaltsev, V.A.; Preul, M.C.; Chen, L. Blood-Brain Barrier, Blood-Brain Tumor Barrier, and Fluorescence-Guided Neurosurgical Oncology: Delivering Optical Labels to Brain Tumors. *Front. Oncol.* **2020**, *10*, 739. [CrossRef] [PubMed]
2. Hadjipanayis, C.G.; Widhalm, G.; Stummer, W. What is the Surgical Benefit of Utilizing 5-Aminolevulinic Acid for Fluorescence-Guided Surgery of Malignant Gliomas? *Neurosurgery* **2015**, *77*, 663–673. [CrossRef] [PubMed]
3. Hohne, J.; Schebesch, K.M.; de Laurentis, C.; Akcakaya, M.O.; Pedersen, C.B.; Brawanski, A.; Poulsen, F.R.; Kiris, T.; Cavallo, C.; Broggi, M.; et al. Fluorescein Sodium in the Surgical Treatment of Recurrent Glioblastoma Multiforme. *World Neurosurg.* **2019**, *125*, e158–e164. [CrossRef] [PubMed]
4. Falco, J.; Cavallo, C.; Vetrano, I.G.; de Laurentis, C.; Siozos, L.; Schiariti, M.; Broggi, M.; Ferroli, P.; Acerbi, F. Fluorescein Application in Cranial and Spinal Tumors Enhancing at Preoperative MRI and Operated With a Dedicated Filter on the Surgical Microscope: Preliminary Results in 279 Patients Enrolled in the FLUOCERTUM Prospective Study. *Front. Surg.* **2019**, *6*, 49. [CrossRef] [PubMed]
5. Smith, E.J.; Gohil, K.; Thompson, C.M.; Naik, A.; Hassaneen, W. Fluorescein-Guided Resection of High Grade Gliomas: A Meta-Analysis. *World Neurosurg.* **2021**, *155*, 181–188.e7. [CrossRef]
6. Acerbi, F.; Broggi, M.; Schebesch, K.M.; Hohne, J.; Cavallo, C.; De Laurentis, C.; Eoli, M.; Anghileri, E.; Servida, M.; Boffano, C.; et al. Fluorescein-Guided Surgery for Resection of High-Grade Gliomas: A Multicentric Prospective Phase II Study (FLUOGLIO). *Clin. Cancer Res.* **2018**, *24*, 52–61. [CrossRef]
7. Erdman, C.M.; Christie, C.; Iqbal, M.O.; Mazzola, C.A.; Tomycz, L. The utilization of sodium fluorescein in pediatric brain stem gliomas: A case report and review of the literature. *Childs Nerv. Syst.* **2021**, *37*, 1753–1758. [CrossRef]
8. Goker, B.; Kiris, T. Sodium fluorescein-guided brain tumor surgery under the YELLOW-560-nm surgical microscope filter in pediatric age group: Feasibility and preliminary results. *Childs Nerv. Syst.* **2019**, *35*, 429–435. [CrossRef]
9. Almojuela, A.; Honey, C.M.; Gomez, A.; Hasen, M.; MacDonald, C.; Kazina, C.; Serletis, D. Using Fluorescein in the Resection of a Pediatric Posterior Fossa Tumor. *Can. J. Neurol. Sci.* **2020**, *47*, 578–580. [CrossRef]
10. Xue, Z.; Kong, L.; Pan, C.C.; Wu, Z.; Zhang, J.T.; Zhang, L.W. Fluorescein-Guided Surgery for Pediatric Brainstem Gliomas: Preliminary Study and Technical Notes. *J. Neurol. Surg. B Skull Base* **2018**, *79*, S340–S346. [CrossRef]

11. Singh, D.K.; Khan, K.A.; Singh, A.K.; Kaif, M.; Yadav, K.; Kumar Singh, R.; Ahmad, F. Fluorescein sodium fluorescence: Role in stereotactic brain biopsy. *Br. J. Neurosurg.* **2021**, 1–4. [CrossRef]

12. Catapano, G.; Sgulo, F.G.; Seneca, V.; Iorio, G.; de Notaris, M.; di Nuzzo, G. Fluorescein-assisted stereotactic needle biopsy of brain tumors: A single-center experience and systematic review. *Neurosurg. Rev.* **2019**, *42*, 309–318. [CrossRef]

13. Nevzati, E.; Chatain, G.P.; Hoffman, J.; Kleinschmidt-DeMasters, B.K.; Lillehei, K.O.; Ormond, D.R. Reliability of fluorescein-assisted stereotactic brain biopsies in predicting conclusive tissue diagnosis. *Acta Neurochir.* **2020**, *162*, 1941–1947. [CrossRef] [PubMed]

14. Louis, D.N.; Ohgaki, H.; Wiestler, O.D.; Cavenee, W.K.; Ellison, D.W.; Figarella-Branger, D.; Perry, A.; Reifenberger, G.; von Deimling, A. *WHO Classification of Tumours of the Central Nervous System*, 4th ed.; IARC: Lyon, France, 2016.

15. WHO Classification of Tumours Editorial Board (Ed.) *WHO Classification of Tumours, Central Nervous System Tumours*, 5th ed.; International Agency for Research on Cancer: Lyon, France, 2021; p. 568.

16. Barry, R.E.; Behrendt, W.A. Studies on the pharmacokinetics of fluorescein and its dilaurate ester under the conditions of the fluorescein dilaurate test. *Arzneimittelforschung* **1985**, *35*, 644–648. [PubMed]

17. Hohne, J.; Acerbi, F.; Falco, J.; Akcakaya, M.O.; Schmidt, N.O.; Kiris, T.; de Laurentis, C.; Ferroli, P.; Broggi, M.; Schebesch, K.M. Lighting Up the Tumor-Fluorescein-Guided Resection of Gangliogliomas. *J. Clin. Med.* **2020**, *9*, 405. [CrossRef] [PubMed]

18. Markosian, C.; Mazzola, C.A.; Tomycz, L.D. Sodium Fluorescein-Guided Gross Total Resection of Pediatric Exophytic Brainstem Glioma: 2-Dimensional Operative Video. *Oper. Neurosurg.* **2021**, *20*, E146–E147. [CrossRef]

19. Packer, R.J. Brain tumors in children. *Arch Neurol.* **1999**, *56*, 421–425. [CrossRef]

20. Dobrovoljac, M.; Hengartner, H.; Boltshauser, E.; Grotzer, M.A. Delay in the diagnosis of paediatric brain tumours. *Eur. J. Pediatr.* **2002**, *161*, 663–667. [CrossRef]

21. Akshulakov, S.K.; Kerimbayev, T.T.; Biryuchkov, M.Y.; Urunbayev, Y.A.; Farhadi, D.S.; Byvaltsev, V.A. Current Trends for Improving Safety of Stereotactic Brain Biopsies: Advanced Optical Methods for Vessel Avoidance and Tumor Detection. *Front. Oncol.* **2019**, *9*, 947. [CrossRef]

22. Gerard, I.J.; Kersten-Oertel, M.; Petrecca, K.; Sirhan, D.; Hall, J.A.; Collins, D.L. Brain shift in neuronavigation of brain tumors: A review. *Med. Image Anal.* **2017**, *35*, 403–420. [CrossRef]

23. Giussani, C.; Trezza, A.; Ricciuti, V.; Di Cristofori, A.; Held, A.; Isella, V.; Massimino, M. Intraoperative MRI versus intraoperative ultrasound in pediatric brain tumor surgery: Is expensive better than cheap? A review of the literature. *Childs Nerv. Syst.* **2022**, *38*, 1445–1454. [CrossRef] [PubMed]

24. Zhang, C.; Boop, F.A.; Ruge, J. The use of 5-aminolevulinic acid in resection of pediatric brain tumors: A critical review. *J. Neurooncol.* **2019**, *141*, 567–573. [CrossRef] [PubMed]

25. Berger, M.S. The fluorescein-guided technique. *Neurosurg. Focus* **2014**, *36*, E6. [CrossRef] [PubMed]

26. Schebesch, K.M.; Brawanski, A.; Hohenberger, C.; Hohne, J. Fluorescein Sodium-Guided Surgery of Malignant Brain Tumors: History, Current Concepts, and Future Project. *Turk Neurosurg.* **2016**, *26*, 185–194. [CrossRef]

27. Hollon, T.; Stummer, W.; Orringer, D.; Suero Molina, E. Surgical Adjuncts to Increase the Extent of Resection: Intraoperative MRI, Fluorescence, and Raman Histology. *Neurosurg Clin. N. Am.* **2019**, *30*, 65–74. [CrossRef]

28. Bander, E.D.; Jones, S.H.; Pisapia, D.; Magge, R.; Fine, H.; Schwartz, T.H.; Ramakrishna, R. Tubular brain tumor biopsy improves diagnostic yield for subcortical lesions. *J. Neurooncol.* **2019**, *141*, 121–129. [CrossRef]

29. Teng, C.W.; Huang, V.; Arguelles, G.R.; Zhou, C.; Cho, S.S.; Harmsen, S.; Lee, J.Y.K. Applications of indocyanine green in brain tumor surgery: Review of clinical evidence and emerging technologies. *Neurosurg. Focus* **2021**, *50*, E4. [CrossRef]

30. Labuschagne, J.J. The Use of 5-Aminolevulinic Acid to Assist Gross Total Resection of Paediatric Posterior Fossa Tumours. *Pediatr. Neurosurg.* **2020**, *55*, 268–279. [CrossRef]

31. Beez, T.; Sarikaya-Seiwert, S.; Steiger, H.J.; Hanggi, D. Fluorescence-guided surgery with 5-aminolevulinic acid for resection of brain tumors in children–a technical report. *Acta Neurochir.* **2014**, *156*, 597–604. [CrossRef]

32. Abdelhafeez, A.; Talbot, L.; Murphy, A.J.; Davidoff, A.M. Indocyanine Green-Guided Pediatric Tumor Resection: Approach, Utility, and Challenges. *Front. Pediatr.* **2021**, *9*, 689612. [CrossRef]

33. Jiang, J.X.; Keating, J.J.; Jesus, E.M.; Judy, R.P.; Madajewski, B.; Venegas, O.; Okusanya, O.T.; Singhal, S. Optimization of the enhanced permeability and retention effect for near-infrared imaging of solid tumors with indocyanine green. *Am. J. Nucl. Med. Mol. Imaging* **2015**, *5*, 390–400. [PubMed]

34. Kwiterovich, K.A.; Maguire, M.G.; Murphy, R.P.; Schachat, A.P.; Bressler, N.M.; Bressler, S.B.; Fine, S.L. Frequency of adverse systemic reactions after fluorescein angiography. Results of a prospective study. *Ophthalmology* **1991**, *98*, 1139–1142. [CrossRef] [PubMed]

35. Fung, T.H.; Muqit, M.M.; Mordant, D.J.; Smith, L.M.; Patel, C.K. Noncontact high-resolution ultra-wide-field oral fluorescein angiography in premature infants with retinopathy of prematurity. *JAMA Ophthalmol.* **2014**, *132*, 108–110. [CrossRef] [PubMed]

36. Diaz, R.J.; Dios, R.R.; Hattab, E.M.; Burrell, K.; Rakopoulos, P.; Sabha, N.; Hawkins, C.; Zadeh, G.; Rutka, J.T.; Cohen-Gadol, A.A. Study of the biodistribution of fluorescein in glioma-infiltrated mouse brain and histopathological correlation of intraoperative findings in high-grade gliomas resected under fluorescein fluorescence guidance. *J. Neurosurg.* **2015**, *122*, 1360–1369. [CrossRef]

37. Shinoda, J.; Yano, H.; Yoshimura, S.; Okumura, A.; Kaku, Y.; Iwama, T.; Sakai, N. Fluorescence-guided resection of glioblastoma multiforme by using high-dose fluorescein sodium. Technical note. *J. Neurosurg.* **2003**, *99*, 597–603. [CrossRef]

38. Acerbi, F.; Broggi, M.; Eoli, M.; Anghileri, E.; Cavallo, C.; Boffano, C.; Cordella, R.; Cuppini, L.; Pollo, B.; Schiariti, M.; et al. Is fluorescein-guided technique able to help in resection of high-grade gliomas? *Neurosurg. Focus* **2014**, *36*, E5. [CrossRef]
39. Acerbi, F.; Cavallo, C.; Broggi, M.; Cordella, R.; Anghileri, E.; Eoli, M.; Schiariti, M.; Broggi, G.; Ferroli, P. Fluorescein-guided surgery for malignant gliomas: A review. *Neurosurg. Rev.* **2014**, *37*, 547–557. [CrossRef]
40. Catapano, G.; Sgulo, F.G.; Seneca, V.; Lepore, G.; Columbano, L.; di Nuzzo, G. Fluorescein-Guided Surgery for High-Grade Glioma Resection: An Intraoperative "Contrast-Enhancer". *World Neurosurg.* **2017**, *104*, 239–247. [CrossRef]
41. Craig, S.E.L.; Wright, J.; Sloan, A.E.; Brady-Kalnay, S.M. Fluorescent-Guided Surgical Resection of Glioma with Targeted Molecular Imaging Agents: A Literature Review. *World Neurosurg.* **2016**, *90*, 154–163. [CrossRef] [PubMed]
42. Li, Y.; Rey-Dios, R.; Roberts, D.W.; Valdes, P.A.; Cohen-Gadol, A.A. Intraoperative fluorescence-guided resection of high-grade gliomas: A comparison of the present techniques and evolution of future strategies. *World Neurosurg.* **2014**, *82*, 175–185. [CrossRef] [PubMed]
43. Liu, J.T.; Meza, D.; Sanai, N. Trends in fluorescence image-guided surgery for gliomas. *Neurosurgery* **2014**, *75*, 61–71. [CrossRef] [PubMed]
44. Chen, D.; Li, X.; Zhu, X.; Wu, L.; Ma, S.; Yan, J.; Yan, D. Diffusion Tensor Imaging with Fluorescein Sodium Staining in the Resection of High-Grade Gliomas in Functional Brain Areas. *World Neurosurg.* **2019**, *124*, e595–e603. [CrossRef] [PubMed]
45. Restelli, F.; Bonomo, G.; Monti, E.; Broggi, G.; Acerbi, F.; Broggi, M. Safeness of sodium fluorescein administration in neurosurgery: Case-report of an erroneous very high-dose administration and review of the literature. *Brain Spine* **2022**, *2*, 101703. [CrossRef]
46. Dilek, O.; Ihsan, A.; Tulay, H. Anaphylactic reaction after fluorescein sodium administration during intracranial surgery. *J. Clin. Neurosci.* **2011**, *18*, 430–431. [CrossRef] [PubMed]
47. Moosbrugger, K.A.; Sheidow, T.G. Evaluation of the side effects and image quality during fluorescein angiography comparing 2 mL and 5 mL sodium fluorescein. *Can J Ophthalmol* **2008**, *43*, 571–575. [CrossRef]
48. Tan, V.A.; Gerez, I.F.; Van Bever, H.P. Prevalence of drug allergy in Singaporean children. *Singap. Med. J.* **2009**, *50*, 1158–1161.
49. Goh, D.Y.; Chew, F.T.; Quek, S.C.; Lee, B.W. Prevalence and severity of asthma, rhinitis, and eczema in Singapore schoolchildren. *Arch. Dis. Child.* **1996**, *74*, 131–135. [CrossRef]
50. O'Goshi, K.; Serup, J. Safety of sodium fluorescein for in vivo study of skin. *Skin Res. Technol.* **2006**, *12*, 155–161. [CrossRef]
51. Kwan, A.S.; Barry, C.; McAllister, I.L.; Constable, I. Fluorescein angiography and adverse drug reactions revisited: The Lions Eye experience. *Clin. Exp. Ophthalmol.* **2006**, *34*, 33–38. [CrossRef]
52. De Laurentis, C.; Bteich, F.; Beuriat, P.-A.; Almeida, L.C.A.; Combet, S.; Mottolese, C.; Vinchon, M.; Szathmari, A.; Di Rocco, F. Sodium fluorescein in pediatric oncological neurosurgery: A pilot study on 50 children. *Childs Nerv. Syst.* **2022**. [CrossRef]
53. Xue, Z.; Kong, L.; Hao, S.; Wang, Y.; Jia, G.; Wu, Z.; Jia, W.; Zhang, J.; Zhang, L. Combined Application of Sodium Fluorescein and Neuronavigation Techniques in the Resection of Brain Gliomas. *Front. Neurol.* **2021**, *12*, 747072. [CrossRef] [PubMed]

Perspective

The Morphospace of Consciousness: Three Kinds of Complexity for Minds and Machines

Xerxes D. Arsiwalla [1,*], Ricard Solé [2,3,4,5], Clément Moulin-Frier [6], Ivan Herreros [1], Martí Sánchez-Fibla [1] and Paul Verschure [7]

1 Department of Information and Communication Technologies, Universitat Pompeu Fabra (UPF), 08018 Barcelona, Spain
2 Complex Systems Lab, Universitat Pompeu Fabra, 08003 Barcelona, Spain
3 Institut de Biologia Evolutiva (CSIC-UPF), 08003 Barcelona, Spain
4 Santa Fe Institute, Santa Fe, NM 87501, USA
5 Institució Catalana de Recerca i Estudis Avançats (ICREA), 08010 Barcelona, Spain
6 Flowers Research Group, Inria Bordeaux, 33405 Talence, France
7 Donders Institute for Brain, Cognition and Behavior, Radboud University, 6525 GD Nijmegen, The Netherlands
* Correspondence: x.d.arsiwalla@gmail.com

Abstract: In this perspective article, we show that a morphospace, based on information-theoretic measures, can be a useful construct for comparing biological agents with artificial intelligence (AI) systems. The axes of this space label three kinds of complexity: (i) autonomic, (ii) computational and (iii) social complexity. On this space, we map biological agents such as bacteria, bees, C. elegans, primates and humans; as well as AI technologies such as deep neural networks, multi-agent bots, social robots, Siri and Watson. A complexity-based conceptualization provides a useful framework for identifying defining features and classes of conscious and intelligent systems. Starting with cognitive and clinical metrics of consciousness that assess awareness and wakefulness, we ask how AI and synthetically engineered life-forms would measure on homologous metrics. We argue that awareness and wakefulness stem from computational and autonomic complexity. Furthermore, tapping insights from cognitive robotics, we examine the functional role of consciousness in the context of evolutionary games. This points to a third kind of complexity for describing consciousness, namely, social complexity. Based on these metrics, our morphospace suggests the possibility of additional types of consciousness other than biological; namely, synthetic, group-based and simulated. This space provides a common conceptual framework for comparing traits and highlighting design principles of minds and machines.

Keywords: consciousness; brain networks; artificial intelligence; synthetic biology; cognitive robotics; complex systems

Citation: Arsiwalla, X.D.; Solé, R.; Moulin-Frier, C.; Herreros, I.; Sánchez-Fibla, M.; Verschure, P. The Morphospace of Consciousness: Three Kinds of Complexity for Minds and Machines. *NeuroSci* **2023**, 4, 79–102. https://doi.org/10.3390/neurosci4020009

Academic Editors: François Ichas and Jonathan Foster

Received: 14 November 2022
Revised: 14 February 2023
Accepted: 20 March 2023
Published: 27 March 2023

1. Introduction

How do we build a taxonomy of consciousness based on evidence from clinical neuroscience, synthetic biology, artificial intelligence (AI) and cognitive robotics? Here, we examine current biologically motivated metrics of consciousness. In view of these metrics, we show how contemporary AI and synthetic systems measure on homologous scales. In what follows, we refer to a phenomenological description of consciousness. In other words, that which can be described in epistemically objective terms, even though aspects of the problem of consciousness may require an ontologically subjective description. Drawing from what is known about the phenomenology of consciousness in biological systems, we build a homologous argument for artificial, collective and simulated systems. For example, in clinical diagnosis of disorders of consciousness, two widely used scales are patient awareness and wakefulness (also referred to as arousal), both of which can be assessed using neurophysiological recordings [1,2]. We extrapolate these metrics to construct a morphospace of consciousness.

The origin of the concept of a morphospace comes from comparative anatomy and paleobiology, where quantitative measures or principal components from clustering methods allow classification in a metric-like space. However, it can also involve a qualitative relation-based approach, as the one we will follow here. A related concept of the so-called *theoretical morphospace*, has also been defined in formal terms, as an *N*-dimensional geometric hyperspace produced by systematically varying the parameter values associated to a given (usually geometric) set of traits [3]. More recently, morphospaces have been used in the study of complex systems, linguistics and biology [4–6]. Hence, a morphospace serves as a useful tool to gain insights on design principles and evolutionary constraints, when looking across a large class of systems (or species) that display complex variations in traits. For the problem of consciousness, we construct this morphospace based on three kinds of complexity. These considerations suggest an embodiment-based taxonomy of consciousness [7].

For practical reasons, many experimental paradigms testing consciousness are designed for humans or higher-order primates (see [8–10] for an overview of the field). In this article, we argue that metrics commonly associated to biological consciousness can be appropriately extrapolated for conceptualizing behaviors of synthetic and artificial intelligence systems. This is insightful not only for understanding parallels between biological and potential synthetic consciousness, but more importantly for unearthing design principles necessary for building biomimetic technology that could potentially acquire consciousness. As evidenced by several historical precedents, bio-inspired design thinking has been at the core of some of the greatest scientific breakthroughs. For instance, early attempts at aviation in the 19th century were inspired by studying flight mechanics in birds and insects (the term aviation itself is derived from the Latin "avis" for "bird"). In fact, biological flight mechanisms are so sophisticated that their biomimetic implementations are still being actively studied within the field of soft robotics [11]. However, it so happened that rather than coming around to mimicking nature exactly, humanity learnt the basic laws of aerodynamics based on observations from nature and looked for other embodiments of those principles. This in fact, led to the invention of the modern aircraft by the Wright brothers in 1903, leading to a completely new way to build machines that fly than those that exactly mimic nature.

Metrics of consciousness are also the right tools to quantitatively study how human and animal intelligence differs from state-of-the-art machine intelligence. Once again, it is instructive to take a historical perspective on human intelligence as laid out by one of the founders of AI, Allen Newell. In 1994, in his seminal work, "Unified Theories of Cognition" [12]. Newell proposed the following thirteen criteria necessary for building human-level cognitive architectures:

- Behave flexibly as a function of the environment
- Exhibit adaptive (rational, goal-oriented) behavior
- Operate in real-time
- Operate in a rich, complex, detailed environment (that is, perceive an immense amount of changing detail, use vast amounts of knowledge, and control a motor system of many degrees of freedom)
- Use symbols and abstractions
- Use language, both natural and artificial
- Learn from the environment and from experience
- Acquire capabilities through development
- Operate autonomously, but within a social community
- Be self-aware and have a sense of self
- Be realizable as a neural system
- Be constructible by an embryological growth process
- Arise through evolution

Current AI architectures still do not meet all these criteria. On the other hand, although Newell did not discuss consciousness back then, the above criteria are equally relevant to the problem of consciousness. Given current advances in our understanding of neural and cognitive mechanisms of consciousness [9], one may well argue that the problem of consciousness supersedes and even subsumes the problem of biological intelligence. While Newell's criteria list signatures that are the consequence of human intelligence, for consciousness it is more useful to have a list of functional criteria that underlie the process of consciousness. Below, we will discuss such functional criteria.

Why are these considerations relevant for understanding the direction of today's AI and the development of new technologies? The field of AI, and particularly neural networks, began as a modest attempt to understand cognition and brain function. It dates back to the 1930s with the first model of neural networks by Nicolas Rashevsky [13], followed by the seminal work of Walter Pitts and Warren McCulloch in 1943 [14] and Frank Rosenblatt's perceptron in 1958 [15]. Eventually, with the use of analytic methods from statistical physics, those simple models paved the way to understanding associative memory and other emergent cognitive phenomena [16]. Even though artificial neural networks did not quite solve the problem of how the brain works, they led to the discovery of brain-inspired computing technologies such as deep learning systems and powerful technologies for computational intelligence such as IBM's Watson. These machines process massive volumes of data and are built for intensive computational tasks that the brain may not even be designed for. Yet, in spite of these computational successes, contemporary AI is still challenged in many tasks that human and animal brains seem to perform effortlessly. For that reason, the next frontier in AI and machine intelligence will be closely tied to our advances in understanding the governing principles of consciousness and its various embodiments. This potentially has a bearing on the development of next-generation biomimetic and sentient technologies. Recent work in this direction can be found in [7,17–19].

2. Biological Consciousness: Insights from Clinical Neuroscience

We begin by reviewing clinical scales used for assessing consciousness in patients with disorders of consciousness. In subsequent sections, we will generalize the complexity measures pertinent to these biological scales and discuss how synthetic systems can be measured on these.

2.1. Clinical Consciousness and its Disorders

In patients with disorders of consciousness ranging from coma, locked-in syndrome to those in vegetative states, levels of consciousness are assessed through a battery of behavioral tests as well as physiological recordings. Cognitive awareness in patients is assessed by testing cognitive functions using behavioral and neurophysiological (fMRI or EEG) protocols [1]. Assessments of wakefulness/arousal in patients are based on metabolic markers (in cases where reporting is not possible) such as glucose uptake in the brain, captured using PET scans. More generally, in [1,2], awareness and wakefulness have been proposed as a two-dimensional operational definition of clinical consciousness, shown in Figure 1 below. While awareness concerns higher and lower-order cognitive functions enabling complex behavior; wakefulness results from biochemical homeostatic mechanisms regulating survival drives and is clinically measured in terms of glucose metabolism in the brain. In fact, in all known organic life forms, biochemical arousal is a necessary precursor supporting the hardware necessary for cognition. In turn, evolution has shaped brains in such a way so as to support the organism's basic survival (using wakefulness/arousal) as well as higher-order drives (using awareness) associated to cooperation and competition in a multi-agent world [20]. Awareness and wakefulness thus taken together, form the clinical markers of consciousness.

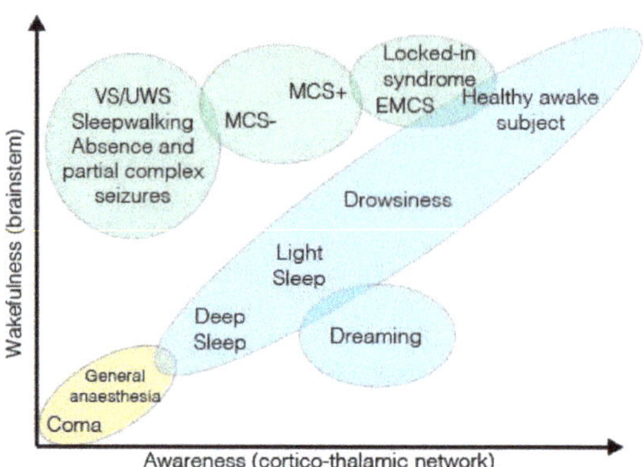

Figure 1. Clinical scales of consciousness. A clustering of disorders of consciousness in humans represented on scales of awareness and wakefulness. Adapted from [2]. In neurophysiological recordings, signatures of awareness have been found in cortico-thalamic activity, whereas wakefulness corresponds to activity in the brainstem and associated systems [1,2]. Abbreviated legends: VS/UWS (vegetative state/unresponsive wakefulness state) [21]; MCS(+/−) (minimally conscious state plus/minus), EMCS (emergence from minimally conscious state) [22].

The clinical definition (or criterion) of consciousness enables a practical classification of closely associated states and disorders of consciousness into clusters on a bivariate scale with awareness and wakefulness on orthogonal axes. Under healthy conditions, these two levels are almost linearly correlated, as in conscious wakefulness (high arousal and high awareness) or in deep sleep (low arousal and low awareness). However, in pathological states, wakefulness without awareness can be observed in the vegetative state [1], while transiently reduced awareness is observed following seizures [23]. Patients in the minimally conscious state show intermittent and limited non-reflexive and purposeful behavior [24,25], whereas patients with hemispatial neglect display reduced awareness of stimuli contralateral to the side where brain damage has occurred [26].

The question is, how can one extrapolate wakefulness and awareness for non-biological systems in order to obtain homologous scales of consciousness that can be applied to artificial systems? As noted above, wakefulness results from autonomous homeostatic mechanisms necessary for the self-preservation of an organism's germ line in a given environment. In other words, it is tied to self-sustaining life processes necessary for basic survival, whereas awareness refers to functionalities pertaining to estimating or predicting states of the world and optimizing the agent's own actions with respect to those states. If biological consciousness as we know it, is supported via a synergistic interaction between metabolic and cognitive processes, then how should this insight be extended to conceive a functional notion of consciousness in synthetic systems? One way of doing so might be generalizing from scales of wakefulness to those of generic autonomic processes; likewise, generalizing from scales of awareness to those of generic computational processes. As such, most autonomic processes are usually considered to be running below the radar of consciousness (or unconscious in certain usages). On the other hand, computational processes provide for mechanistic descriptions for many neural and cognitive functions associated to consciousness. However, as evident from the examples above (including those of disorders of consciousness), biological forms of consciousness seem to require both types of processes. For that reason, we will consider both, autonomic and computational processes when formulating homologous scales of consciousness in synthetic systems.

2.2. Candidate Measures in Brain and Behavioral Studies

Given the above discussion of clinical scales of consciousness (based on wakefulness and awareness), in Section 5, we will attempt to identify the homologues of these biological scales for artificial systems. As a precursor to that discussion, in this subsection we will review prominent candidate measures of autonomy and computation in brain and behavioral sciences.

Measures of autonomy and computation, including information processing performed by cognitive agents, have been discussed in various psychometric [27] and neurophysiological studies [28]. However, generalizing these measures to artificial systems and comparing those values to biological systems is certainly not so straightforward (due to completely different processing substrates as well as differing comparative benchmarks). Nonetheless, biological/cognitive measures of autonomy and computation suggest a first step in this direction. For example, [27] introduced an "Index of Autonomous Functioning", tested on healthy human subjects (via psychometric questionnaires). This index aims to assess psychological ownership, interest-taking and susceptibility to external controls. This is similar to the concept of volition (or agency), introduced in the cognitive neurosciences [29,30], which seeks to determine the neural correlates of self-regulation, referring to actions regulated by internal drives rather than exclusively driven by external contingencies.

Psychometric attempts to quantify awareness have been discussed in [31] in the context of a unified psychological theory of self-functioning. In consciousness research, a measure of awareness that has gained a lot of traction is integrated information [32] (often denoted as Φ). This is an information-theoretic complexity measure. It was first introduced as a measure that might be useful for realistic neural data. Based on mutual information, Φ has been touted as a correlate of consciousness [32]. Integrated information is loosely defined as the quantity of information generated by a network as a whole, due to its causal dynamical interactions, over and above the information generated independently by the disjoint sum of its parts. As a complexity measure, Φ seeks to operationalize the intuition that complexity arises from simultaneous integration and differentiation of the network's structure and dynamics, thus enabling the emergence of the system's collective states. The interplay between integration and differentiation generates information that is highly diversified yet integrated, creating patterns of high complexity. Following initial proposals [32–34], several approaches have been developed to compute integrated information [35–50].

Notably the measure discussed in [38] will be relevant for our discussion as it develops methods for large-scale network computations of integrated information, applied to the human brain's connectome network. The human connectome network consists of structural connectivity of white matter fiber tracts in the cerebral cortex, extracted using diffusion spectrum imaging and tractography [51,52] (see [53–58] for visualization of neurodynamical data and model dynamics on this network). Compared to a randomly re-wired network, it was seen that the particular topology of the human brain generates greater information complexity for all allowed couplings associated to the network's attractor states, as well as its non-stationary dynamical states [38].

Φ as described above, is not specific to biological systems and can also be applied to artificial dynamical systems. In Section 5 we will exploit the applicability of Φ and use it as a generic measure of computational complexity for artificial systems.

3. Synthetic Consciousness? Insights from Synthetic Biology and Artificial Intelligence

Let us now look at the evidence in synthetic biology and AI to see how these systems qualitatively compare to biological systems. Oftentimes, our methods for probing biological systems can be limited due to natural design constraints. On the other hand, the potential for exploring synthetic counterparts provides a unique opportunity to probe the nature of life and intelligence processes. It has been suggested that artificial simulations, in silico implementations and engineered alternatives are in fact, much needed for understanding the origins of evolutionary dynamics, including cognitive transitions [59]. What can be learned in relation to consciousness from artificial agents?

Within the context of non-cognitive phenomena, synthetic biology provides a valuable example of the classes of relevant questions that can be answered. Examples are the possibility of creating living systems from non-living chemistry, generating multicellular assemblies, creating synthetic organoids or even artificial immune systems. Here advanced genetic engineering techniques along with a systems view of biology has been able to move beyond standard design principles provided by evolution. Examples of this are new genetic codes with extra genetic letters in the alphabet that have been designed and successfully inherited [60], synthetic protocells with replicative potential [61] and even whole synthetic chromosomes that have defined a novel bacterium species [62]. Ongoing research has also revealed the potential for creating cognitive networks of interacting microorganisms capable of displaying collective intelligence [63].

Of course, the criteria for consciousness, as stated in sections above, are not even remotely satisfied by any of these synthetic systems. They either have some limited form of intelligence or life but not yet both. Nevertheless, there have been some noteworthy recent developments in these areas. In this context, AlphaGo's 2016 feat in beating the top human Go champion was remarkable for a couple of reasons. Unlike Chess, possible combinations in Go run into the millions and when played using a timer, any brute-force algorithm trying to scan the entire search space would simply run out of computational capacity or time. Hence, an efficient pattern recognition algorithm was crucial to the development of AlphaGo, where using deep reinforcement learning, the system was trained on a large number of games [64]. Most interestingly, it played counterintuitive moves that shocked the best human players and the sole game of the series that Lee Sedol, the human champion, won out of five, was only possible after he himself adopted a brilliant counterintuitive strategy. Subsequent AI systems such as AlphaGo Zero, AlphaZero and MuZero have gone even further. While AlphaGo learnt the game by playing thousands of matches with amateur and professional players, AlphaGo Zero learnt by playing against itself over and over again, starting from completely random play, while reinforcing successful sequence of plays through the weights of its deep neural networks [64,65]. This aspect of playing itself is akin to training via social interactions as will be described below. Then we have AlphaZero, which is a single system that taught itself from scratch how to master the games of chess, shogi and Go [66]. And MuZero takes these ideas one step further. It matches the performance of AlphaZero on Go, chess and shogi, while also mastering a range of visually complex Atari games, all without being told the rules of any game [67].

Thus, AI such as AlphaGo and its successors do demonstrate a rather broad form of domain intelligence (that is within a game or across games). In contrast, most forms of biological problem-solving capabilities span across domains (related to ecologically-realistic constraints). Moreover, one would agree that AlphaGo is not equipped with any form of wakefulness mechanisms coupled to its computational capabilities [68].

The same can be said for other state-of-the-art AI systems including deep convolutional neural networks, or deep recurrent networks. Both these latter architectures were inspired from Hubel and Weisel's groundbreaking work on the coding properties of the visual system, which led to the realization of a hierarchical processing architecture [69]. Today deep convolutional networks are widely used for image classification [70] and recurrent neural networks for speech recognition [71], among countless other applications. Recent developments have advanced this by virtue of larger data sets and more computational power. For example, there have been attempts to build biologically-plausible models of learning in the visual cortex using recurrent neural networks [72]. In summary, deep architectures have made remarkable progress in domain-specific AI.

However, asking whether AI can be conscious in exactly the same way that a human is, is similar to asking whether a submarine can swim. Even if it did so, it might well do so differently. If the goal of a system is to learn and solve complex tasks close to human performance or better, current machines are already doing that in specific domains [73–79]. However, these machines are still far from learning and solving problems in generic domains and more importantly, in ways that would couple its problem-solving capabilities to its autonomous survival drives. On the other end of things, neither have any of the synthetic life systems discussed above been used to build architectures with complex computing or cognitive capabilities. Nevertheless, this does suggest that a future synthesis between artificial life forms and AI could be evaluated using homologous scales of consciousness to the ones currently applicable to biological beings. This plausible form of synthetic consciousness, if based on a life form with different survival drives and mechanisms, along with non-human forms of intelligence or computation, would also likely lead to non-human behavioral outcomes.

In summary, these phenomenological considerations suggest that autonomic and computational complexity provide the necessary abstractions to wakefulness and awareness, which can be applied to a wide spectrum of synthetic agents in terms of their underlying mechanistic processes. In the next section, we will make the case for a third kind of complexity, necessary to build the morphospace of consciousness, namely, social complexity.

4. The Function of Consciousness: Insights from Evolutionary Game Theory and Cognitive Robotics

Reviewing insights from evolution and cognitive robotics, this section looks at the functional role of consciousness [20,80–83]. The biological substrates of consciousness presumably evolved through natural selection driven by social co-operation and competition. This can be framed in the context of evolutionary game-theory. In [80,84] it was suggested that rather than being thought of as a problem, consciousness could rather be seen as a solution to the problem of autonomous goal-oriented action, when faced with a world filled with other agents. This was formulated as the *H5W problem*.

4.1. The H5W Problem

What does an agent operating in a social world need to do in order to optimize its fitness? It needs to perceive the world, to act, and through time, to understand the consequences of its actions so it can start to reason about its goals and how to achieve them. This requires building a representation of the world grounded on the agent's own sensorimotor history and use that to reason and act. The agent will witness a scene of agents, including itself, and objects interacting in various manners, times and places. This comprises the six fundamental problems that the agent is faced with, together referred to as the H5W problem [80,84]: In order to act in the physical world an agent needs to determine a behavioral procedure to achieve a goal state; that is, it has to answer the HOW of action. In turn this requires the agent to: (a) Define the motivation for action in terms of its needs, drives and goals, that is, the WHY of action; (b) Determine knowledge of objects it needs to act upon and their affordances in the world, pertaining to the above goals, that is, the WHAT of action; (c) Determine the location of these objects, the spatial configuration of the task domain and the location of the self, that is, the WHERE of action; (d) Determine the sequencing and timing of action relative to dynamics of the world and self, that is, the WHEN of action; and (e) Estimate hidden mental states of other agents when action requires cooperation or competition, that is, the WHO of action.

While the first four of the above questions suffice for generating simple goal-oriented behaviors, the last of the Ws (the WHO) is of particular significance as it involves intentionality, in the sense of estimating the future course of action of other agents based on their social behaviors and psychological states. However, because mental states of other agents that are predictive of their actions are hidden, these can at best be inferred from incomplete sensory data such as location, posture, vocalization, social salience, etc. As a result the acting agent faces the challenge to univocally assess, in a deluge of sensory data those exteroceptive and interoceptive states that are relevant to ongoing and future action and therefore has to deal with the ensuing credit assignment problem in order to optimize its own actions. Furthermore, this results in a reciprocity of behavioral dynamics, where the agent is now acting on a social and dynamical world that is in turn acting upon itself. It was proposed in [84] that consciousness is associated to the ability of an agent to maintain a transient and autonomous memory of the virtualized agent-environment interaction, that captures the hidden states of the external world, in particular, the intentional states of other agents and the norms that they implicitly convey through their actions.

4.2. Evolutionary Game Theory

From the above, we surmise that the function of consciousness is to enable an acting agent to solve its H5W problem while being engaged in social cooperation and competition with other agents in its eviroment, who are trying to solve their own H5W problem in a world with limited resources [80,81]. This brings our discussion to the setting of evolutionary game theory.

First, consider a scenario with only a small number of other agents. Then any given agent might use statistical learning approaches to learn and classify behaviors of the few others agents in that game. For example, multiple robots interacting to learn naming conventions of perceptual aspects of the world [85]. At the least, this requires embodiment so that agents can interpret perceptual cues presented by other agents (for example, by pointing at objects) [86]. Another example is the emergence of signaling languages in sender-receiver games based on replicator dynamics described by David Lewis in 1969 in his seminal work, Convention [87,88]. These are all examples where social norms are acquired in the process of iterative multi-agent interactions, and can thus be investigated in the setting of multi-agent game theory using evolutionary algorithms.

Note however, that most game-theoretic strategies involving statistical learning are computationally feasible only when a limited number of players are involved. They are often sub-optimal in the event of an explosion in the number of players (see [89] for an overview of these limits). Likewise, in a social environment comprising a large number of agents trying to solve the H5W problem, machine learning strategies for reward and punishment valuations may soon become computationally unfeasible for an agent's processing capacities and memory storage. Therefore, for a large population to sustain itself in an evolutionary game involving complex forms of cooperation and competition would require strategies other than merely data-driven statistical learning. One such strategy involves modeling and inferring intentional states of the self and that of other agents. Emotion-driven flight or fight responses depend on such intentional inferences and so do higher-order psychological drives. The mechanisms of consciousness enable such strategies, whereas, contemporary AI systems such as AlphaGo do not possess such capabilities.

The importance of the role that sociality plays in surviving a multi-agent world suggests a possible function of consciousness: it is a mechanism that enables agents to learn and acquire complex social game-theoretic strategies based on emotional cues. From an evolutionary perspective, social behaviors result from generations of cooperation–competition games, with natural selection filtering out unfavorable strategies. Presumably, winning strategies were eventually encoded as anatomical mechanisms, such as emotional responses. The complexity of these behaviors depends on the ability of an agent to make complex social inferences.

While evolutionary game theory itself does not hinge on consciousness (and there are plenty of examples of emergent behaviors acquired in iterative multi-agent games involving reciprocating agents driven by cultural cues, where the presence or need for consciousness does not arise [90]), nor is consciousness the end-product of all evolutionary games; the key point we wish to emphasize here is that the mechanisms of biological consciousness, which allow organisms to have highly flexible autonomous action and cognitive processing capabilities, provide a competitive advantage to agents operating in complex social environments. Based on fossil records, the evolutionary and genetic origins of consciousness have been traced back to the Cambrian Period over 500 million years ago [91], when early vertebrates with somatotopically-organized neural representations acquired sensory capabilities (with vision being postulated as the first conscious sense [91]). These early markers of sensory consciousness enabled agents to navigate complex social scenarios (without having to rely exclusively on extensive computational resources).

All of this discussion suggests a third dimension in the morphospace of consciousness (see Figures 2 and 3 below), namely, social complexity, which serves as a measure of an agent's social intelligence.

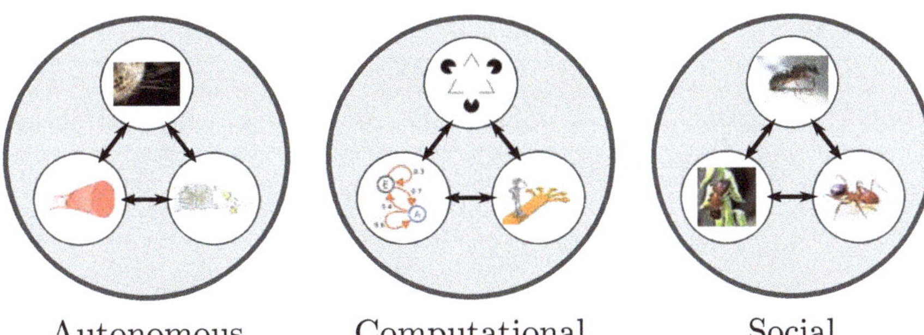

Autonomous   Computational   Social

Figure 2. Schematic representation of autonomic, computational and social complexity. Each complexity measure is illustrated as a whole (the large circles) constituted of its parts (the inner circles), their interactions (the arrows) and the emerging properties resulting from these interactions (the inner space within the large circles, in light grey). Autonomic complexity (left) refers to the collective phenomena resulting from the interactions between typical components of reactive behavior such as sensors (illustrated by whiskers in the top inner circle), actuators (illustrated by a muscle in the bottom-left inner circle) and low-level sensorimotor coupling (illustrated by a spinal cord in the bottom-right inner circle). Computational complexity is associated to higher-level cognitive processes such as visual perception (top inner circle), planning (bottom-left inner circle) or decision making (bottom-right inner circle). Social complexity is associated to interactions between individuals of a population, such as a queen ant (top inner circle), a worker ant (bottom-left inner circle) and a soldier ant (bottom-right inner circle).

5. Three Kinds of Complexity to Characterize Consciousness

In this section we discuss the construction of the aforementioned morphospace as well as candidate complexity measures to parametrize it.

5.1. Why Distinguish Between Complexity?

The phenomenology of consciousness draws upon a variety of empirical disciplines including cognitive and clinical neuroscience, synthetic biology, artificial intelligence, evolutionary biology and robotics. The theoretical challenge is then to find a formal explanatory framework that provides an abstraction of the phenomenology across substrates. One attempt at doing so has been through complexity measures. However, based on the evidence discussed above, a universal complexity measure may be insufficient to parse through the types of process and functional specifications supporting consciousness. This point has also been mentioned in [92], albeit from purely clinical considerations. Hence, in this work, we make the case for at least three kinds of complexity, based on process types. These are autonomic complexity, computational complexity and social complexity (see Figure 2 and Table 1 below). Table 1 lists the respective building blocks, systems-level realizations, and associated emergent phenomena for each of these complexity kinds.

A common type of complexity measure, that is often discussed in neuroscientific and consciousness-related paradigms, is a whole-versus-parts measure. Here, a system's complexity C is defined by how much an integrated whole outdoes the sum of its independent parts in terms of an information processing metric. Generally, $C = \mathcal{I}_{substrate} - \sum_{\{parts\}} \mathcal{I}_{part}$, where $\mathcal{I}$ refers to an appropriate type of information. For instance, when $\mathcal{I}$ is the conditional entropy, C yields the measure Φ of integrated information theory and its many derivatives. Besides whole-versus-parts measures, there are a host of others which capture different aspects of information processing in complex systems. Below, we discuss the relevance of each of these measures with respect to the three kinds of complexity proposed in this article and how they may be collectively used for labelling states of consciousness.

Autonomic complexity $C_{Autonomic}$ measures the complexity of processes enabling the system to act autonomously in its environment. In eukaryotes, autonomic action refers to arousal mechanisms resulting from coordinated nervous system activity; in prokaryotes, this refers to reactive behaviors such as chemotaxis, stress responses to temperature, toxins, mechanical damage, etc., all of these resulting from coordinated cellular signaling processes; in robotics, autonomic systems refer to homeostatic mechanisms driving reactive behaviors. Hence, autonomic complexity quantifies information processing by the collective dynamics of the systems driving autonomous behaviors.

Table 1. Presented in the table below is a classification of three kinds of complexity relevant for charting a taxonomy of consciousness, namely, autonomic, computational and social complexity. This classification is based on the respective building blocks or substrates of each complexity kind, the systems-level realizations of these substrates, and their associated emergent phenomena.

	$\mathcal{C}_{Autonomic}$	$\mathcal{C}_{Computational}$	$\mathcal{C}_{Social}$
Building Blocks	Sensors, Actuators	Neurons, Transistors	Individual Agents
Systems-Level Realizations	Prokaryotes, Autonomic Nervous System, Bots	Cognitive Systems, Brains, Microprocessors	Population of Agents, Social Organizations
Emergent Phenomena	Self-Regulated Real-Time Behavior	Problem Solving Capabilities	Signaling Conventions, Language, Social Norms, Arts, Science, Culture

On the other hand, computational complexity $\mathcal{C}_{Computational}$ refers to the ability of an agent to integrate information over space and time across computational or cognitive tasks. In higher biological forms, this complexity is typically associated to neural processes; in artificial computational systems, it refers to microprocessor signaling. The distinction between $\mathcal{C}_{Computational}$ and $\mathcal{C}_{Autonomic}$ is specified by the tasks that they refer to, rather than substrates. $\mathcal{C}_{Autonomic}$ refers to autonomous control loops, whereas $\mathcal{C}_{Computational}$ refers to computational and inferential mechanisms.

In whole-versus-parts terminology, social complexity $\mathcal{C}_{Social}$ would refer to information generated by a population as a whole, during the course of social interactions, over the information generated additively by individual agents of that population. Unlike $\mathcal{C}_{Autonomic}$ or $\mathcal{C}_{Computational}$, $\mathcal{C}_{Social}$ is not assigned to an individual, but rather to a specific population (its own species) within which the individual has been interacting. As discussed earlier, by way of social games, these interactions are believed to have contributed to the emergence of the agent's consciousness on an evolutionary time-scale. Note that $\mathcal{C}_{Social}$ as defined here, does not refer to group consciousness (we will discuss that in following sections), rather it quantifies the environmental complexity due to a population of agents (this in turn, applies selection pressures on individual agents).

5.2. Candidate Complexity Measures

This section takes an overview on candidate complexity measures for quantifying each of the three complexity kinds discussed above. Table 2 below provides a summary of these measures.

Let us begin with autonomic complexity. Besides integrated information, which may also be customized to the components of the autonomic system, other measures that have proven more practical for capturing the complexity of autonomous processes in systems are morphological computation [93,94], synergistic information [95,96] and the index of autonomous functioning [27,97,98]. The first of these is particularly useful for systems with high morphological dexterity such as in biology and soft robotics. It captures the extent to which a system's morphological properties are used to delegate and distribute its informational processing capabilities towards the goal of autonomous action. Synergistic information refers to information provided by the simultaneous knowledge of multiple variables, that is not available from any of the individual variables by themselves. In [96], this was used to show how an agent's cue sensors jointly carry cue information with the agent's interneurons (in fact, in this example, this measure quantifies the coupling between systems referring to $\mathcal{C}_{Autonomic}$ and $\mathcal{C}_{Computational}$). The index of autonomous functioning has extensively been used in human behavioral studies to quantify regulation of action by the self. All these studies, particularly [98] emphasize the necessity of autonomy (and hence, autonomic complexity) for systems realizing goal-oriented behaviors.

Now let us turn our attention to measures capturing systems and processes referring to computational complexity $C_{Computational}$. These have been studied extensively in neuroscience and AI. In the context of consciousness research, the most prominent among these is the measure of integrated information, Φ. However, there have been several candidates for this measure and its many approximations. The earliest version of IIT was based on a measure called neural density [32] (see also [33,34]). Subsequently, version 2 of the theory, IIT v2, defined Φ in terms of a Kullback-Leibler divergence, which was used as a relative entropy measure to quantify the information generated by the whole over the sum of its parts [43,99]. The current version of the theory, dubbed IIT 3.0, uses the Earth Mover's Distance (EMD) [46]. Despite its conceptual appeal, the algorithm proposed by IIT has been computationally intractable for realistic biological or artificial systems. This is where either related or approximate integrated information measures have been useful. Examples of related measures include stochastic interaction (also called total information flow) [49], stochastic integrated information [35,38,44] and geometric measures of integrated information [42]. Examples of various empirical approximations to Φ include the 'Perturbational Complexity Index' based on Lempel-Ziv compression [100] and causal connectivity based on Granger causality [101] (see [40] for a review of theoretical and empirical consciousness measures).

Besides integrated information, other information-theoretic measures that have been used in cognitive and computational neuroscience are mutual information and specific information, both of which have been used in neural coding paradigms [96]. From partial information decomposition methods, one has measures of redundant information, unique information and synergistic information [95,96]. These measures are relevant in situations where multiple sources potentially carry information about a measurement outcome or cue variable. Synergistic information refers to the property of multiple random variables cooperating to predict, or reduce the uncertainty of, a single target variable. In general cases, these are quite difficult to compute. In the case of two and three source variables, a formulation of these measures can be found in [95,96]. Yet another class of information-theoretic measures applicable to computational systems is that which describes the dynamics of information processing. Examples of these information-theoretic measures are transfer entropy, information gain and information transfer (discussed in [96]). The last two of these are of practical relevance. They have been tested on an artificial cognitive agent with a brain, body and environment [96]. This study shows that information-theoretic analysis reveals important insights on how task-relevant information flows through the embodied agent and is combined into a categorization decision. Furthermore, a dynamical systems analysis reveals the key geometrical and temporal interrelationships underlying the above categorization task performed by the agent.

Finally, let us discuss social complexity. There are two broad measures that have been discussed in the literature for this kind of complexity. One is, not surprisingly, integrated information, discussed in [102]. The other is the collective intelligence factor, discussed in [103,104], which refers to how well groups perform on a diverse set of group problem-solving tasks. The primary influences on a group's collective intelligence were identified to as the following: (a) the group composition (e.g., the members' skills, diversity, and intelligence) and (b) the group interaction (e.g., structures, processes, and norms of the group). Other related works that discuss collective intelligence include [105–108]. All of these cited measures seek to capture social complexity of groups with respect to complex tasks involving the group as a whole. Below we shall see that this kind of complexity is important for the morphospace of consciousness where the groups refer to species of animals or kinds of technologies.

Table 2. A summary of complexity measures that have been tested on various autonomic, computational and social systems.

Complexity Kind	Complexity Measures
$\mathcal{C}_{Autonomic}$	Index of Autonomous Functioning Synergistic Information Morphological Computation
$\mathcal{C}_{Computational}$	Integrated Information (v1, v2, v3, geometric) Stochastic Information/Total Information Flow Mutual & Specific Information Redundant & Unique Information Synergistic Information Transfer Entropy & Information Transfer
$\mathcal{C}_{Social}$	Collective Intelligence Factor Integrated Information v2

5.3. Constructing the Morphospace

Using these definitions for the three complexity kinds, we construct the morphospace of consciousness in Figure 3. While this space is only a first attempt at constructing a common framework for biological and artificial agents, the precise coordinates of various systems within this morphospace might change due to the rapid pace of new and developing technologies, but we expect the relative locations of each example to remain the same. We start with the human brain, which is taken as the benchmark in this space, defining a limit case located at one upper corner with highest scores on all the three axes. The human brain can perform computational tasks across a variety of domains such as making logical inference, planning an optimal path in a complex environment or dealing with recursive problems and hence leads with respect to computational complexity due to these cross-domain capabilities. On the social axis, human social interactions have resulted in the emergence of language, music, art, culture or socio-political systems. Other biological entities such as non-human primates [109,110] or social insects would score lower on the social and computational axis than humans. Additionally, other species of vertebrates such as some types of birds and cephalopods have been shown to exhibit complex behavior and possess sophisticated nervous systems. These two groups have actually been enormously useful in the search for animal consciousness [111,112].

Current AI systems such as IBM Watson [113], AlphaGo [64], DQNs [114] and Siri [115] are powerful computing systems over a narrow set of domains, but in their current form they do not show general-purpose functionality, that is, the capacity to independently interact with the world and successfully perform multiple tasks in different domains [116], or as proposed by Allen Newell, the capability with which anything can become a task [117]. These AI systems are still clustered high on the computational axis, but lower than humans (due to their domain-specificity). Also they score low on autonomic and social complexity. Synthetic forms of life such as protocells show some levels of autonomous functioning, reacting to chemicals and stressors, but currently show minimal capabilities for computation or inference and minimal interactions with other agents [118].

Interest in the field of multi-agent robotics has led to the rise of machines where emergent collective behaviors, e.g., coordination (KiloBot [119], Multi-Agent Deep Network [120]) or shared semantic conventions (Talking Heads [86]) self-organize out of multi-agent interactions. These systems are designed to display simple forms of navigation, object-detection, etc., while interacting with other agents performing the same task. However, they show lower social and autonomic complexity than most biological agents. Being embodied systems, they currently score lower on computational complexity than heavy-powered AI systems such as Watson or AlphaGo.

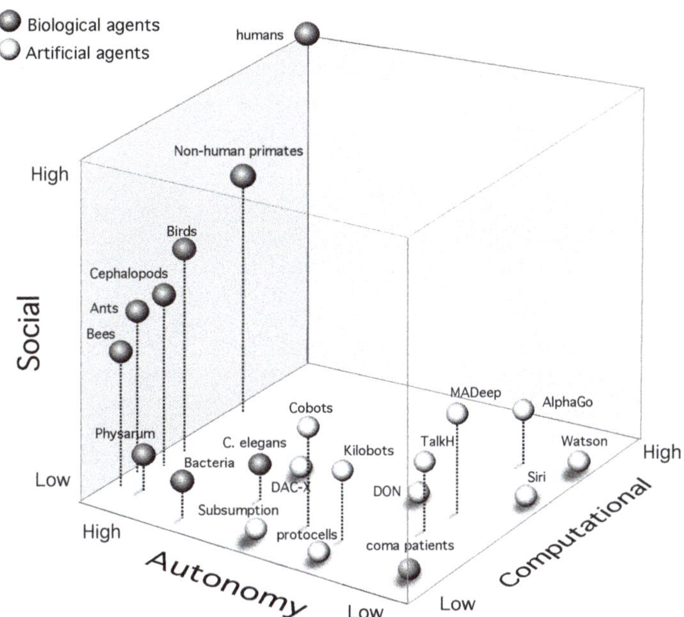

Figure 3. Morphospace of consciousness. Autonomous, computational and social complexity constitute the three axes of the consciousness morphospace. Human consciousness is used as a reference in one corner of the space. Current AI implementations cluster together in the high computation, low autonomy and low social complexity regime, while multi-agent cognitive robotics cluster around low computational, but moderate autonomous and social complexities. Abbreviated legends: MADeep (multi-agent deep reinforcement system) [120]; TalkH (talking heads) [86]; DQN (deep Q-learning) [114]; DAC-X (distributed adaptive control) [121], CoBot (cockroach robot) [122], Kilobot (swarm robot) [119], Subsumption (mobile robot architecture) [123].

An important use of the morphospace within evolutionary biology is related to the actual occupation of this space by different solutions. Notice that in the morphospace in Figure 3, a large part of the space is left vacant. A similar observation was made in [5] in the context of the morphospace of synthetic organs and organoids. In both cases, such an observation points towards new classes of artificial life and intelligence. Most present-day artificial systems (both, synthetic biology and AI), depicted in the morphospace, remain in the lower part of the cube. This is indicative of the currently minimal role played by social context in the development of these systems. On the other hand, in natural systems, social interactions have played an important role in shaping the minds of the organisms (those close to the left wall in Figure 3, involving high autonomy and sociality). Complex organisms equipped with brains and exhibiting cooperative behavior have evolved to live together with others. This is because social synergies increase the resilience of the group to many environmental and predatory challenges.

5.4. Relation to General Intelligence

How do our discussions on consciousness relate to theories of general intelligence? The idea that consciousness resides in select regions of a morphospace, that is constructed from function-specific types of complexity, has implications for any theory of general intelligence. The dimensions of our morphospace implicitly entail (or rather subsume) distinct types of intelligence. In cognitive psychology, manifestations of human intelligence have been discussed in the context of Howard Gardner's theory of multiple intelligences [124]. Here we want to understand how the dimensions of our morphospace help group different types of intelligences. This works as follows (We thank Carlos E. Perez for bringing this point to our attention). A discussion about how Gardner's intelligence types may be realized in machines using deep learning can be found in his recent book [125]). The autonomic axis reflects adaptive intelligence found in biological organisms. This encapsulates Gardner's kinesthetic, musical and spatial intelligence (some of these also require computational complexity). The computational axis refers to recognition, planning and decision-making capabilities that we find in computers as well as in humans. These are tasks involving logical deduction or inference. Hence, this complexity refers to those types of intelligences that require computational capabilities, such as logical reasoning, linguistic intelligence, etc. The third axis of the morphospace, social complexity, relates to social capabilities required for interacting with other agents. This refers to interpersonal and introspective intelligence, in Gardner's terms. These types of intelligences are also associated to the evolution of language, social conventions and culture. Then there are also other types of intelligences described in Gardner's theory such as naturalistic and pedagogical intelligence, which involve a composition of social and computational complexity.

As described above, the defining dimensions of our morphospace account for all of the multiple types of intelligence proposed by Gardner. Taking these intelligence (or their associated complexity) types into account, while building artificially intelligent machines, elucidates the wide spectrum of problems that future AI could potentially address. In the light of both, Gardner's theory and Newell's criteria, our morphospace, in fact, suggests that consciousness as we know it, subsumes a specific form of integrated multiple intelligence. Note that one ought to be careful to *not* claim that consciousness *'is'* general intelligence. Following William James, in cognitive psychology, consciousness is seen as a process that enables action for survival purposes [126]. We claim that this process, enabling action, constitutes mechanisms and phenomenology that realize an integration of specific types of intelligences and their associated complexities in such a way so as to meet survival goals. On the other hand, intelligence by itself can be thought of as any task-specific capability (or a process realizing that capability), that is not necessarily tied to existential pressures [127]. However, currently we have yet to fully understand how several of the intelligence types mentioned above, especially the non-computational ones [128], can be functionally realized in machines, let alone understanding the mechanisms that lead to integration of types. Nonetheless, given the myriad of recent advances in human-machine interactions, a complexity-based conceptualization of consciousness provides a practical and quantitative framework for studying ways in which interactions with machines might enhance our joint complexities and competences.

The outlook of these complexity kinds with respect to general intelligence is that systems and processes referring to computational intelligence will bring about new cures in medicine, new scientific understanding, and more efficient and less wasteful processes. Machines with autonomous intelligence capabilities will bring about greater conveniences such as self-driving automobiles, robotic care-takers in the workplace and in the home, and intuitive user interfaces. The third kind, systems with social intelligence, will be beneficial with regards to advertising to the masses, promoting global causes and managing social unrest.

5.5. Other Embodiments of Consciousness in the Morphospace

What other forms of consciousness does our morphospace suggest? Because there is no precise definition or consensus to benchmark consciousness even in biological life forms, the best one can do at the moment is to pursue a comparative functional approach as has been followed here. The three complexity axes on the morphospace encapsulate processes necessary to support functions that consciousness serves. Moreover, from the earlier discussion on multiple intelligence types being distributed across the morphospace, one may ask what forms of systems might reside in distinct regions of this space. Higher biological life-forms, those which are generally believed to possess some degree consciousness (in terms of reportable behaviors), tend to cluster closer to one corner of the cube in Figure 3. The other corners are suggestive of agents or systems with different embodiments and functionalities. Below we identify these embodiments and the form of consciousness or intelligence that they might potentially refer to (again, purely on functional grounds).

To illustrate these embodiments, it is instructive to represent the morphospace as a Boolean graph, where vertices are labelled by their corresponding cartesian co-ordinates in the cubic morphospace and edges refer to one of the three complexity axes along the cube (Figure 4). The $(1, 1, 1)$ vertex corresponds to human consciousness. The $(1, 0, 0)$ vertex and $(0, 1, 0)$ vertex correspond to present day synthetic biological systems and AI technologies, respectively. Neither of them are considered conscious. Examples of current technologies near the $(0, 0, 1)$ vertex would be highly interactive reactive systems. Even these are not what one would consider conscious. The $(1, 1, 0)$ vertex corresponds to an agent that is highly autonomous and computational, but lacking social drives. Evolutionarily, such agents would be disfavored. Technologically, they offer somewhat similar utilities as agents on either the $(1, 0, 0)$ or $(0, 1, 0)$ vertex. For our purposes, the interesting vertices are $(1, 0, 1)$ and $(0, 1, 1)$. This is where future intelligent technologies or potentially new forms of conscious systems may be found. Below we identify three system embodiments, corresponding to potential forms of consciousness, that occupy these vertices.

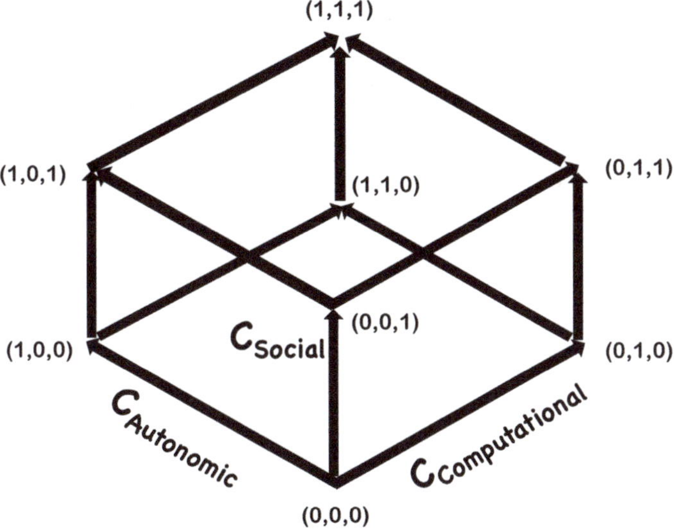

Figure 4. A Boolean Graphical Representation of the Morphospace.

5.5.1. Synthetic Consciousness

In Section 3 above, we alluded to two kinds of synthetic systems: synthetic biological systems and AI technologies. The question is, what augmentations would these systems need in order to be able to solve the H5W problem (of Section 4), with respect to their conspecifics? When that criterion is met, these systems could be considered to possess basic forms of consciousness on functional grounds. For certain, this would require them to have sufficient social complexity. Systems located near the $(1, 0, 1)$ vertex would show behaviors similar to some biological life-forms. Systems located near the $(0, 1, 1)$ vertex would be where current efforts in AGI (artificial general intelligence) are trying to get. Of course, at present, we do not know of any objective tests to ascertain if these systems on the $(1, 0, 1)$ and $(0, 1, 1)$ vertices may have first-person experiences (this criticism holds also for many biological species that one would otherwise argue as possessing phenomenological traits of consciousness). The comparison of these to a form of consciousness is made on behavioral grounds. Alternatively, one might as well refer to these systems as a form of synthetic general intelligence.

5.5.2. Group Consciousness

In a sense, biological consciousness itself can be thought of as a collective phenomenon where individual cells making up an organism are themselves not considered to be conscious (with respect to the three complexity measures defining the morphospace), even though the organism as a whole is. But what happens when the system itself is not localized? We postulate group consciousness as an extension of the above idea to composite or distributed systems that display levels of computational, autonomic and social complexity that are sufficient to answer the H5W problem. Note that, as per this specification of group consciousness, the group itself is treated as one entity. Hence, social complexity now refers to the interactions of this group with other similar groups.

This bears some resemblance to the notion of collective intelligence, which is a widely studied phenomenon in complex systems ranging from ant colonies [107], to a swarm of robots (the Kilobot in [119] and the CoRobot in [122]), to social networks [129]. However, these are usually not thought of as conscious systems. As a whole, they are not considered as autonomic forms with survival drives that compete or cooperate with other similar agents. However, these distinctions begin to get blurred during transient epochs when collective survival comes under threat. For example, when a bee colony comes under attack by hornets, collectively it demonstrates a prototypical survival drive, similar to lower-order organisms. Other examples of such behaviors have also been studied in the context of group interactions in humans, where social sensitivity, cooperation and diversity have been shown to correlate with the collective intelligence of the group [103]. Following this, the notion of collective intentionality has been discussed in [108]. More recently, [102] have applied integrated information Φ to group interactions, suggesting a new kind of group consciousness. While it is known that Φ in adapting agents increases with fitness [130], one can ask a similar question for an entire group: what processes (evolutionary games, learning, etc.) enable an increase in all three complexity types for an entire group such that it can solve the H5W problem while cooperating or competing with other groups?

For these reasons, this type of system, if conscious, in terms of being able to solve the H5W problem with respect to its conspecifics (other groups), will cluster around the $(1, 0, 1)$ vertex of the morphospace.

5.5.3. Simulated Consciousness

Our discussions on complexity kinds also suggest yet another potential type of consciousness, namely, simulated consciousness, wherein embodied virtual agents in a simulated reality interact with other virtual agents, while satisfying the complexity bounds that enable them to answer the H5W questions within the simulation. In this case, consciousness is strictly confined to the simulated environment. The agents cannot perceive or communicate with entities outside of the simulation but satisfy all the criteria we have discussed above within the simulation. How these embodied virtual agents could acquire consciousness is not yet known. Presumably by evolving across multiple generations of agents that adapt and learn to optimize fitness conditions. It is also not clear what precise traits or mechanisms would have to be coded into the simulation (as initializations or priors) in order to enable consciousness to evolve. The point here is simply that the same criteria that we have identified with consciousness in biological agents in the physical world, could in principle be admitted by agents within a simulation and confined to their interactions within that simulation. This has parallels to the notion of "Machine Consciousness" discussed in [131], which proposes that neural processes leading to consciousness might be realizable as a machine simulation (it even goes further to claim that computer systems might someday be able to emulate consciousness). At the moment, these are all open challenges in AI and consciousness research. Examples of studies discussing embodied virtual agents can be found in the work of [132,133]. More recent implementations of embodied virtual agents have been using gaming technology, such as the Minecraft platform [134,135].

Simulations, if conscious in the functional sense—that is, being able to solve the H5W problem with respect to its conspecifics (other groups) within the simulation—will cluster around the $(0, 1, 1)$ vertex of the morphospace.

6. Discussion

The objective of this article was to bring together diverse ideas from neuroscience, AI, synthetic biology, evolutionary theory and robotics in order to identify measures and mechanisms that relate to the problem of consciousness. Synergies between these disciplines have started to converge towards a systematic science of consciousness. Following through with these developments, we have attempted to generalize the applicability of current clinical scales of consciousness to artificial agents. In particular, starting from clinical measures of consciousness that calibrate awareness and wakefulness in patients, we have investigated how contemporary AI agents and synthetically engineered organisms compare on homologous measures. An abstraction of processes involving awareness and wakefulness can be generically associated to forms of computational and/or autonomic complexity.

Furthermore, based on insights from evolutionary game theory, we have discussed the function that consciousness serves in nature, and argued that the mechanisms of consciousness arose as an evolutionary game-theoretic strategy. This was why we introduced a third kind of complexity to describe consciousness, namely, social complexity. Social interactions play a crucial role in driving and regulating adaptive responses through behavioral feedback in both natural and artificial systems [20]. In [80,84], it has been suggested that complex social interactions may have evolutionarily served as a trigger for consciousness. For these reasons, social complexity is crucial for constructing a morphospace of consciousness.

A morphospace is a useful construct to study systems-level properties of complex systems based on information-theoretic complexity measures. The three kinds of complexity specified here, capture functional characteristics of biological as well as synthetic complex systems. Using these scales, we have shown how biological organisms including bacteria, bees, C. elegans, primates and humans compare to current AI systems such as deep networks, multi-agent systems, social robots, intelligent assistants such as Siri and computational systems such as IBM's Watson. Put together, the above three kinds complexity help characterize both, biological and artificial agents in a common framework.

Besides consciousness as we know it (in biology), distinct regions in the morphospace suggest other plausible manifestations of consciousness (based on functional criteria), namely, synthetic, group and simulated consciousness, each based on a distinct embodiment. However, what is far from clear is whether there exist specific thresholds in the values of each complexity, that an agent must surpass in order to attain a level of consciousness. Certainly, from developmental biology we know that both humans (and many higher-order animals) undergo extensive periods of cognitive and social learning, concurrent to physiological development, from infancy to maturation. These phases of physiological, cognitive and social training are necessary for the development of autonomic and cognitive abilities leading to levels of consciousness attained by brains.

Even though we may still be far from understanding most of the engineering principles required to build conscious machines, a complexity-based comparison between biological and artificial systems reveals interesting insights. For example, current AI systems using deep learning tend to cluster along the computational complexity axis of the morphospace, whereas synthetically engineered life forms group closer along the autonomic complexity axis. On the other hand, biologically conscious agents are distributed in regions of the morphospace corresponding to relatively high complexity along all three of the axes (which suggests necessary, if not sufficient, conditions for biological consciousness). In terms of Newell's criteria, mentioned in the introduction, excluding those criteria that refer exclusively to human-specific traits (language, symbolic reasoning), the remaining ones are completely satisfiable by any agent located in the high complexity region of all three axes of the morphospace. In contrast, current AI or synthetic systems do not check out on this list. Though in 1994 Newell was not explicitly referring to consciousness, it is remarkable to note how those ideas to formulate theories of cognition and intelligence seem to reconcile with current ideas of consciousness. One could summarize the crux of Newell's criteria as referring to agents displaying autonomous adaptive behaviors with cross-domain problem-solving capabilities, which can be decomposed to the kinds of complexity discussed here.

This perspective on consciousness opens several possibilities for future work. For instance, it may be interesting to further refine the morphospace described here. In particular, computational complexity itself may involve several sub-types involving learning, adaptation, acquiring sensorimotor representations, etc, all of which are relevant for cognitive robotics. Another question arising out of our discussion is whether the emergence of consciousness in a multi-agent social environment can be identified as a Nash equilibrium of a cooperation–competition game. In a game where, say, two species attain consciousness, the population pay-offs in cooperation and competition between them are likely to reach one of possible equilibria due to the recursive nature of social inferences, when an agent attempts to infer the inferences of other agent about its own intentions. Multi-agent models might offer a viable approach to test ideas such as these.

Furthermore, a conceptualization of a morphospace of consciousness offers an interesting comparative perspective on leading candidate theories of consciousness. The main contenders in this case are: (i) Integrated Information Theory (IIT) [10], (ii) Global Workspace Theory (GWT) [8], (iii) Predictive Processing Theories (PPT) [136], (iv) Higher Order Theories (HOT) [137], and (v) Orchestrated Objective Reduction Theory (Orch-OR) [138]. These are also the major theoretical paradigms of consciousness currently being pitted against each other as part of the 'Structured Adversarial Collaboration Projects' initiative being supported by the Templeton World Charity Foundation. The ultimate goal (and test) of any theory of consciousness is to satisfactorily explain the so-called "hard problem of consciousness", that is, 'How and why first-person phenomenal experiences arise, and what the nature of qualia may be?' [139]. Our intention, in this work, is not to directly address any of those fundamental questions or propose a new theory of consciousness. Rather, we have investigated a taxonomy of conscious and artificial agents based on complexity, with the objective of highlighting design constraints shared across minds and machines. These constraints may help fine-tune future iterations of the above candidate proposals of consciousness.

Given the morphospace of consciousness, we can now ask the following questoin: What kinds of complexity could the above-mentioned candidate theories of consciousness admit? The first four of these, for the most part, associate consciousness to computational complexity. IIT, with its information-theoretic Φ measure, says little about autonomic or social processes, deferring consciousness to computational mechanisms with high integrated information. GWT explicitly proposes conscious access as a kind of computation [140]. PPT operates within the framework of predictive coding and Bayesian inference. These models are grounded in sensorimotor interactions and, to a certain extent, also involve autonomic processes (see also [141]). In HOT, phenomenal consciousness is postulated to be a higher-order representation of perceptual or quasi-perceptual contents, that is, thoughts or perceptions about first-order mental states. Orch-OR, on the other hand, explicitly states that consciousness is a non-computational process (one that cannot be algorithmically implemented in the Turing sense). This theory associates "proto-consciousness" to an orchestrated objective reduction of the quantum wave-function in dendritic microtubuli. Of the three complexity kinds, processes postulated in Orch-OR belong to the autonomic class and manifest at the molecular level. It is also worth mentioning other non-computational processes such as stochastic dynamics [142] or non-Darwinian mechanisms [143,144] that are relevant to molecular and systems-level biology, but have not yet been fully exploited in the context of consciousness research.

What we have learned here, from synthesizing cross-disciplinary evidence about brains and machines into a unified framework of a morphospace, is that consciousness (at least, as we know it in biology) is supported by processes of at least three kinds of complexity and that these processes are closely intertwined with each other. This poses a challenge for all of the above candidate theories of consciousness. Hence, these theories have to explain how the core mechanisms they associate to consciousness unfold into all three complexity kinds. In the absence of that, the proposed theories at best only describe individuated components of the full problem and may require building bridges with each other in order to reconcile how autonomic, computational and social processes collectively give rise to consciousness.

Related to the above point, the morphospace, described here, suggests a taxonomy of complexity into three kinds of systems-level processes. In a sense, these correspond to the brain ($\mathcal{C}_{Computational}$), body ($\mathcal{C}_{Autonomic}$) and environment ($\mathcal{C}_{Social}$). While this taxonomy was derived from functional arguments (the H5W problem), this correspondence to the brain, body and environment suggests architectural constraints on conscious agents, namely, that such architectures include computational, embodied, situated and social modalities. However, this by itself is not surprising; there is extensive work in the cognitive science literature studying each of these paradigms, some of which also concern the easy problem versus hard problem dichotomy [139]. The point we emphasize in this work is that a morphospace forces one to think of these functions and design constraints in a common framework. It is also a challenge for all existing theories of consciousness to show how the axioms they propose to address the "hard" problem reconcile with the integration of the "easy" problems via interactions between the brain, body and environment. This is the issue that a morphospace of consciousness brings to the forefront.

As a final remark, note that the taxonomy of the three complexity kinds discussed here, also shows up in current AI architectures where physics and psychology engines provide priors to hierarchical Bayesian networks used for meta-learning or learning to learn [145]. The physics engine in this case can be identified to processes mostly along the autonomic axis. The psychology engine largely accounts for social processes. Computational and reasoning processes are implemented on Bayesian inference engines. Of course, these engines do not operate mutually exclusively, being closely coupled and co-ordinated with each other. This observation lends evidence to the claim that a morphospace, as we have described here, serves as a useful construct for identifying general design principles and constraints in our theories and architectures of biological as well as artificial intelligence.

Author Contributions: Manuscript conceptualized by X.D.A. with substantial inputs from R.S. All authors contributed to subsequent development of ideas and ensuing discussions. X.D.A. wrote the original draft with significant contributions from R.S. All authors contributed to reviewing and editing large swaths of the draft article leading to the final version. All authors have read and agreed to the published version of the manuscript.

Funding: This research was funded by the European Research Council's CDAC project: "The Role of Consciousness in Adaptive Behavior: A Combined Empirical, Computational and Robot based Approach" (ERC-2013- ADG 341196).

Institutional Review Board Statement: Not applicable.

Informed Consent Statement: Not applicable.

Data Availability Statement: No new data created or used in this study.

Acknowledgments: We thank Riccardo Zucca and Sytse Wierenga for help with graphics. This work has been supported by the European Research Council's CDAC project: "The Role of Consciousness in Adaptive Behavior: A Combined Empirical, Computational and Robot based Approach" (ERC-2013- ADG 341196). RS acknowledges the support of the Secretaria d'Universitats i Recerca del Departament d'Economia i Coneixement de la Generalitat de Catalunya, the Botin Foundation, by Banco Santander through its Santander Universities Global Division and by the Santa Fe Institute.

Conflicts of Interest: The authors declare no conflict of interest. The funders had no role in the design of the study; in the collection, analyses, or interpretation of data; in the writing of the manuscript; or in the decision to publish the results.

References

1. Laureys, S.; Owen, A.M.; Schiff, N.D. Brain function in coma, vegetative state, and related disorders. *Lancet Neurol.* **2004**, *3*, 537–546. [CrossRef] [PubMed]
2. Laureys, S. The neural correlate of (un) awareness: Lessons from the vegetative state. *Trends Cogn. Sci.* **2005**, *9*, 556–559. [CrossRef] [PubMed]
3. McGhee, G.R. *Theoretical Morphology: The Concept and Its Applications*; Columbia University Press: New York, NY, USA, 1999.
4. Avena-Koenigsberger, A.; Goñi, J.; Solé, R.; Sporns, O. Network morphospace. *J. R. Soc. Interface* **2015**, *12*, 20140881. [CrossRef] [PubMed]
5. Ollé-Vila, A.; Duran-Nebreda, S.; Conde-Pueyo, N.; Montañez, R.; Solé, R. A morphospace for synthetic organs and organoids: The possible and the actual. *Integr. Biol.* **2016**, *8*, 485–503. [CrossRef]
6. Seoane, L.F.; Solé, R. The morphospace of language networks. *Sci. Rep.* **2018**, *8*, 10465. [CrossRef]
7. Arsiwalla, X.D.; Herreros, I.; Verschure, P. On Three Categories of Conscious Machines. In Proceedings of the Biomimetic and Biohybrid Systems: 5th International Conference, Living Machines 2016, Edinburgh, UK, 19–22 July 2016; Lepora, N.F., Mura, A., Mangan, M., Verschure, P.F., Desmulliez, M., Prescott, T.J., Eds.; Springer International Publishing: Cham, Switzerland, 2016; pp. 389–392. [CrossRef]
8. Baars, B.J. Global workspace theory of consciousness: Toward a cognitive neuroscience of human experience. *Prog. Brain Res.* **2005**, *150*, 45–53.
9. Koch, C.; Massimini, M.; Boly, M.; Tononi, G. Neural correlates of consciousness: Progress and problems. *Nat. Rev. Neurosci.* **2016**, *17*, 307–321. [CrossRef]
10. Tononi, G.; Boly, M.; Massimini, M.; Koch, C. Integrated information theory: From consciousness to its physical substrate. *Nat. Rev. Neurosci.* **2016**, *17*, 450–461. [CrossRef]
11. Mischiati, M.; Lin, H.T.; Herold, P.; Imler, E.; Olberg, R.; Leonardo, A. Internal models direct dragonfly interception steering. *Nature* **2015**, *517*, 333–338. [CrossRef]
12. Newell, A. *Unified Theories of Cognition*; Harvard University Press: Cambridge, MA, USA, 1994.
13. Rashevsky, N. Outline of a physico-mathematical theory of excitation and inhibition. *Protoplasma* **1933**, *20*, 42–56. [CrossRef]
14. McCulloch, W.S.; Pitts, W. A logical calculus of the ideas immanent in nervous activity. *Bull. Math. Biophys.* **1943**, *5*, 115–133. [CrossRef]
15. Rosenblatt, F. The perceptron: A probabilistic model for information storage and organization in the brain. *Psychol. Rev.* **1958**, *65*, 386. [CrossRef]
16. Hopfield, J.J. Neural networks and physical systems with emergent collective computational abilities. *Proc. Natl. Acad. Sci. USA* **1982**, *79*, 2554–2558. [CrossRef]
17. Sole, R. Rise of the Humanbot. *arXiv* **2017**, arXiv:1705.05935.
18. Arsiwalla, X.D.; Freire, I.T.; Vouloutsi, V.; Verschure, P. Latent Morality in Algorithms and Machines. In Proceedings of the Conference on Biomimetic and Biohybrid Systems, Nara, Japan, 9–12 July 2019; Springer: Cham, Switzerland, 2019; pp. 309–315.
19. Freire, I.T.; Urikh, D.; Arsiwalla, X.D.; Verschure, P.F. Machine Morality: From Harm-Avoidance to Human-Robot Cooperation. In Proceedings of the Conference on Biomimetic and Biohybrid Systems, Freiburg, Germany, 28–30 July 2020; Springer: Cham, Switzerland, 2020; pp. 116–127.
20. Verschure, P.F.; Pennartz, C.M.; Pezzulo, G. The why, what, where, when and how of goal-directed choice: Neuronal and computational principles. *Phil. Trans. R. Soc. B* **2014**, *369*, 20130483. [CrossRef] [PubMed]
21. Laureys, S.; Celesia, G.G.; Cohadon, F.; Lavrijsen, J.; León-Carrión, J.; Sannita, W.G.; Sazbon, L.; Schmutzhard, E.; von Wild, K.R.; Zeman, A.; et al. Unresponsive wakefulness syndrome: A new name for the vegetative state or apallic syndrome. *BMC Med.* **2010**, *8*, 68. [CrossRef]
22. Bruno, M.A.; Vanhaudenhuyse, A.; Thibaut, A.; Moonen, G.; Laureys, S. From unresponsive wakefulness to minimally conscious PLUS and functional locked-in syndromes: Recent advances in our understanding of disorders of consciousness. *J. Neurol.* **2011**, *258*, 1373–1384. [CrossRef]
23. Blumenfeld, H. Impaired consciousness in epilepsy. *Lancet Neurol.* **2012**, *11*, 814–826. [CrossRef]
24. Giacino, J.T.; Ashwal, S.; Childs, N.; Cranford, R.; Jennett, B.; Katz, D.I.; Kelly, J.P.; Rosenberg, J.H.; Whyte, J.; Zafonte, R.; et al. The minimally conscious state definition and diagnostic criteria. *Neurology* **2002**, *58*, 349–353. [CrossRef]
25. Giacino, J.T. The vegetative and minimally conscious states: Consensus-based criteria for establishing diagnosis and prognosis. *NeuroRehabilitation* **2004**, *19*, 293–298. [CrossRef]
26. Parton, A.; Malhotra, P.; Husain, M. Hemispatial neglect. *J. Neurol. Neurosurg. Psychiatry* **2004**, *75*, 13–21. [PubMed]
27. Weinstein, N.; Przybylski, A.K.; Ryan, R.M. The index of autonomous functioning: Development of a scale of human autonomy. *J. Res. Personal.* **2012**, *46*, 397–413. [CrossRef]
28. Wibral, M.; Vicente, R.; Lindner, M. Transfer Entropy in Neuroscience. In *Directed Information Measures in Neuroscience*; Wibral, M., Vicente, R., Lizier, J.T., Eds.; Springer: Berlin/Heidelberg, Germany, 2014; pp. 3–36. [CrossRef]

29. Haggard, P. Human volition: Towards a neuroscience of will. *Nat. Rev. Neurosci.* **2008**, *9*, 934–946. [CrossRef]
30. Lopez-Sola, E.; Moreno-Bote, R.; Arsiwalla, X.D. Sense of agency for mental actions: Insights from a belief-based action-effect paradigm. *Conscious. Cogn.* **2021**, *96*, 103225. [CrossRef] [PubMed]
31. Deci, E.L.; Ryan, R.M. The "what" and "why" of goal pursuits: Human needs and the self-determination of behavior. *Psychol. Inq.* **2000**, *11*, 227–268. [CrossRef]
32. Tononi, G.; Sporns, O.; Edelman, G.M. A measure for brain complexity: Relating functional segregation and integration in the nervous system. *Proc. Natl. Acad. Sci. USA* **1994**, *91*, 5033–5037. [CrossRef]
33. Tononi, G. An information integration theory of consciousness. *BMC Neurosci.* **2004**, *5*, 42. [CrossRef]
34. Tononi, G.; Sporns, O. Measuring information integration. *BMC Neurosci.* **2003**, *4*, 31. [CrossRef]
35. Arsiwalla, X.D.; Verschure, P.F.M.J. Integrated information for large complex networks. In Proceedings of the 2013 International Joint Conference on Neural Networks (IJCNN), Dallas, TX, USA, 4–9 August 2013; pp. 1–7. [CrossRef]
36. Arsiwalla, X.D.; Verschure, P. Computing Information Integration in Brain Networks. In *Advances in Network Science, Proceedings of the 12th International Conference and School, NetSci-X 2016, Wroclaw, Poland, 11–13 January 2016*; Wierzbicki, A., Brandes, U., Schweitzer, F., Pedreschi, D., Eds.; Springer International Publishing: Cham, Switzerland, 2016; pp. 136–146. [CrossRef]
37. Arsiwalla, X.D.; Verschure, P.F.M.J., High Integrated Information in Complex Networks Near Criticality. In *Artificial Neural Networks and Machine Learning – ICANN 2016, Proceedings of the 25th International Conference on Artificial Neural Networks, Barcelona, Spain, 6–9 September 2016; Part I*; Villa, A.E.; Masulli, P.; Pons Rivero, A.J., Eds.; Springer International Publishing: Cham, Switzerland, 2016; pp. 184–191. [CrossRef]
38. Arsiwalla, X.D.; Verschure, P.F. The global dynamical complexity of the human brain network. *Appl. Netw. Sci.* **2016**, *1*, 16. [CrossRef]
39. Arsiwalla, X.D.; Verschure, P. Why the Brain Might Operate Near the Edge of Criticality. In Proceedings of the International Conference on Artificial Neural Networks, Alghero, Italy, 11–14 September 2017; Springer: Cham, Switzerland, 2017; pp. 326–333.
40. Arsiwalla, X.D.; Verschure, P. Measuring the Complexity of Consciousness. *Front. Neurosci.* **2018**, *12*, 424. [CrossRef]
41. Arsiwalla, X.D.; Pacheco, D.; Principe, A.; Rocamora, R.; Verschure, P. A Temporal Estimate of Integrated Information for Intracranial Functional Connectivity. In Proceedings of the International Conference on Artificial Neural Networks, Rhodes, Greece, 4–7 October 2018; Springer: Cham, Switzerland, 2018; pp. 403–412.
42. Ay, N. Information geometry on complexity and stochastic interaction. *Entropy* **2015**, *17*, 2432–2458. [CrossRef]
43. Balduzzi, D.; Tononi, G. Integrated information in discrete dynamical systems: Motivation and theoretical framework. *PLoS Comput. Biol.* **2008**, *4*, e1000091. [CrossRef] [PubMed]
44. Barrett, A.B.; Seth, A.K. Practical measures of integrated information for time-series data. *PLoS Comput. Biol.* **2011**, *7*, e1001052. [CrossRef] [PubMed]
45. Griffith, V. A Principled Infotheoretic/phi-like Measure. *arXiv* **2014**, arXiv:1401.0978.
46. Oizumi, M.; Albantakis, L.; Tononi, G. From the phenomenology to the mechanisms of consciousness: Integrated information theory 3.0. *PLoS Comput. Biol.* **2014**, *10*, e1003588. [CrossRef] [PubMed]
47. Petersen, K.; Wilson, B. Dynamical intricacy and average sample complexity. *arXiv* **2015**, arXiv:1512.01143.
48. Tegmark, M. Improved Measures of Integrated Information. *arXiv* **2016**, arXiv:1601.02626.
49. Wennekers, T.; Ay, N. Stochastic interaction in associative nets. *Neurocomputing* **2005**, *65*, 387–392. [CrossRef]
50. Sarasso, S.; Casali, A.G.; Casarotto, S.; Rosanova, M.; Sinigaglia, C.; Massimini, M. Consciousness and complexity: A consilience of evidence. *Neurosci. Conscious.* **2021**, *7*, 1–24. [CrossRef]
51. Hagmann, P.; Cammoun, L.; Gigandet, X.; Meuli, R.; Honey, C.J.; Wedeen, V.J.; Sporns, O. Mapping the Structural Core of Human Cerebral Cortex. *PLoS Biol.* **2008**, *6*, 15. [CrossRef]
52. Honey, C.; Sporns, O.; Cammoun, L.; Gigandet, X.; Thiran, J.P.; Meuli, R.; Hagmann, P. Predicting human resting-state functional connectivity from structural connectivity. *Proc. Natl. Acad. Sci. USA* **2009**, *106*, 2035–2040. [CrossRef] [PubMed]
53. Arsiwalla, X.D.; Betella, A.; Bueno, E.M.; Omedas, P.; Zucca, R.; Verschure, P.F. The Dynamic Connectome: A Tool For Large-Scale 3D Reconstruction Of Brain Activity In Real-Time. In Proceedings of the ECMS, Ålesund, Norway, 27–30 May, 2013; pp. 865–869.
54. Betella, A.; Martínez, E.; Zucca, R.; Arsiwalla, X.D.; Omedas, P.; Wierenga, S.; Mura, A.; Wagner, J.; Lingenfelser, F.; André, E.; et al. Advanced Interfaces to Stem the Data Deluge in Mixed Reality: Placing Human (Un)Consciousness in the Loop. In Proceedings of the ACM SIGGRAPH 2013 Posters, Hong Kong, China, 19–22 November 2013; ACM: New York, NY, USA, 2013; SIGGRAPH '13, p. 68:1. [CrossRef]
55. Betella, A.; Cetnarski, R.; Zucca, R.; Arsiwalla, X.D.; Martínez, E.; Omedas, P.; Mura, A.; Verschure, P.F.M.J. BrainX3: Embodied Exploration of Neural Data. In Proceedings of the Proceedings of the 2014 Virtual Reality International Conference, Laval, France, 9–11 April 2014; ACM: New York, NY, USA, 2014; VRIC '14, pp. 37:1–37:4. [CrossRef]
56. Omedas, P.; Betella, A.; Zucca, R.; Arsiwalla, X.D.; Pacheco, D.; Wagner, J.; Lingenfelser, F.; Andre, E.; Mazzei, D.; Lanatá, A.; et al. XIM-Engine: A Software Framework to Support the Development of Interactive Applications That Uses Conscious and Unconscious Reactions in Immersive Mixed Reality. In Proceedings of the Proceedings of the 2014 Virtual Reality International Conference, Laval, France, 9–11 April 2014; ACM: New York, NY, USA, 2014; VRIC '14, pp. 26:1–26:4. [CrossRef]
57. Arsiwalla, X.D.; Zucca, R.; Betella, A.; Martinez, E.; Dalmazzo, D.; Omedas, P.; Deco, G.; Verschure, P. Network Dynamics with BrainX3: A Large-Scale Simulation of the Human Brain Network with Real-Time Interaction. *Front. Neuroinformatics* **2015**, *9*, 2. [CrossRef]

58. Arsiwalla, X.D.; Dalmazzo, D.; Zucca, R.; Betella, A.; Brandi, S.; Martinez, E.; Omedas, P.; Verschure, P. Connectomics to Semantomics: Addressing the Brain's Big Data Challenge. *Procedia Comput. Sci.* **2015**, *53*, 48–55. [CrossRef]

59. Solé, R. Synthetic transitions: Towards a new synthesis. *Philos. Trans. R. Soc. B* **2016**, *371*, 20150438. [CrossRef]

60. Malyshev, D.A.; Dhami, K.; Lavergne, T.; Chen, T.; Dai, N.; Foster, J.M.; Corrêa, I.R.; Romesberg, F.E. A semi-synthetic organism with an expanded genetic alphabet. *Nature* **2014**, *509*, 385–388. [CrossRef]

61. Solé, R.V.; Munteanu, A.; Rodriguez-Caso, C.; Macia, J. Synthetic protocell biology: From reproduction to computation. *Philos. Trans. R. Soc. Lond. B Biol. Sci.* **2007**, *362*, 1727–1739. [CrossRef]

62. Hutchison, C.A.; Chuang, R.Y.; Noskov, V.N.; Assad-Garcia, N.; Deerinck, T.J.; Ellisman, M.H.; Gill, J.; Kannan, K.; Karas, B.J.; Ma, L.; et al. Design and synthesis of a minimal bacterial genome. *Science* **2016**, *351*, aad6253. [CrossRef]

63. Solé, R.; Amor, D.R.; Duran-Nebreda, S.; Conde-Pueyo, N.; Carbonell-Ballestero, M.; Montañez, R. Synthetic collective intelligence. *Biosystems* **2016**, *148*, 47–61. [CrossRef]

64. Silver, D.; Huang, A.; Maddison, C.J.; Guez, A.; Sifre, L.; van den Driessche, G.; Schrittwieser, J.; Antonoglou, I.; Panneershelvam, V.; Lanctot, M.; et al. Mastering the game of Go with deep neural networks and tree search. *Nature* **2016**, *529*, 484–489. [CrossRef]

65. Tian, Y.; Ma, J.; Gong, Q.; Sengupta, S.; Chen, Z.; Pinkerton, J.; Zitnick, C.L. Elf opengo: An analysis and open reimplementation of alphazero. *arXiv* **2019**, arXiv:1902.04522.

66. Silver, D.; Hubert, T.; Schrittwieser, J.; Antonoglou, I.; Lai, M.; Guez, A.; Lanctot, M.; Sifre, L.; Kumaran, D.; Graepel, T.; et al. A general reinforcement learning algorithm that masters chess, shogi, and Go through self-play. *Science* **2018**, *362*, 1140–1144. [CrossRef] [PubMed]

67. Schrittwieser, J.; Antonoglou, I.; Hubert, T.; Simonyan, K.; Sifre, L.; Schmitt, S.; Guez, A.; Lockhart, E.; Hassabis, D.; Graepel, T.; et al. Mastering atari, go, chess and shogi by planning with a learned model. *Nature* **2020**, *588*, 604–609. [CrossRef] [PubMed]

68. Marcus, G. Innateness, alphazero, and artificial intelligence. *arXiv* **2018**, arXiv:1801.05667.

69. Hubel, D.H.; Wiesel, T.N. Receptive fields, binocular interaction and functional architecture in the cat's visual cortex. *J. Physiol.* **1962**, *160*, 106–154. [CrossRef]

70. Ciresan, D.C.; Meier, U.; Masci, J.; Maria Gambardella, L.; Schmidhuber, J. Flexible, high performance convolutional neural networks for image classification. In Proceedings of the IJCAI Proceedings-International Joint Conference on Artificial Intelligence, Barcelona, Spain, 16–22 July 2011; Volume 22, p. 1237.

71. Sak, H.; Senior, A.W.; Beaufays, F. Long short-term memory recurrent neural network architectures for large scale acoustic modeling. In Proceedings of the Interspeech, Singapore, 14–18 September 2014; pp. 338–342.

72. Liao, Q.; Poggio, T. Bridging the gaps between residual learning, recurrent neural networks and visual cortex. *arXiv* **2016**, arXiv:1604.03640.

73. Moulin-Frier, C.; Arsiwalla, X.D.; Puigbò, J.Y.; Sanchez-Fibla, M.; Duff, A.; Verschure, P.F. Top-down and bottom-up interactions between low-level reactive control and symbolic rule learning in embodied agents. In Proceedings of the CoCo@ NIPS Conference, Barcelona, Spain, 9 December 2016.

74. Sánchez-Fibla, M.; Moulin-Frier, C.; Arsiwalla, X.; Verschure, P. Social Sensorimotor Contingencies: Towards Theory of Mind in Synthetic Agents. In *Recent Advances in Artificial Intelligence Research and Development*; IOS Press: Amsterdam, The Netherlands, 2017; pp. 251–256.

75. Freire, I.T.; Moulin-Frier, C.; Sanchez-Fibla, M.; Arsiwalla, X.D.; Verschure, P. Modeling the formation of social conventions in multi-agent populations. *arXiv* **2018**, arXiv:1802.06108.

76. Freire, I.T.; Puigbò, J.Y.; Arsiwalla, X.D.; Verschure, P.F. Modeling the Opponent's Action Using Control-Based Reinforcement Learning. In Proceedings of the Conference on Biomimetic and Biohybrid Systems, Paris, France, 17–20 July 2018; Springer: Cham, Switzerland, 2018, pp. 179–186.

77. Freire, I.T.; Arsiwalla, X.D.; Puigbò, J.Y.; Verschure, P. Modeling theory of mind in multi-agent games using adaptive feedback control. *arXiv* **2019**, arXiv:1905.13225.

78. Freire, I.T.; Moulin-Frier, C.; Sanchez-Fibla, M.; Arsiwalla, X.D.; Verschure, P.F. Modeling the formation of social conventions from embodied real-time interactions. *PLoS ONE* **2020**, *15*, e0234434. [CrossRef]

79. Demirel, B.; Moulin-Frier, C.; Arsiwalla, X.D.; Verschure, P.F.; Sánchez-Fibla, M. Distinguishing Self, Other, and Autonomy From Visual Feedback: A Combined Correlation and Acceleration Transfer Analysis. *Front. Hum. Neurosci.* **2021**, *15*, 560657. [CrossRef]

80. Arsiwalla, X.D.; Herreros, I.; Moulin-Frier, C.; Sanchez, M.; Verschure, P.F., Is Consciousness a Control Process? In *Artificial Intelligence Research and Development*; Nebot, A., Binefa, X., Lopez de Mantaras, R., Eds.; IOS Press: Amsterdam, The Netherlands, 2016; pp. 233–238. [CrossRef]

81. Arsiwalla, X.D.; Herreros, I.; Moulin-Frier, C.; Verschure, P. Consciousness as an Evolutionary Game-Theoretic Strategy. In Proceedings of the Conference on Biomimetic and Biohybrid Systems, Stanford, CA, USA, 26–28 July 2017; Springer: Cham, Switzerland 2017; pp. 509–514.

82. Herreros, I.; Arsiwalla, X.; Verschure, P. A forward model at Purkinje cell synapses facilitates cerebellar anticipatory control. In Proceedings of the Advances in Neural Information Processing Systems, Barcelona, Spain, 5–10 December 2016; pp. 3828–3836.

83. Moulin-Frier, C.; Puigbò, J.Y.; Arsiwalla, X.D.; Sanchez-Fibla, M.; Verschure, P.F. Embodied Artificial Intelligence through Distributed Adaptive Control: An Integrated Framework. *arXiv* **2017**, arXiv:1704.01407.

84. Verschure, P.F. Synthetic consciousness: The distributed adaptive control perspective. *Phil. Trans. R. Soc. B* **2016**, *371*, 20150448. [CrossRef] [PubMed]

85. Steels, L. Evolving grounded communication for robots. *Trends Cogn. Sci.* **2003**, *7*, 308–312. [CrossRef] [PubMed]

86. Steels, L.; Hild, M. *Language Grounding in Robots*; Springer Science & Business Media: New York, NY USA, 2012.

87. Lewis, D. *Convention: A Philosophical Study*; Wiley-Blackwell: Cambridge, MA, USA, 1969.

88. Lewis, D. *Convention: A Philosophical Study*; John Wiley & Sons: Hoboken, NJ, USA, 2008.

89. Hofbauer, J.; Huttegger, S.M. Feasibility of communication in binary signaling games. *J. Theor. Biol.* **2008**, *254*, 843–849. [CrossRef] [PubMed]

90. Alexander, J.M. The evolutionary foundations of strong reciprocity. *Anal. Krit.* **2005**, *27*, 106–112. [CrossRef]

91. Feinberg, T.E.; Mallatt, J. The evolutionary and genetic origins of consciousness in the Cambrian Period over 500 million years ago. *Front. Psychol.* **2013**, *4*, 667. [CrossRef] [PubMed]

92. Bayne, T.; Hohwy, J.; Owen, A.M. Are there levels of consciousness? *Trends Cogn. Sci.* **2016**, *20*, 405–413. [CrossRef]

93. Zahedi, K.; Ay, N. Quantifying morphological computation. *Entropy* **2013**, *15*, 1887–1915. [CrossRef]

94. Füchslin, R.M.; Dzyakanchuk, A.; Flumini, D.; Hauser, H.; Hunt, K.J.; Luchsinger, R.H.; Reller, B.; Scheidegger, S.; Walker, R. Morphological computation and morphological control: Steps toward a formal theory and applications. *Artif. Life* **2013**, *19*, 9–34. [CrossRef]

95. Griffith, V.; Koch, C. Quantifying synergistic mutual information. In *Guided Self-Organization: Inception*; Springer: Berlin/Heidelberg, Germany, 2014; pp. 159–190.

96. Beer, R.D.; Williams, P.L. Information processing and dynamics in minimally cognitive agents. *Cogn. Sci.* **2015**, *39*, 1–38. [CrossRef]

97. Deci, E.L.; Ryan, R.M. The support of autonomy and the control of behavior. *J. Personal. Soc. Psychol.* **1987**, *53*, 1024. [CrossRef]

98. Ryan, R.M.; Deci, E.L. Autonomy Is No Illusion: Self-Determination Theory and the Empirical Study of Authenticity, Awareness, and Will. In *Handbook of Experimental Existential Psychology*; Guilford Press: New York, NY, USA, 2004; pp. 449–479.

99. Balduzzi, D.; Tononi, G. Qualia: The geometry of integrated information. *PLoS Comput. Biol.* **2009**, *5*, e1000462. [CrossRef] [PubMed]

100. Casali, A.G.; Gosseries, O.; Rosanova, M.; Boly, M.; Sarasso, S.; Casali, K.R.; Casarotto, S.; Bruno, M.A.; Laureys, S.; Tononi, G.; et al. A theoretically based index of consciousness independent of sensory processing and behavior. *Sci. Transl. Med.* **2013**, *5*, 198ra105. [CrossRef] [PubMed]

101. Barrett, A.B.; Murphy, M.; Bruno, M.A.; Noirhomme, Q.; Boly, M.; Laureys, S.; Seth, A.K. Granger causality analysis of steady-state electroencephalographic signals during propofol-induced anaesthesia. *PLoS ONE* **2012**, *7*, e29072. [CrossRef]

102. Engel, D.; Malone, T.W. Integrated Information as a Metric for Group Interaction: Analyzing Human and Computer Groups Using a Technique Developed to Measure Consciousness. *arXiv* **2017**, arXiv:1702.02462.

103. Woolley, A.W.; Chabris, C.F.; Pentland, A.; Hashmi, N.; Malone, T.W. Evidence for a collective intelligence factor in the performance of human groups. *Science* **2010**, *330*, 686–688. [CrossRef] [PubMed]

104. Woolley, A.W.; Aggarwal, I.; Malone, T.W. Collective intelligence and group performance. *Curr. Dir. Psychol. Sci.* **2015**, *24*, 420–424. [CrossRef]

105. Wolpert, D.H.; Tumer, K. An introduction to collective intelligence. *arXiv* **1999**, arXiv:cs/9908014.

106. Matarić, M.J. From Local Interactions to Collective Intelligence. In *Prerational Intelligence: Adaptive Behavior and Intelligent Systems Without Symbols and Logic, Volume 1, Volume 2, Prerational Intelligence: Interdisciplinary Perspectives on the Behavior of Natural and Artificial Systems, Volume 3* ; Springer: Dordrecht, The Netherlands 2000; pp. 988–998.

107. Dorigo, M.; Birattari, M. Swarm Intelligence. *Scholarpedia* **2007**, *2*, 1462. [CrossRef]

108. Huebner, B. *Macrocognition: A Theory of Distributed Minds and Collective intentionality*; Oxford University Press: Oxford, UK, 2013.

109. Borjon, J.I.; Takahashi, D.Y.; Cervantes, D.C.; Ghazanfar, A.A. Arousal Dynamics Drive Vocal Production in Marmoset Monkeys. *J. Neurophysiol.* **2016**, *116*, 753–764. [CrossRef]

110. de Waal, F.B. Apes know what others believe. *Science* **2016**, *354*, 39–40. [CrossRef]

111. Edelman, D.B.; Seth, A.K. Animal consciousness: A synthetic approach. *Trends Neurosci.* **2009**, *32*, 476–484. [CrossRef]

112. Emery, N.J.; Clayton, N.S. The mentality of crows: Convergent evolution of intelligence in corvids and apes. *Science* **2004**, *306*, 1903–1907. [CrossRef] [PubMed]

113. High, R. *The Era of Cognitive Systems: An Inside Look at IBM Watson and How It Works*; Redbooks; IBM Corporation: New York, NY, USA, 2012.

114. Mnih, V.; Kavukcuoglu, K.; Silver, D.; Rusu, A.A.; Veness, J.; Bellemare, M.G.; Graves, A.; Riedmiller, M.; Fidjeland, A.K.; Ostrovski, G.; et al. Human-level control through deep reinforcement learning. *Nature* **2015**, *518*, 529–533. [CrossRef] [PubMed]

115. Aron, J. How innovative is Apple's new voice assistant, Siri? *New Sci.* **2011**, *212*, 24. [CrossRef]

116. Legg, S.; Hutter, M. A Collection of Definitions of Intelligence. *Front. Artif. Intell. Appl.* **2007**, *157*, 17.

117. Newell, A. You Can't Play 20 Questions with Nature and Win: Projective Comments on the Papers of this Symposium. In *Visual Information Processing*; Academic Press: New York, NY, USA, 1973, pp. 283–308.

118. Kurihara, K.; Okura, Y.; Matsuo, M.; Toyota, T.; Suzuki, K.; Sugawara, T. A recursive vesicle-based model protocell with a primitive model cell cycle. *Nat. Commun.* **2015**, *6*, 8352. [CrossRef] [PubMed]

119. Rubenstein, M.; Cornejo, A.; Nagpal, R. Programmable self-assembly in a thousand-robot swarm. *Science* **2014**, *345*, 795–799. [CrossRef]

120. Tampuu, A.; Matiisen, T.; Kodelja, D.; Kuzovkin, I.; Korjus, K.; Aru, J.; Aru, J.; Vicente, R. Multiagent cooperation and competition with deep reinforcement learning. *PLoS ONE* **2017**, *12*, e0172395. [CrossRef]

121. Maffei, G.; Santos-Pata, D.; Marcos, E.; Sánchez-Fibla, M.; Verschure, P.F. An embodied biologically constrained model of foraging: From classical and operant conditioning to adaptive real-world behavior in DAC-X. *Neural Netw.* **2015**, *72*, 88–108. [CrossRef]

122. Halloy, J.; Sempo, G.; Caprari, G.; Rivault, C.; Asadpour, M.; Tâche, F.; Said, I.; Durier, V.; Canonge, S.; Amé, J.M.; et al. Social integration of robots into groups of cockroaches to control self-organized choices. *Science* **2007**, *318*, 1155–1158. [CrossRef]

123. Brooks, R. A Robust Layered Control System for a Mobile Robot. *IEEE J. Robot. Autom.* **1986**, *2*, 14–23. [CrossRef]

124. Gardner, H. *Frames of Mind: The Theory of Multiple Intelligences*; Hachette UK: London, UK, 2011.

125. Perez, C.E. *The Deep Learning AI Playbook: Strategy for Disruptive Artificial Intelligence*; Intuition Machine: San Francisco, CA, USA, 2017.

126. James, W. The Stream of Consciousness. *Psychology* **1892**, 151–175.

127. Arsiwalla, X.D.; Signorelli, C.M.; Puigbo, J.Y.; Freire, I.T.; Verschure, P. What is the Physics of Intelligence? In *Frontiers in Artificial Intelligence and Applications, Proceedings of the 21st International Conference of the Catalan Association for Artificial Intelligence, Catalonia, Spain, 19–21 October 2018*; IOS Press: Amsterdam, The Netherlands, 2018; Volume 308.

128. Arsiwalla, X.D.; Signorelli, C.M.; Puigbo, J.Y.; Freire, I.T.; Verschure, P.F. Are Brains Computers, Emulators or Simulators? In Proceedings of the Conference on Biomimetic and Biohybrid Systems, Paris, France, 17–20 July 2018; Springer: Berlin/Heidelberg, Germany, 2018; pp. 11–15.

129. Goleman, D. *Social Intelligence*; Random House: New York, NY, USA, 2007.

130. Edlund, J.A.; Chaumont, N.; Hintze, A.; Koch, C.; Tononi, G.; Adami, C. Integrated information increases with fitness in the evolution of animats. *PLoS Comput. Biol.* **2011**, *7*, e1002236. [CrossRef] [PubMed]

131. Reggia, J.A. The rise of machine consciousness: Studying consciousness with computational models. *Neural Networks* **2013**, *44*, 112–131. [CrossRef]

132. Cassell, J. *Embodied Conversational Agents*; MIT Press: Cambridge, MA, USA, 2000.

133. Burden, D.J. Deploying embodied AI into virtual worlds. *Knowl.-Based Syst.* **2009**, *22*, 540–544. [CrossRef]

134. Aluru, K.; Tellex, S.; Oberlin, J.; MacGlashan, J. Minecraft as an experimental world for AI in robotics. In Proceedings of the AAAI Fall Symposium, Arlington, VA, USA, 12–14 November 2015.

135. Johnson, M.; Hofmann, K.; Hutton, T.; Bignell, D. The malmo platform for artificial intelligence experimentation. In Proceedings of the International joint conference on artificial intelligence (IJCAI), New York, NY, USA, 9–15 July 2016; p. 4246.

136. Clark, A. Whatever next? Predictive brains, situated agents, and the future of cognitive science. *Behav. Brain Sci.* **2013**, *36*, 181–204. [CrossRef] [PubMed]

137. Lau, H.; Rosenthal, D. Empirical support for higher-order theories of conscious awareness. *Trends Cogn. Sci.* **2011**, *15*, 365–373. [CrossRef]

138. Hameroff, S.; Penrose, R. Consciousness in the universe: A review of the Orch OR theory. *Phys. Life Rev.* **2014**, *11*, 39–78. [CrossRef]

139. Chalmers, D.J. Facing up to the problem of consciousness. *J. Conscious. Stud.* **1995**, *2*, 200–219.

140. Dehaene, S.; Lau, H.; Kouider, S. What is consciousness, and could machines have it? *Science* **2017**, *358*, 486–492. [CrossRef]

141. Arsiwalla, X.D.; Moreno Bote, R.; Verschure, P. Beyond neural coding? Lessons from perceptual control theory. *Behav. Brain Sci.* **2019**, *42*, e217. [CrossRef]

142. Goertzel, B. *Chaotic logic: Language, Thought, and Reality from the Perspective of Complex Systems Science*; Springer Science & Business Media: New York, NY, USA, 2013; Volume 9.

143. King, J.L.; Jukes, T.H. Non-Darwinian Evolution: Most evolutionary change in proteins may be due to neutral mutations and genetic drift. *Science* **1969**, *164*, 788–798. [CrossRef] [PubMed]

144. Killeen, P.R. The non-Darwinian evolution of behavers and behaviors. *Behav. Process.* **2019**, *161*, 45–53. [CrossRef] [PubMed]

145. Lake, B.M.; Ullman, T.D.; Tenenbaum, J.B.; Gershman, S.J. Building machines that learn and think like people. *Behav. Brain Sci.* **2017**, *40*, e253. [CrossRef] [PubMed]

MDPI

St. Alban-Anlage 66

4052 Basel

Switzerland

Tel. +41 61 683 77 34

Fax +41 61 302 89 18

www.mdpi.com

NeuroSci Editorial Office

E-mail: neurosci@mdpi.com

www.mdpi.com/journal/neurosci